AF412378

Standard Stereograms for Materials Science

P T Clarke

National Physical Laboratory

Department of Industry

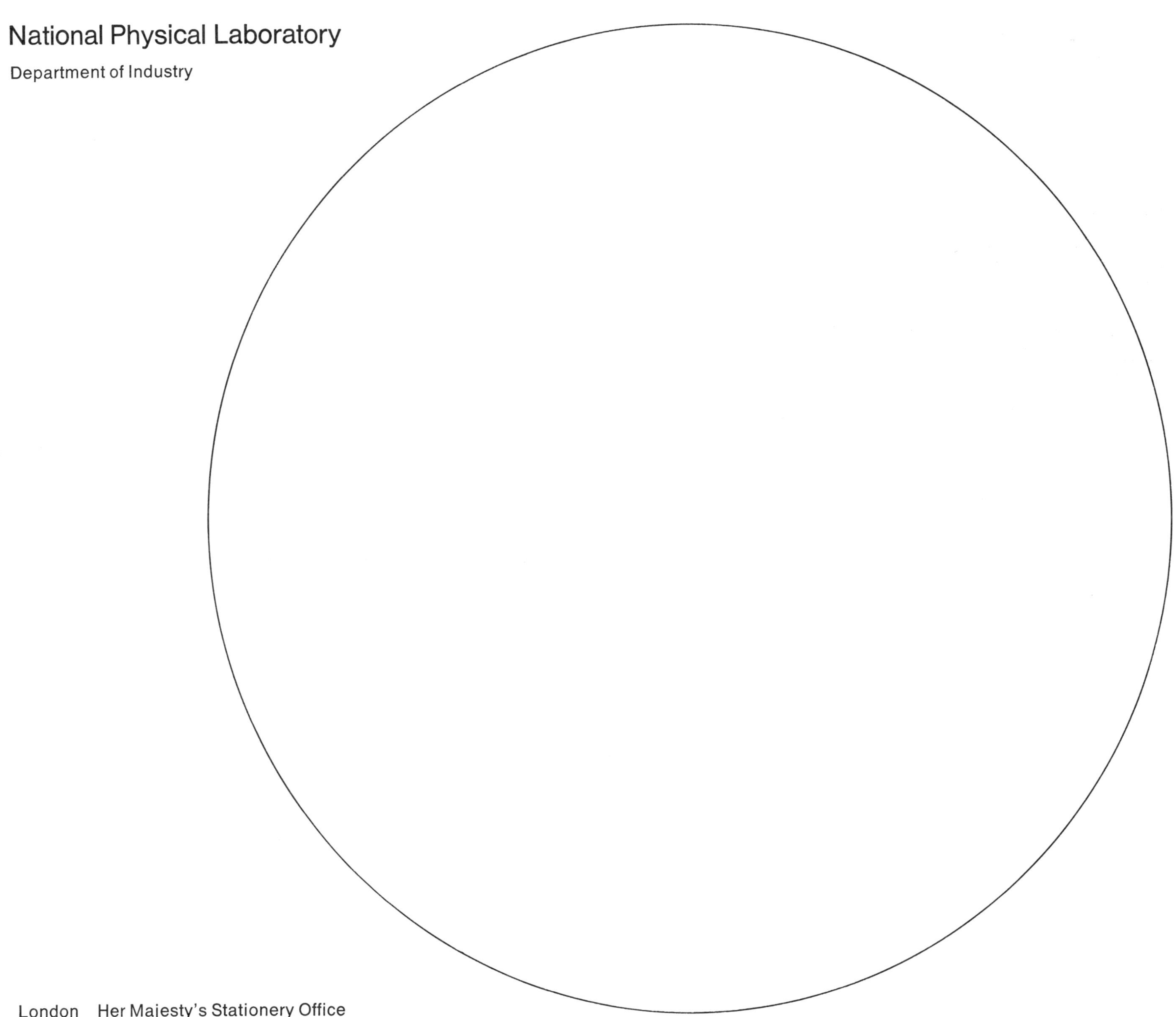

London   Her Majesty's Stationery Office

© *Crown copyright 1976*
First published 1976

HER MAJESTY'S STATIONERY OFFICE

Government Bookshops

49 High Holborn, London WC1V 6HB
13a Castle Street, Edinburgh EH2 3AR
41 The Hayes, Cardiff CF1 1JW
Brazennose Street, Manchester M60 8AS
Southey House, Wine Street, Bristol BS1 2BQ
258 Broad Street, Birmingham B1 2HE
80 Chichester Street, Belfast BT1 4JY

Government Publications are also available
through booksellers

Printed in England for Her Majesty's Stationery Office by
Bemrose & Sons Ltd.
Dd 506563   K10   5/76

ISBN 0 11 480038 3*

. . . and thanne wol the verrey lyne
meridional of thyn Astrolabie lye evene
south, and the est lyne wole lye est, and
the west lyne west, and north lyne
north, so that thou werke softly and
avisely in the couching; and thus hastow
the 4 quarters of the firmament. And for
the more declaracioun, lo here the
figure.

Geoffrey Chaucer

A Treatise on the Astrolabe (circa 1391)
Part II exercise 29

Astrolabe signed Erasmus Habermel fe 1585
Diameter 27.25 cm

National Maritime Museum, London

Introduction

The purpose of this volume is to provide an accurate set of standard stereograms which may be used to facilitate diffraction calculations in the field of materials science. It is assumed that users of these diagrams are already acquainted with the techniques of their use. There are several textbooks available to explain how to use stereograms (e.g. Smaill, 1972; Johari and Thomas, 1969; and Andrews, Dyson and Keown, 1971), but in order to apply the standard methods of calculation it is necessary to have a standard projection of the material of interest. In this book I have attempted to supply sufficient diagrams to cover all materials of major economic importance in the cubic, hexagonal, and tetragonal crystal systems, and to provide at least a basic set for some commonly occurring compounds in the lower crystal systems. I have tried to give a sufficiently close coverage of the main range of axial ratios for hexagonal and tetragonal crystals to provide a projection suitable for experimental purposes for any substance. Actual materials have been used to provide this coverage rather than hypothetical substances at standard arbitrary intervals of the axial ratio. If a material in question is not in this collection then there should be one with an axial ratio sufficiently close to serve for all usual calculations with experimental data.

When using standard stereograms it is frequently very convenient to be able to draw directly onto the diagram. To expedite this mode of working this volume is of a type designed to facilitate the photo-copying of individual diagrams for such purposes. The diameter of the stereograms, 200mm, was also chosen as being the largest, and hence most accurate, of the usual sizes to fit standard A4 paper and so permit the use of the most common types of copying machine.

The standard Wulff and Polar nets supplied with this volume have been redrawn by computer techniques for use with these diagrams. These will also be available separately.

There are particularly large sets of diagrams for the cubic and ideal close packed hexagonal cases. Ideally, similarly large numbers of projections are needed for all other crystal symmetries. Unfortunately no other packing arrangement of atoms is of anywhere near such general applicability. Other arrangements are therefore represented by the three projections of greatest symmetry, or projections down the three principal axes. To provide a comprehensive coverage of the lower symmetry classes would however be prohibitively expensive because whilst for tetragonal and hexagonal crystals there is only one parameter required to characterize the distribution of poles, namely the axial ratio, c/a, there are two for orthorhombic, three for monoclinic and five for triclinic. Fortunately, most materials of technological interest belong to one of the first three crystal classes.

The present volume shows a distinct bias towards ferrous metallurgy. This reflects the economic importance of this field of research which has resulted historically in a concentration of effort in that area. I hope to be able to do something to redress this imbalance in a subsequent volume and to provide improved coverage of the lower symmetry materials.

Hexagonal Indices

Three index notation, hkl rather than $hkil$, is used for all hexagonal crystals in this volume. The reasons are simple. Firstly, the i index is redundant, since $i = -(h+k)$. It is merely a mnemonic to reduce the calculation of equivalent indices to a simple permutation, so nothing is lost by omitting it. Secondly, by omitting it, it is possible to print 20 per cent more poles in the diagrams without overlap of the indices.

For some compounds both hexagonal and rhombohedral cells have been used to prepare stereograms. In these cases the corresponding diagrams are not exactly equivalent in that only some of the poles displayed are common to both diagrams. This is because the smaller rhombohedral cell exhibits symmetry aspects of the structure which are achieved in the diffraction patterns of the corresponding hexagonal cell by selection rules resulting from the internal rhombohedral symmetry. These selection rules are ignored when calculating a stereogram which displays directions without regard to internal symmetry requirements, which in any case do not apply to all materials with similar axial ratios.

Limits and accuracy of the diagrams

Nearly all of the stereograms in this volume contain over 300 poles. The limit of the indices displayed is chosen as a lower limit on the interplanar spacing. This is especially important for crystals with notably non-orthogonal axes, since the choice of a limitation based on a maximum value for the indices leaves notable gaps in the diagrams. This is particularly objectionable in diagrams of hexagonal materials where the use of index limitations can give rise to a diagram with only monoclinic symmetry. However, the full symmetry is preserved by using a minimum interplanar spacing to limit the number of poles.

The computations for these stereograms were carried out by a KDF9 computer with a program written in ALGOL. The mathematics used is essentially that described previously (Stokes, Keown, and Dyson, 1968), with a few modifications to suit local conditions (Clarke and Woolf, 1970). The diagrams were drawn on an incremental plotter with a step of 0.1 mm, so the nominal accuracy should have been 0.05 mm, however, problems arising from dimensional instability of the paper used limited the accuracy of the original diagrams to about 0.2 mm in an original diameter of 300 mm. This is still better than the accuracy of any Wulff nets presently available. It is to be expected that the exigences of time and the inescapable imperfections of optical and mechanical systems will result in a final accuracy of about 0.4 mm referred to the diameter of a stereogram or 0.2 mm referred to the centre. To some extent these errors can be minimized by comparison with the standard charts printed at the beginning of this book, since these charts have undergone the same treatment as the stereograms and have therefore suffered similar dimensional changes. In any case, the accuracy of these diagrams will still be far better than that of the vast majority of experimental data.

Acknowledgements

I should like to thank Mrs A Woolf for her help in writing the computer program, and for assisting in the onerous task of preparing printable versions of these diagrams. I should also like to thank Dr B R Heap for the use of his graphical output routines, without which this book could not have been attempted.

P T Clarke

References

K W Andrews, D J Dyson and S R Keown
Interpretation of Electron Diffraction Patterns, (2nd edition)
London, Hilger, 1971

P T Clarke and Anne Woolf
The Construction of Stereographic Projections and Ideal Laue
Photographs by Computer Program
NPL Materials Applications Report No. 9, 1970

O Johari and G Thomas
The Stereographic Projection and its Applications
New York, Wiley, 1969

J S Smaill
Metallurgical Stereographic Projections
London, Hilger, 1972

G K Stokes, S R Keown and D J Dyson
The Construction of Stereographic Projections by Computer
J. app. Crystallogr., 1968, **1,** 68-70

Contents

Hexagonal projections continued

Page	Axial ratio c/a	Material
129	3.4900	Y phase
132	4.9058	Alpha silicon carbide SiC
135	5.8260	Molybdenum disulphide MoS_2
138	7.2203	Samarium Sm:
141		Rhombohedral [111] Alpha 23.21

Tetragonal projections
Standard projections are down the axes
[001], [100], and [110]
Unless otherwise stated all materials listed below are represented
by the standard projections

Page	Axial ratio c/a	Material
142	0.3004	Potassium niobium oxide $K_4Nb_6O_{17}$
145	0.4370	Niobium dioxide NbO_2
148	0.5200	Sigma phase
151	0.5456	Tin Sn
154	0.6440	Titanium dioxide: rutile TiO_2
157	0.7454	Lead oxide Pb_3O_4
160	0.8315	Iron boride Fe_2B
163	0.9012	Zircon $ZrSiO_4$
166	1.1000	Iron nitride $Fe_{16}N_2$
169	1.2144	Beta spodumene $LiAlSi_2O_6$
172	1.3320	Delta' plutonium Pu
175	1.3952	Low christobalite SiO_2
178	1.5215	Indium In
181	1.5984	Thallium Tl 001 axis only
182	1.6401	Calcium carbide CaC_2
185	1.7556	Apophyllite $Ca_4KFSi_8O_{20}$
188	1.8628	Iron silicide boride Fe_5SiB_2
191	1.9487	Chromium carbide Cr_5B_3
194	2.4350	Z phase
197	2.5130	Titanium dioxide: anatase TiO_2
200	3.0000	Gamma ferric oxide: maghemite Fe_2O_3

Orthorhombic projections
Standard projections are down the axes
[001], [010] and [100]
All materials listed below are represented by the standard
projections

Page	Axial ratios			Material
203	0.311	0.568	1	Gamma plutonium Pu
206	0.485	1	0.844	Alpha uranium U
209	1	0.616	0.696	M_aC_b Fe_2MoC
212	0.671	0.755	1	Cementite Fe_3C
215	0.975	1	0.751	Sillimanite Al_2SiO_5

Monoclinic projections
Standard projections are down the axes
[001], [010] and [100]

Page	Axial ratios			beta	Material
218	0.564	0.440	1	101.82	Alpha plutonium Pu

Triclinic projections
Standard projections are down the axes
[001], [010] and [100]

Page	Axial ratios			alpha	beta	gamma	Material
221	0.998	1	0.633	98.92	91.22	85.75	Anorthite

Wulff net at 2° intervals

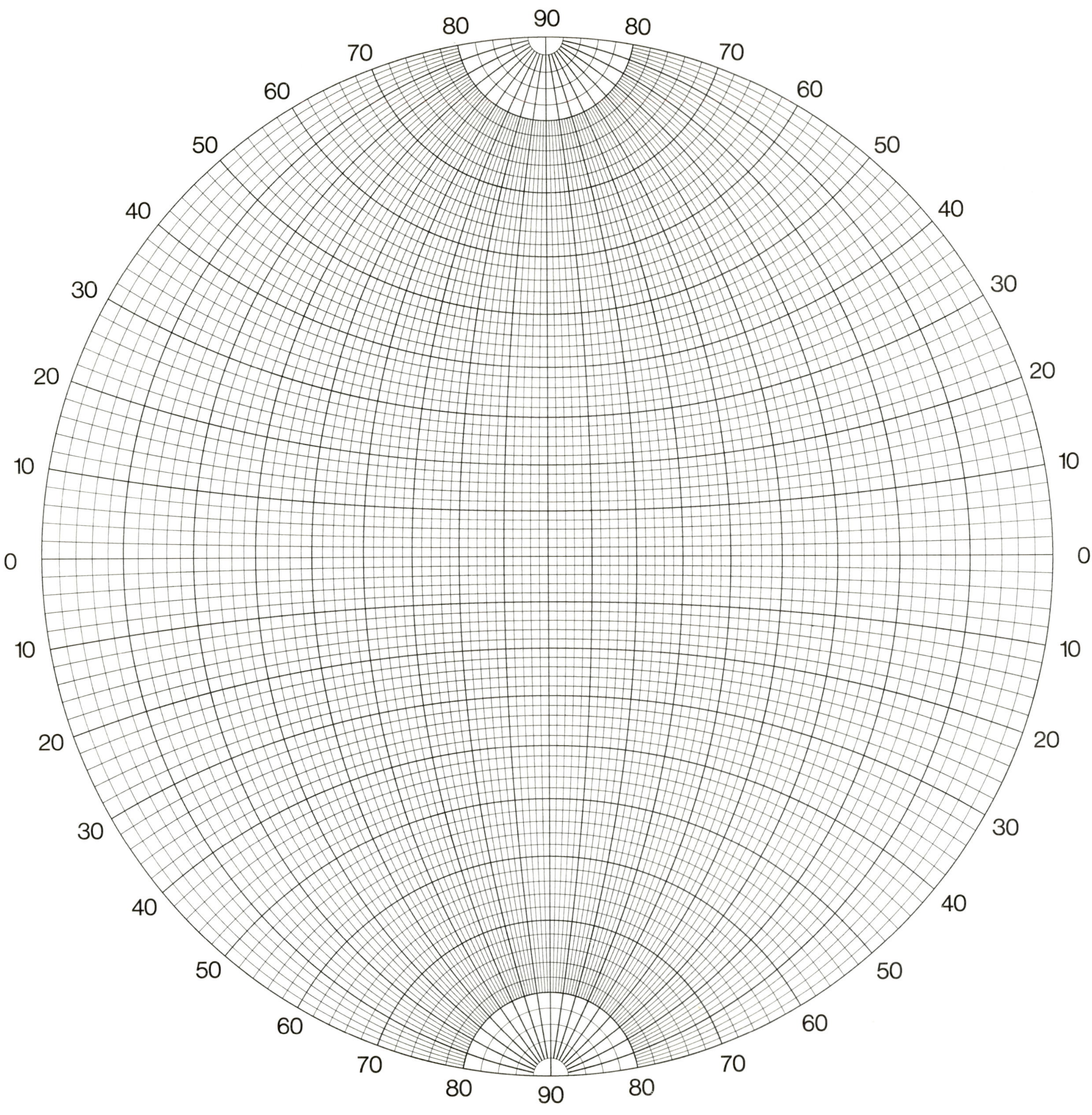

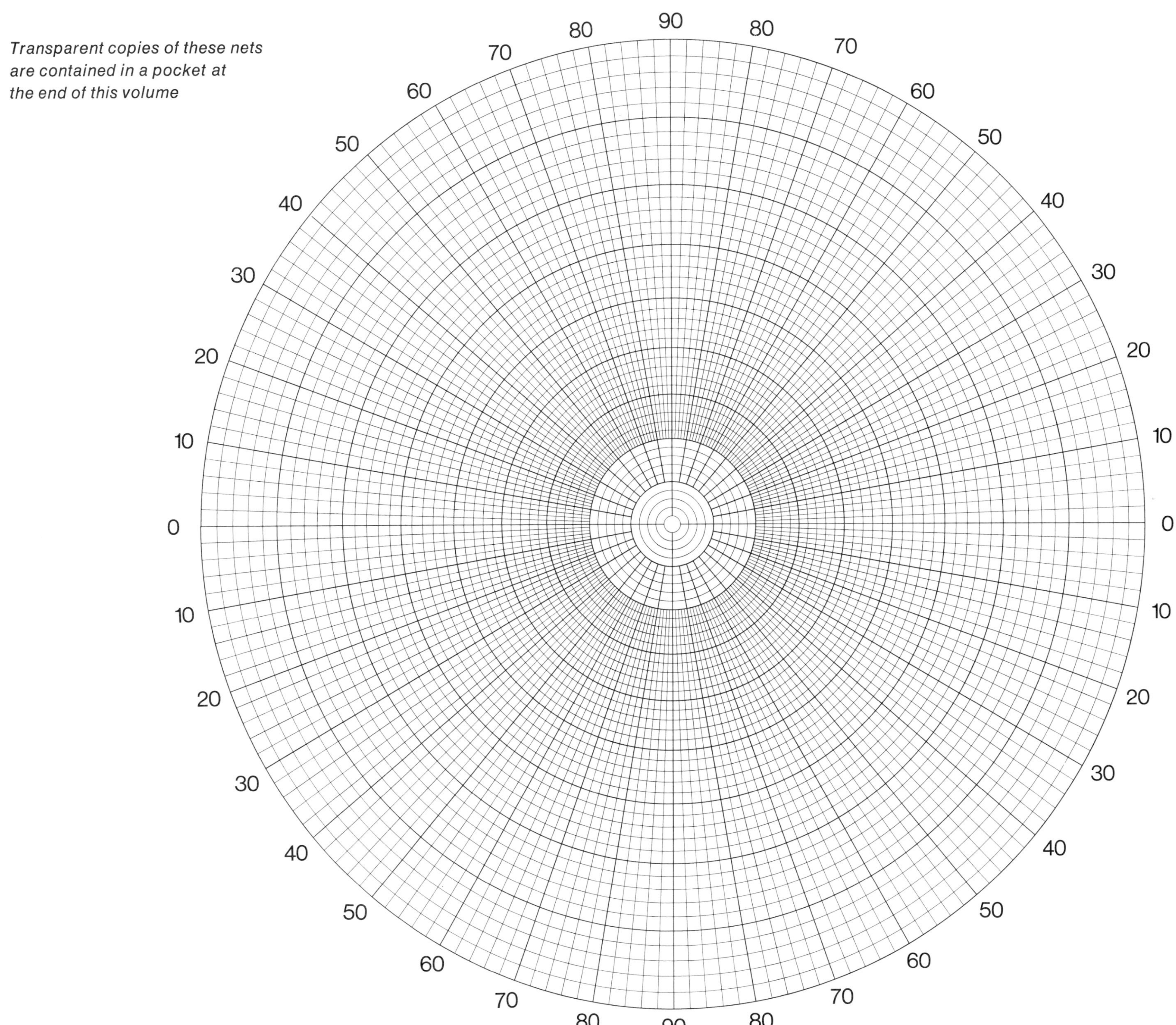
Transparent copies of these nets
are contained in a pocket at
the end of this volume

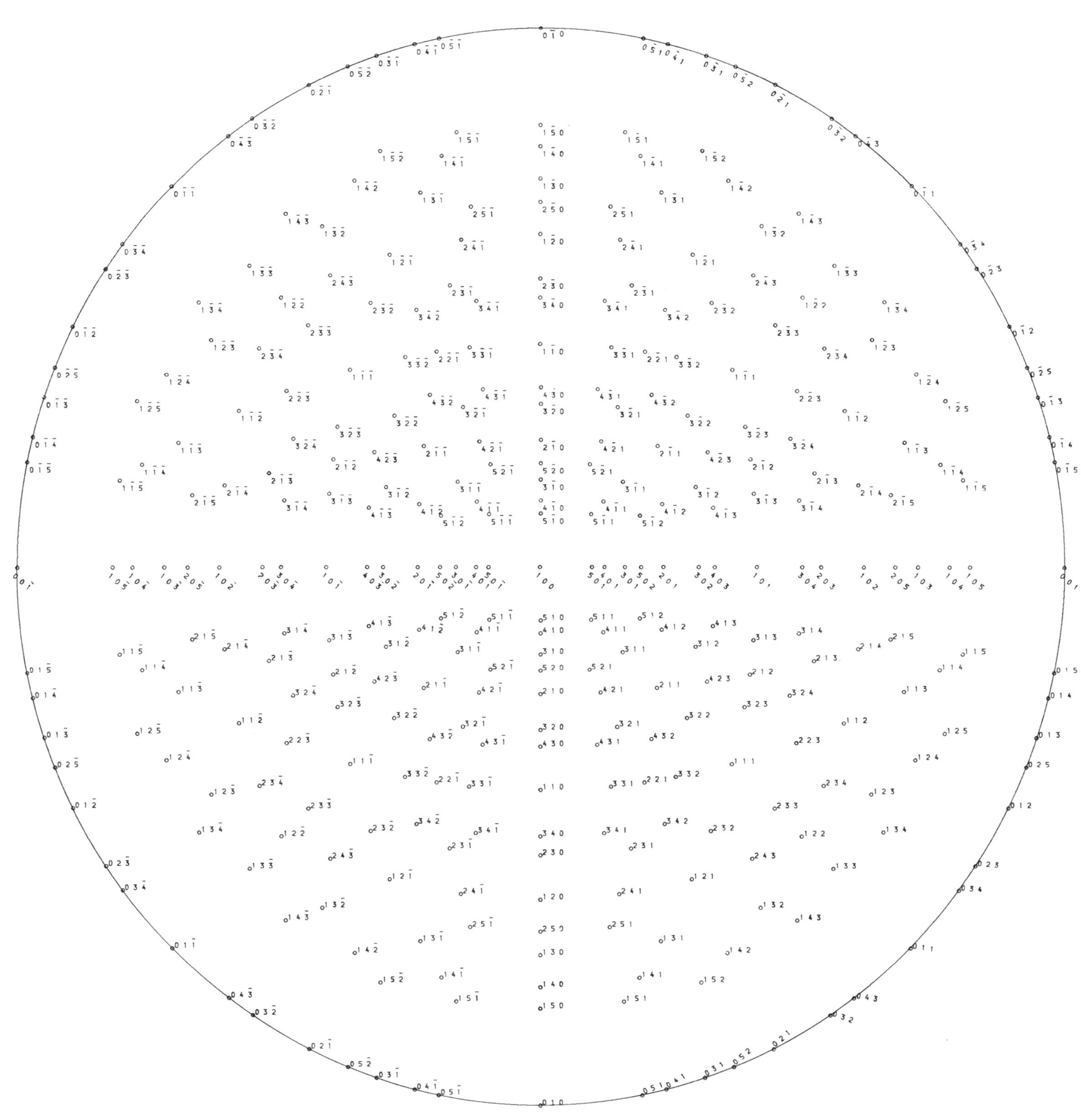

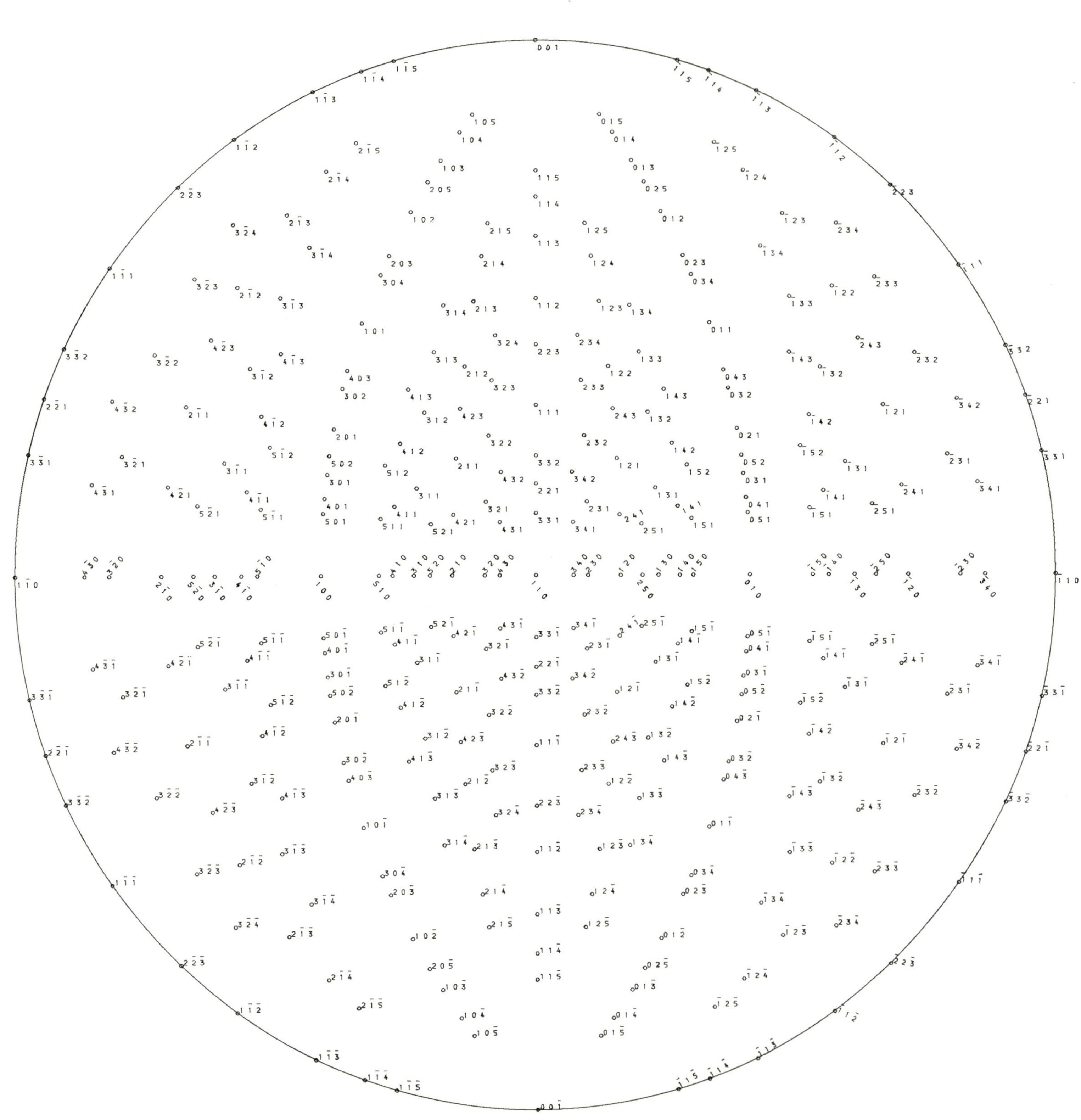

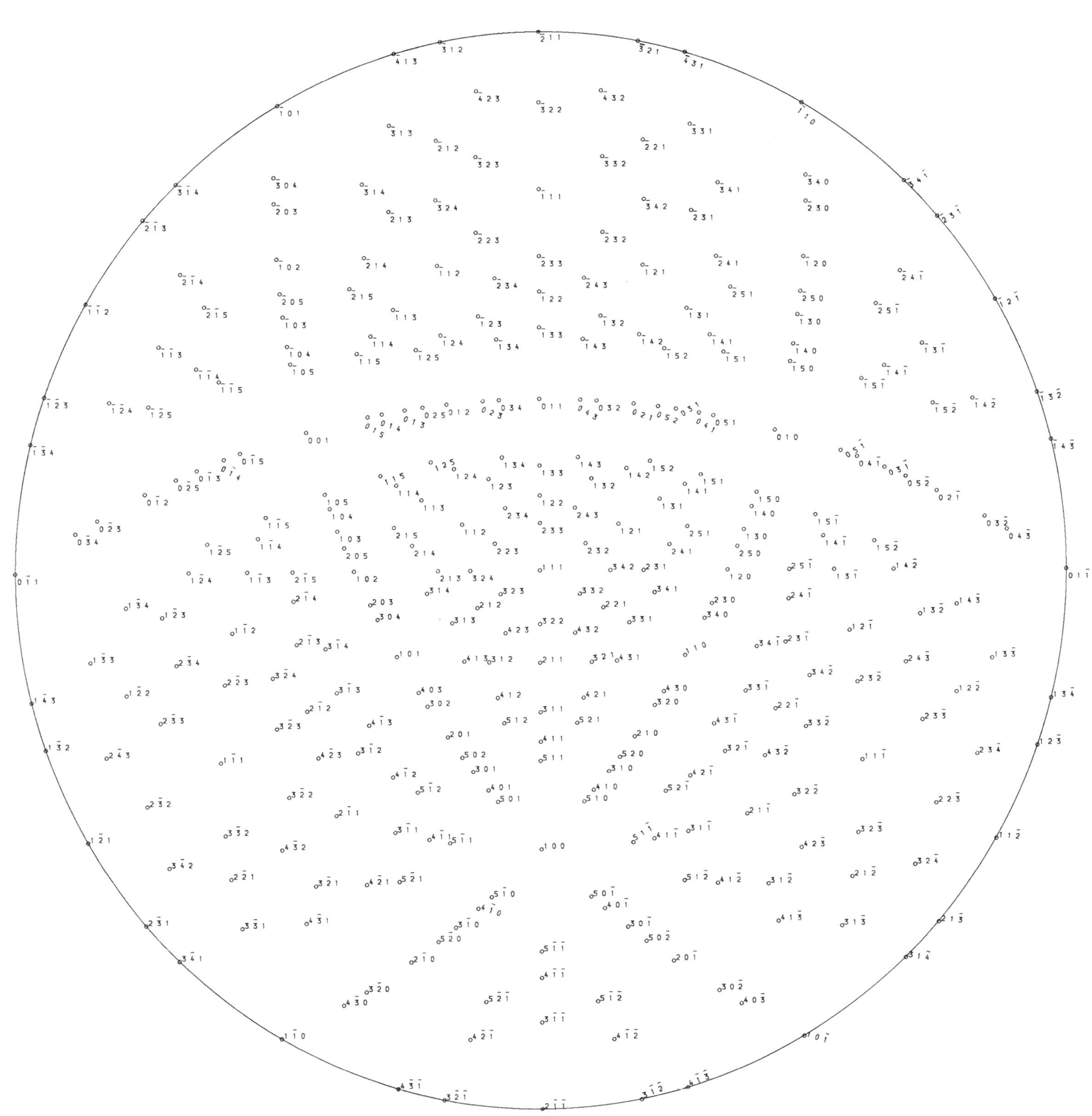

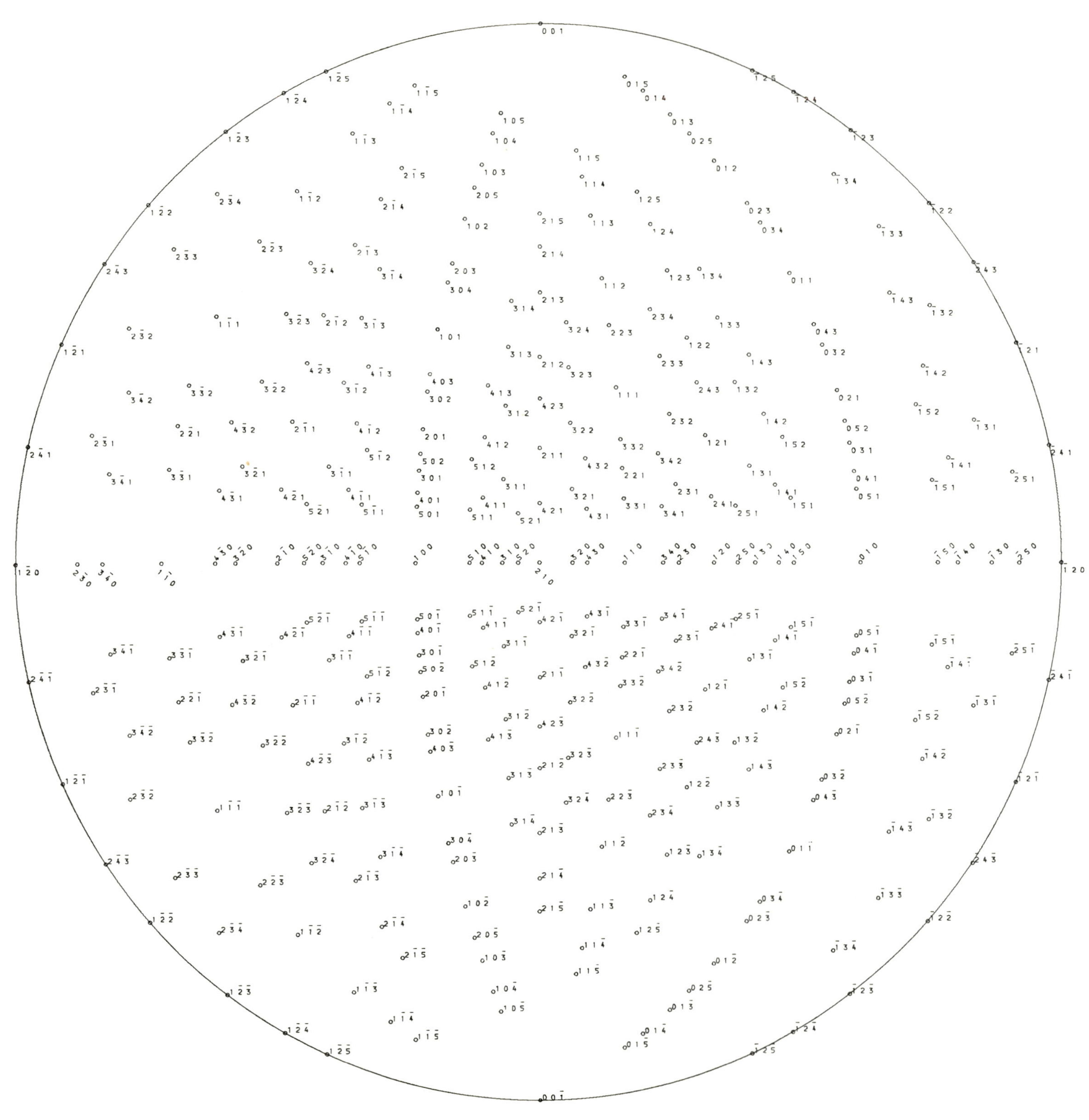

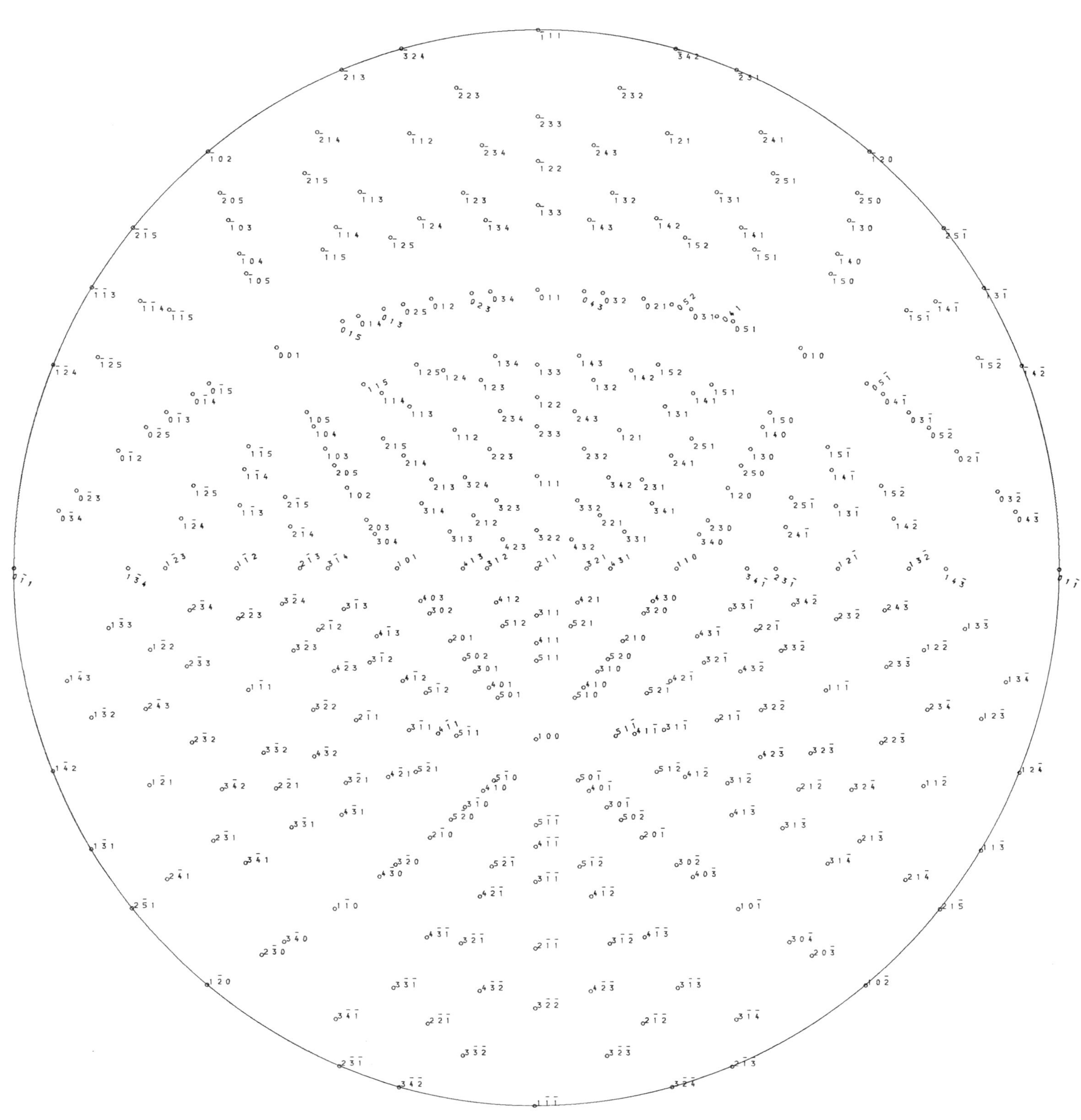

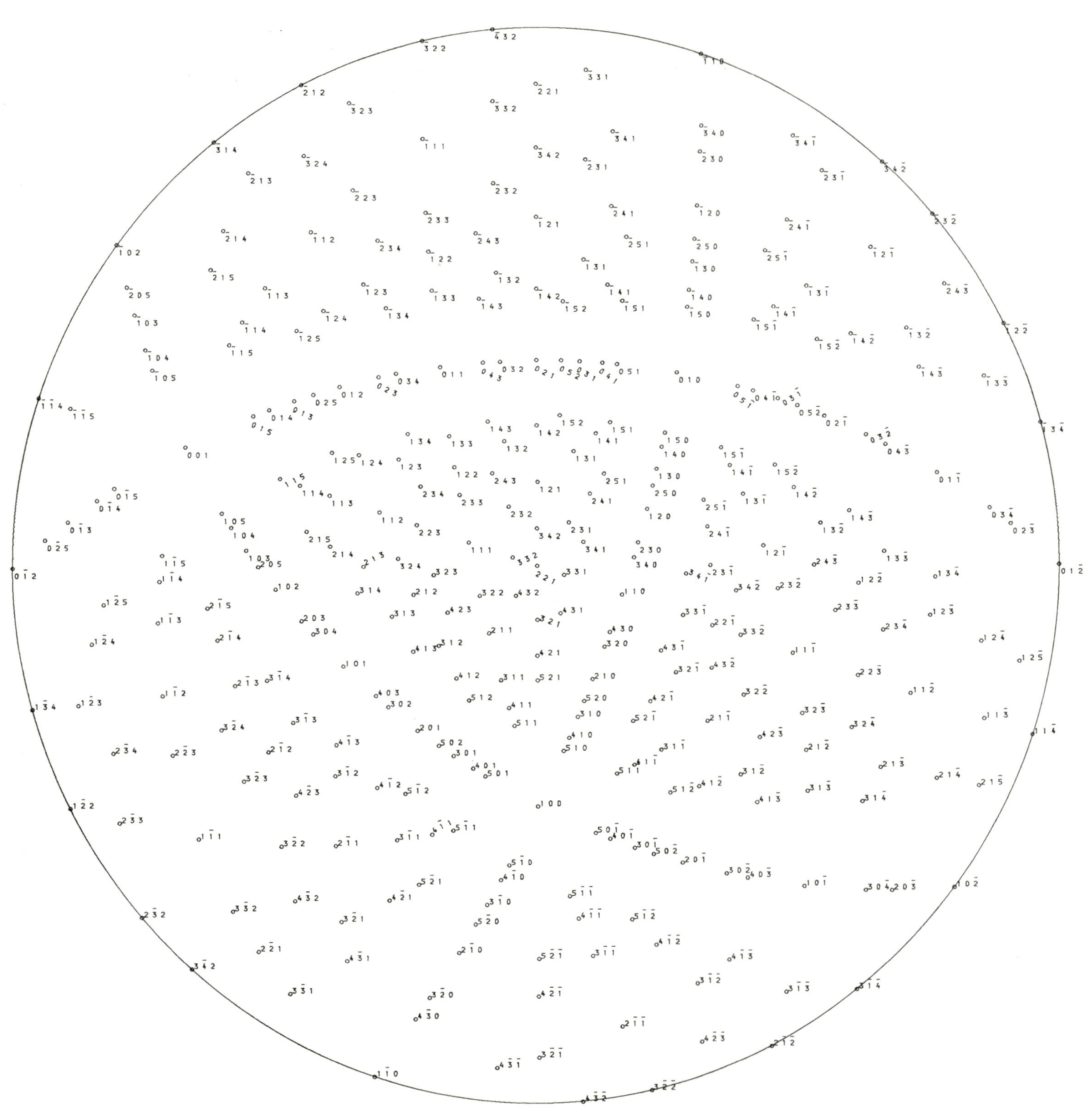

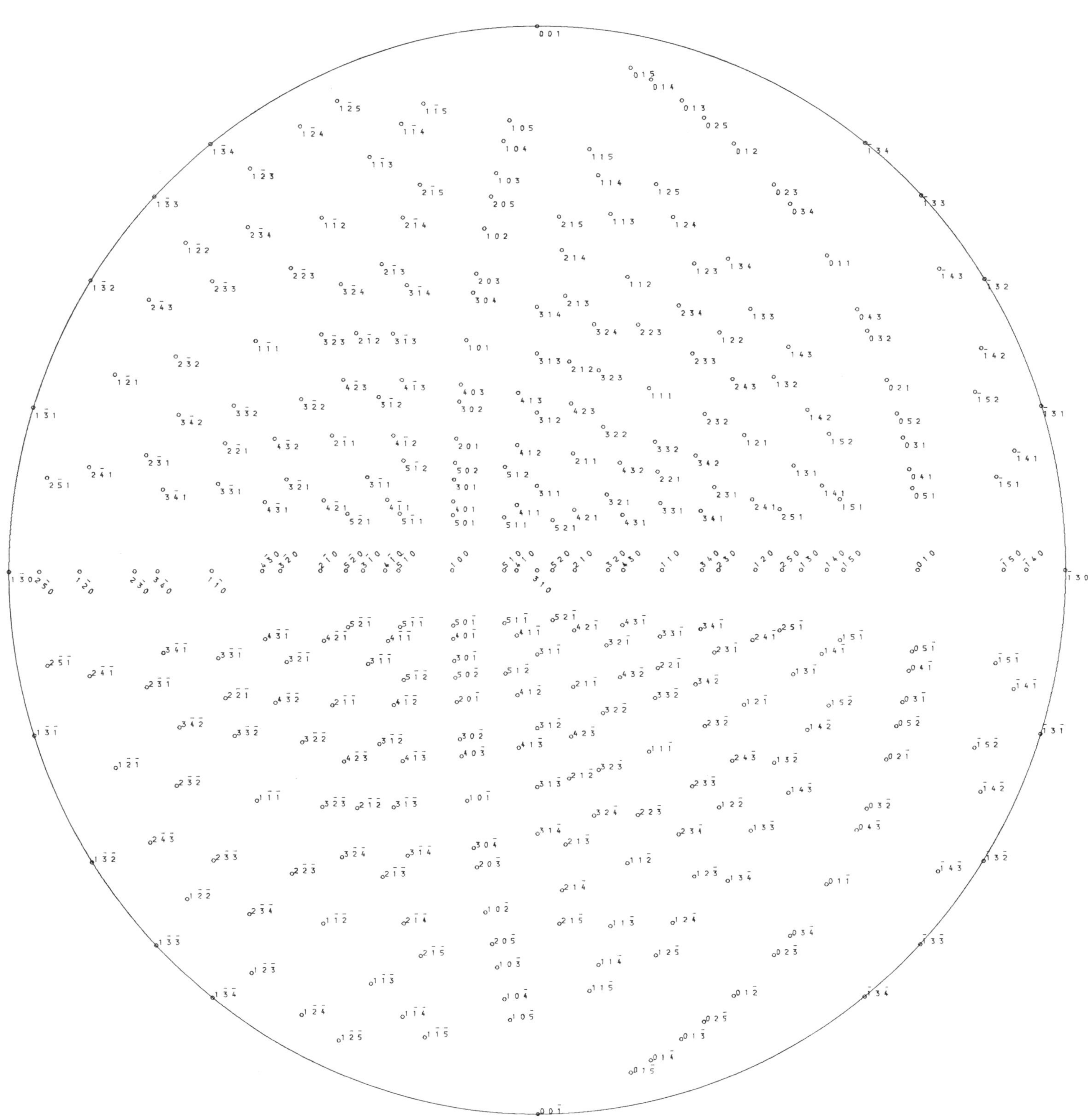

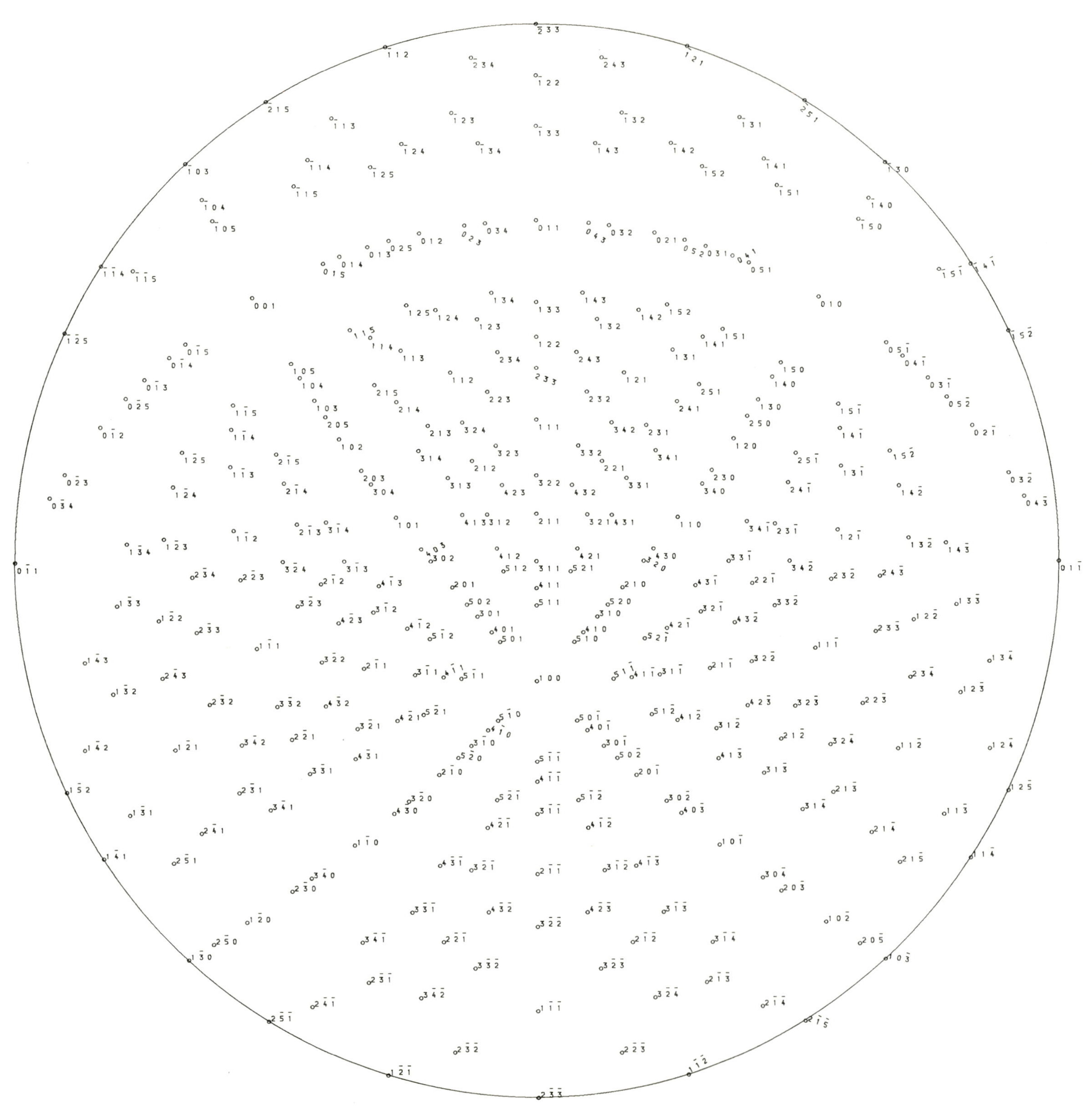

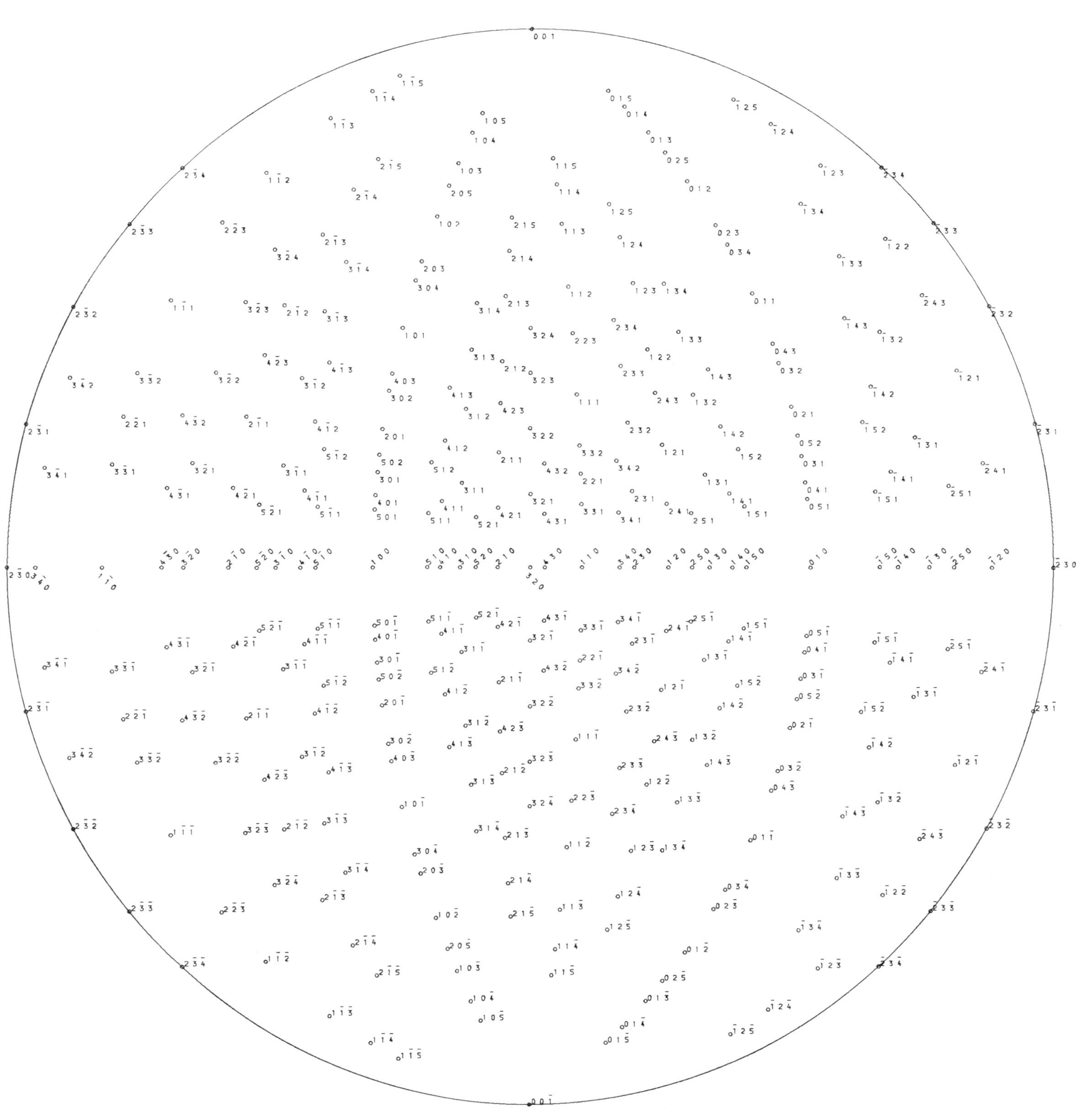

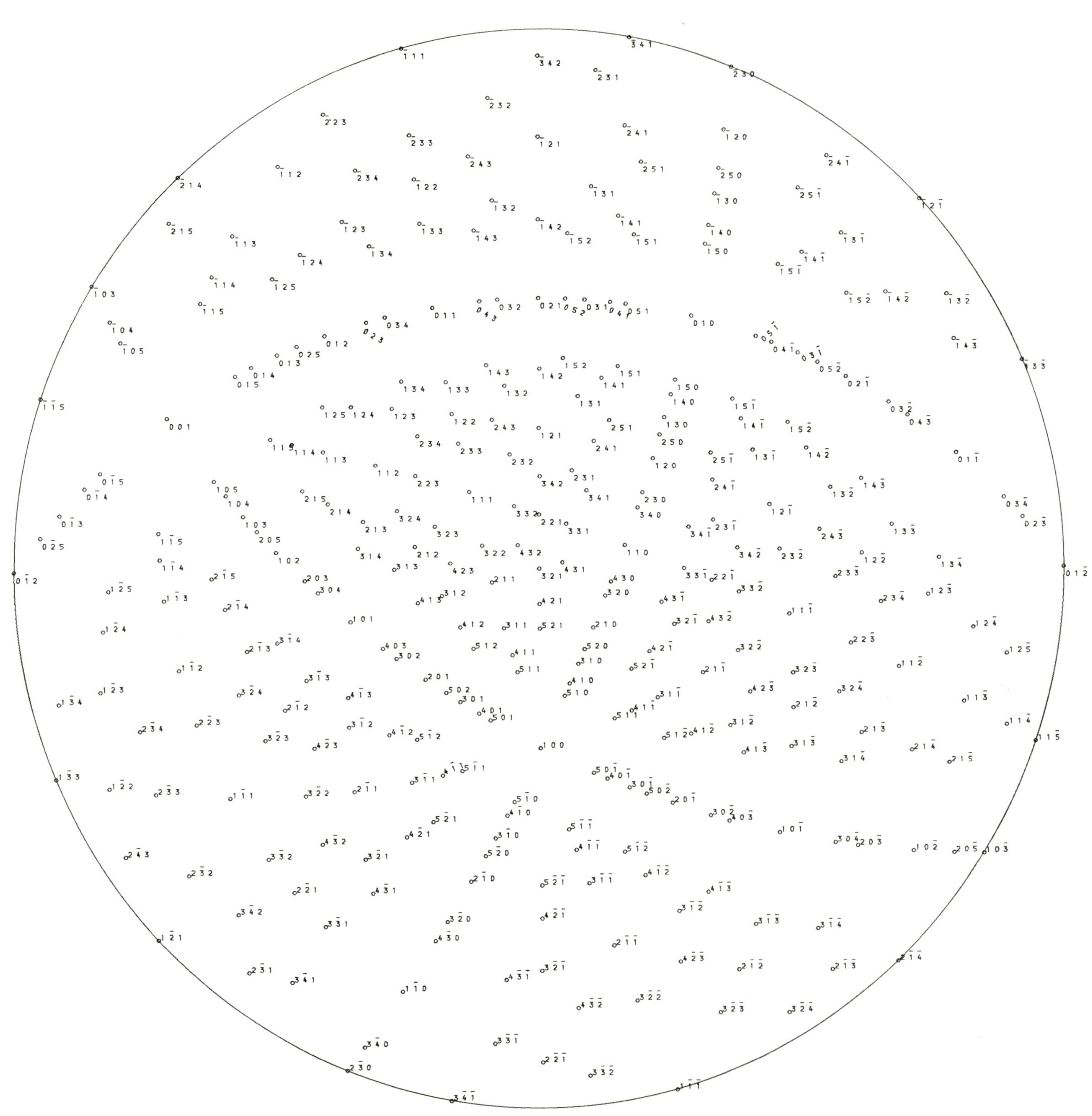

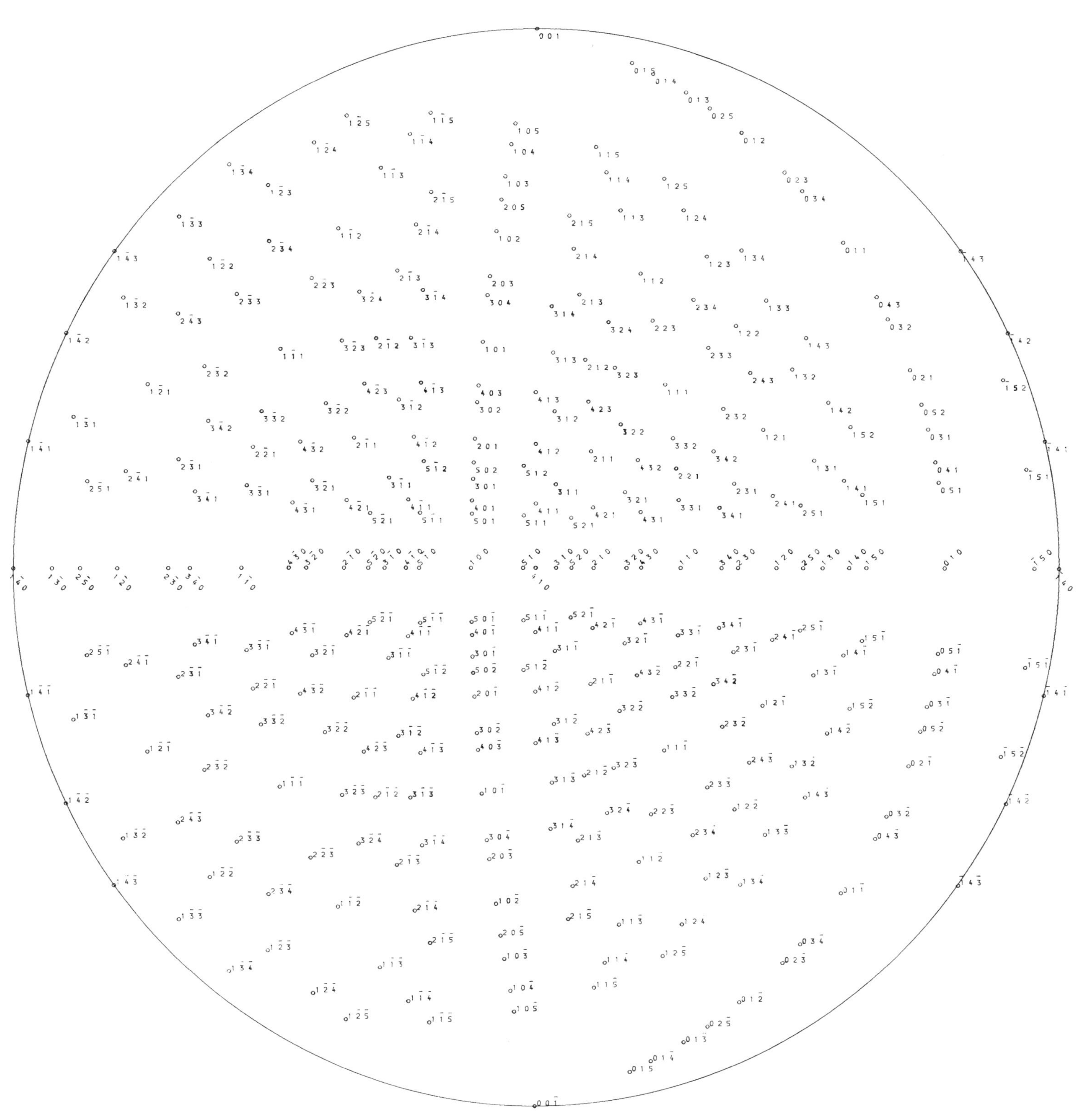

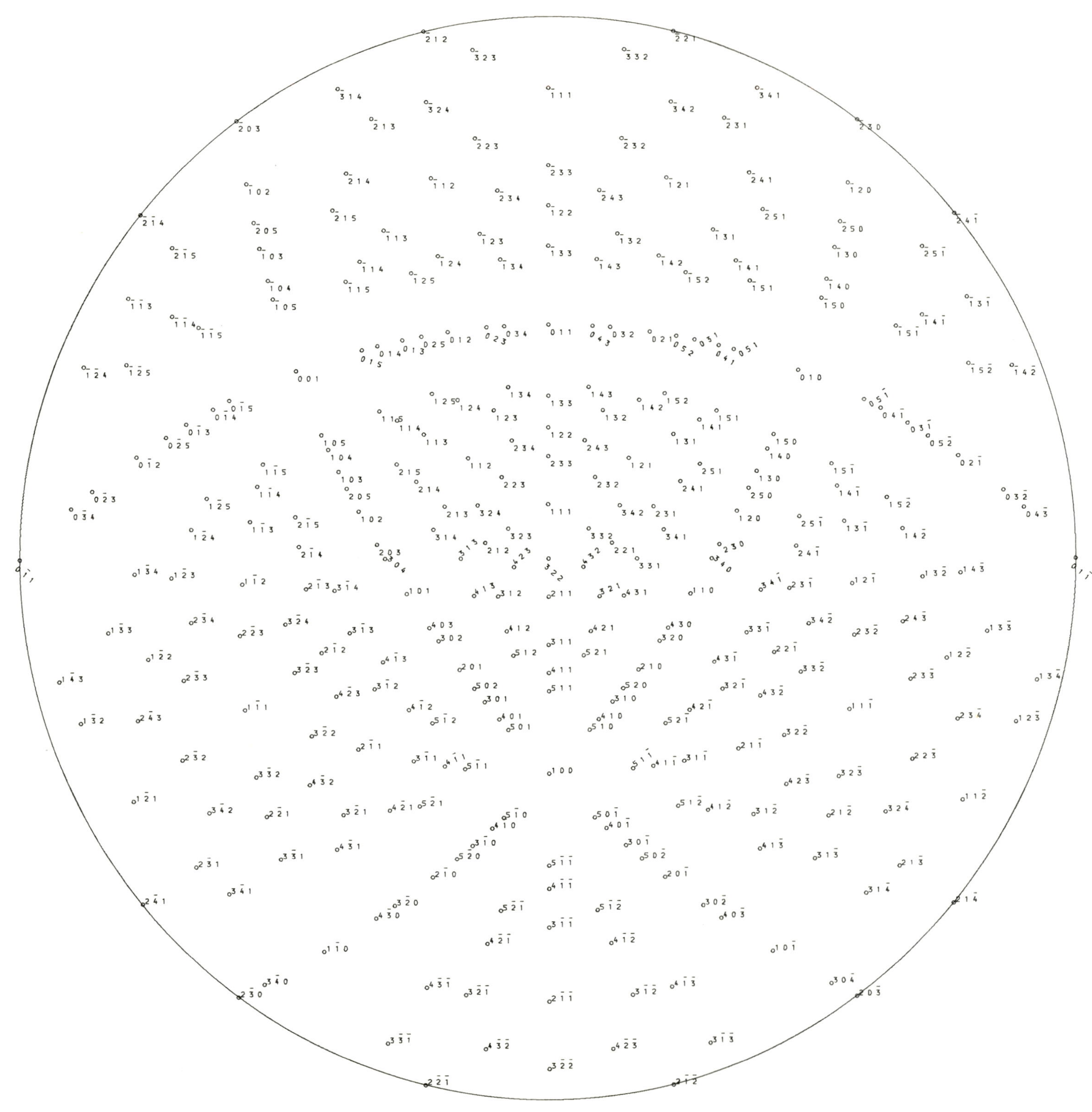

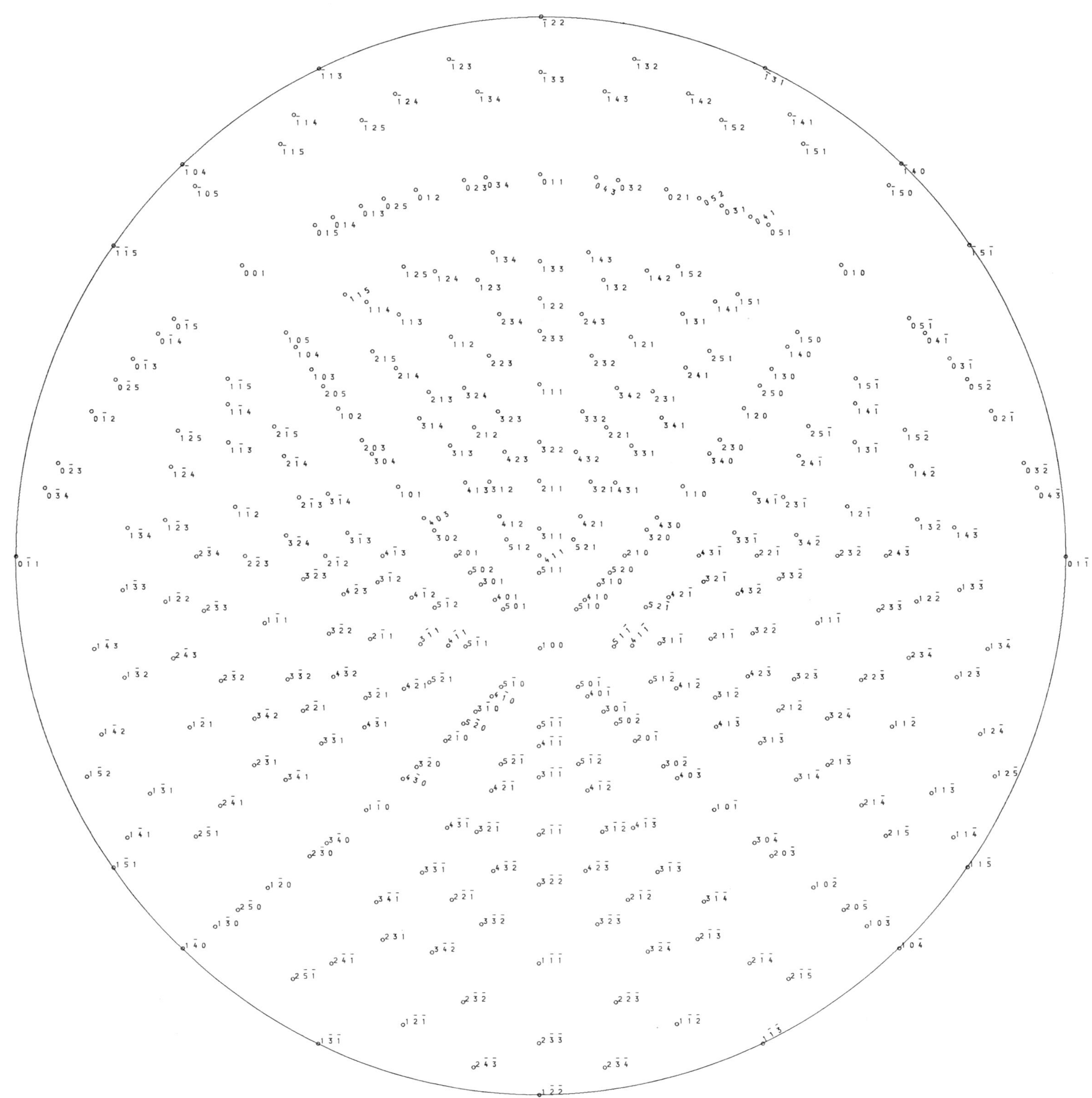

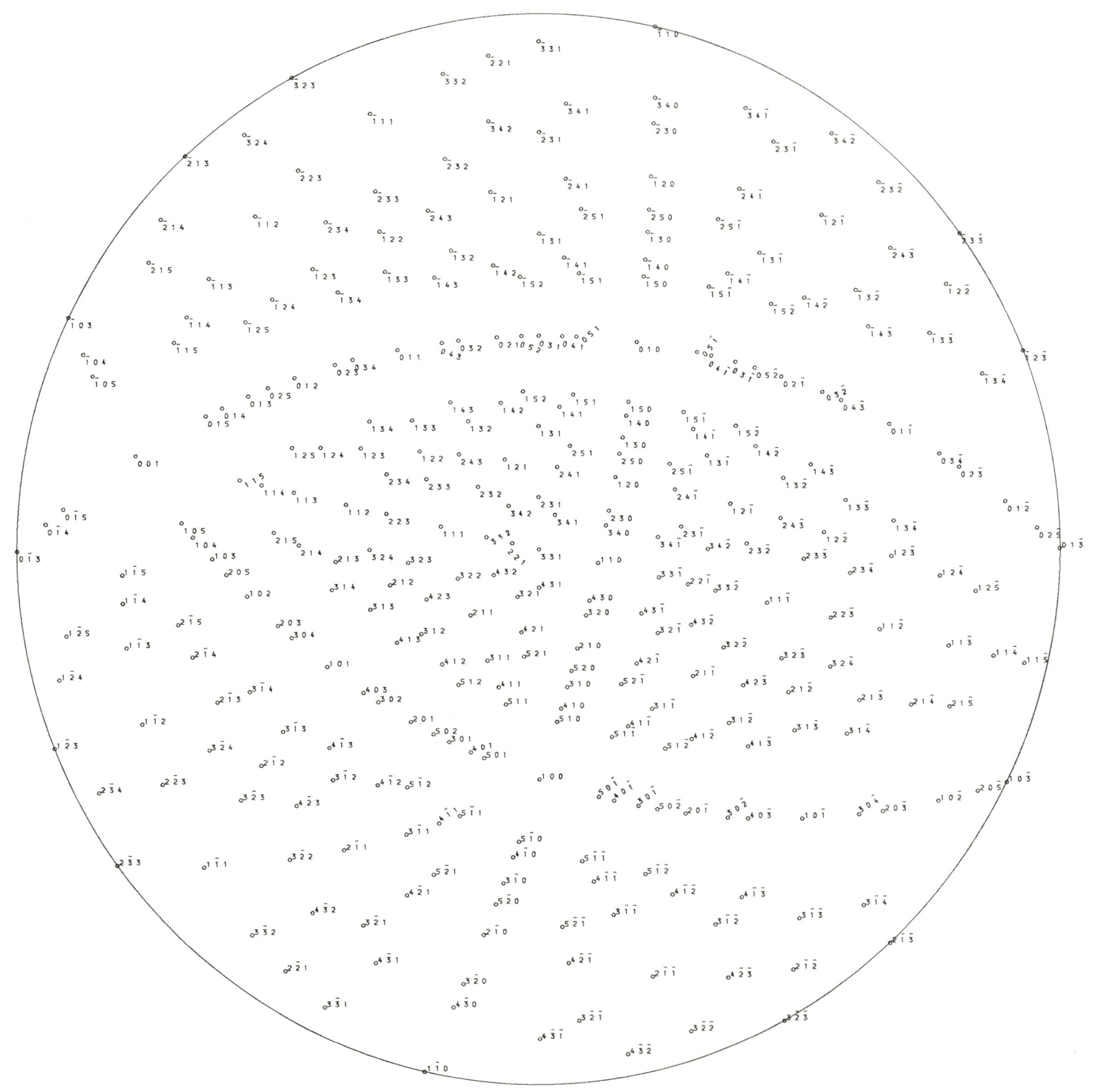

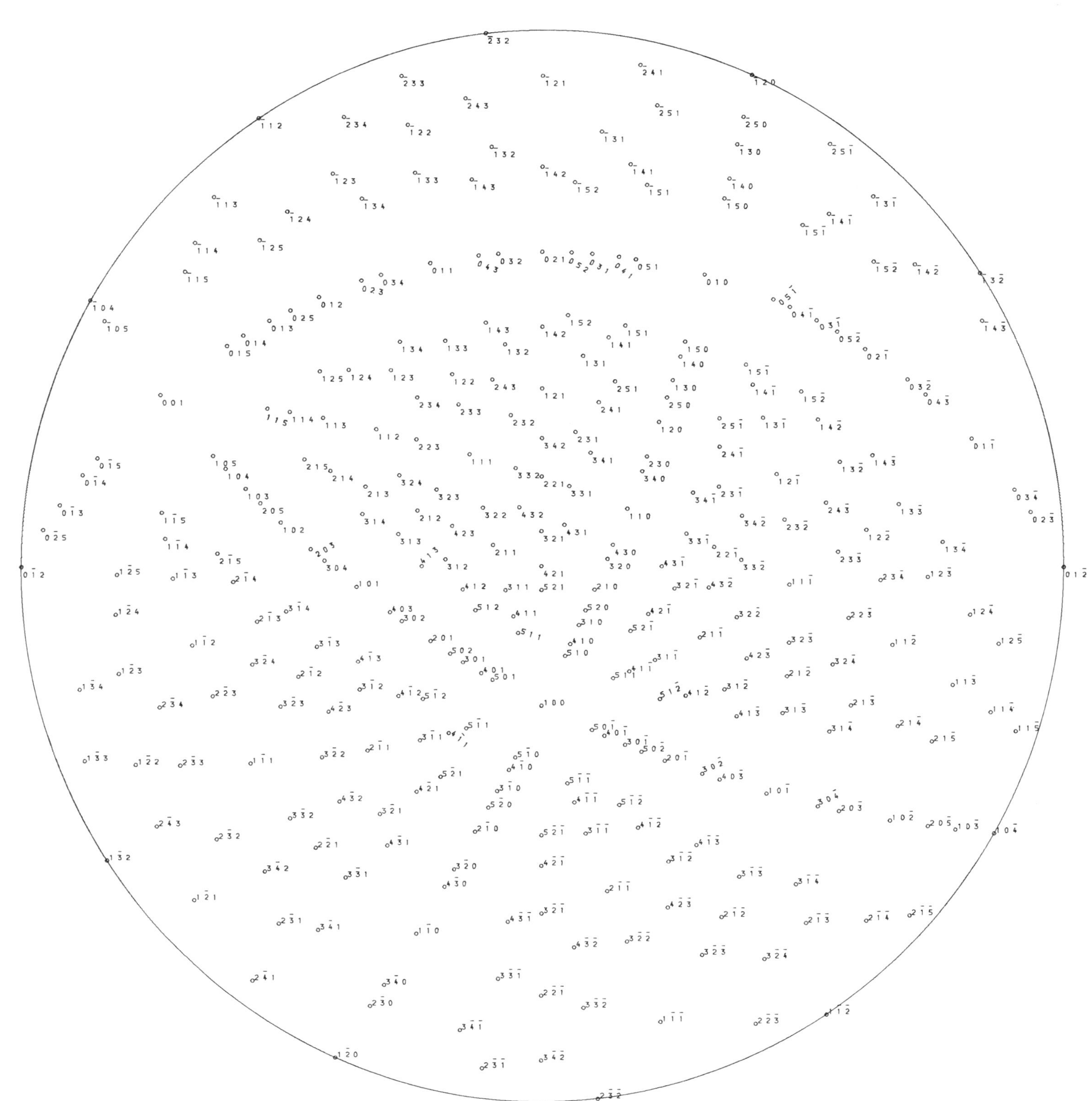

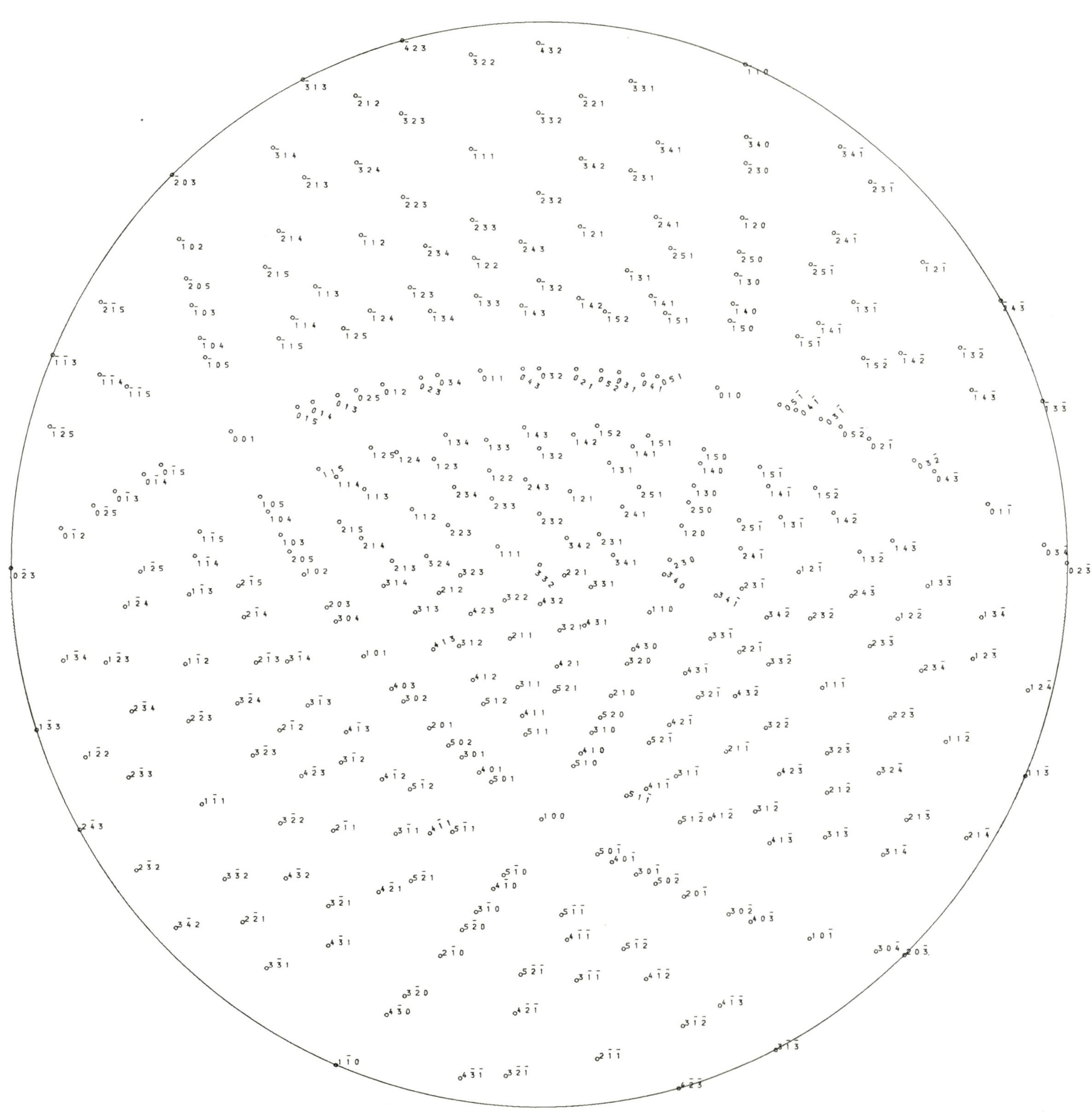

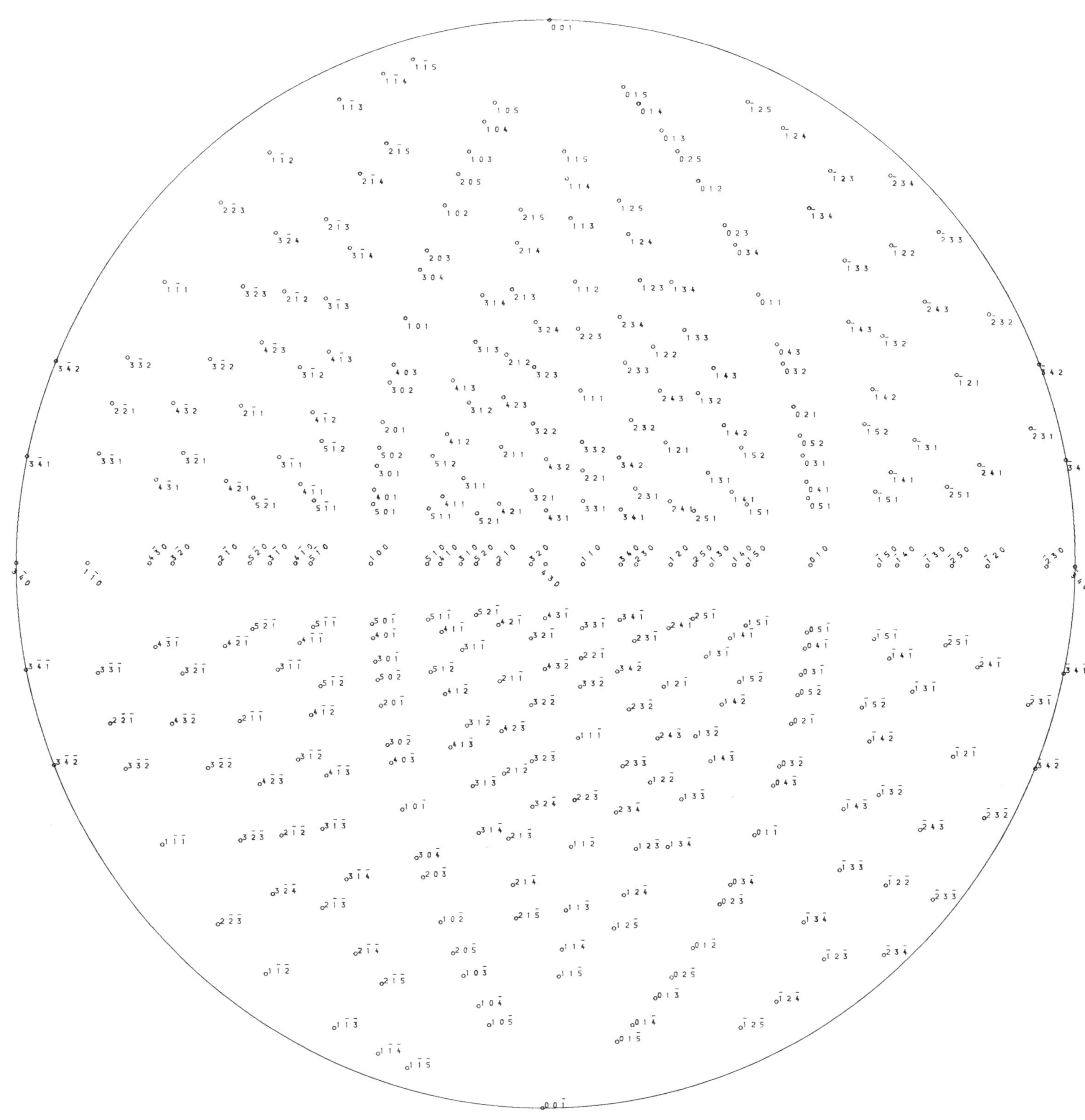

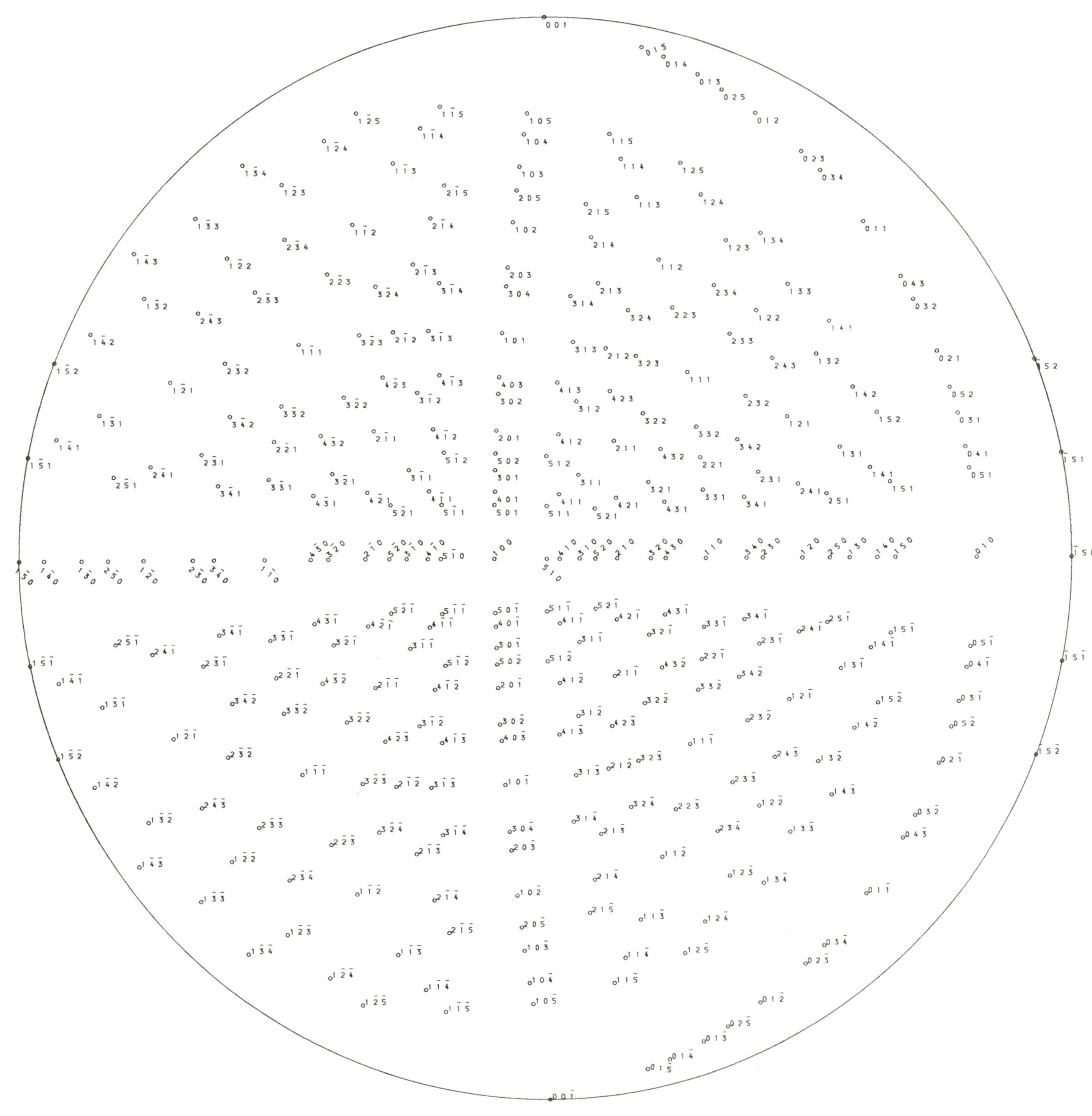

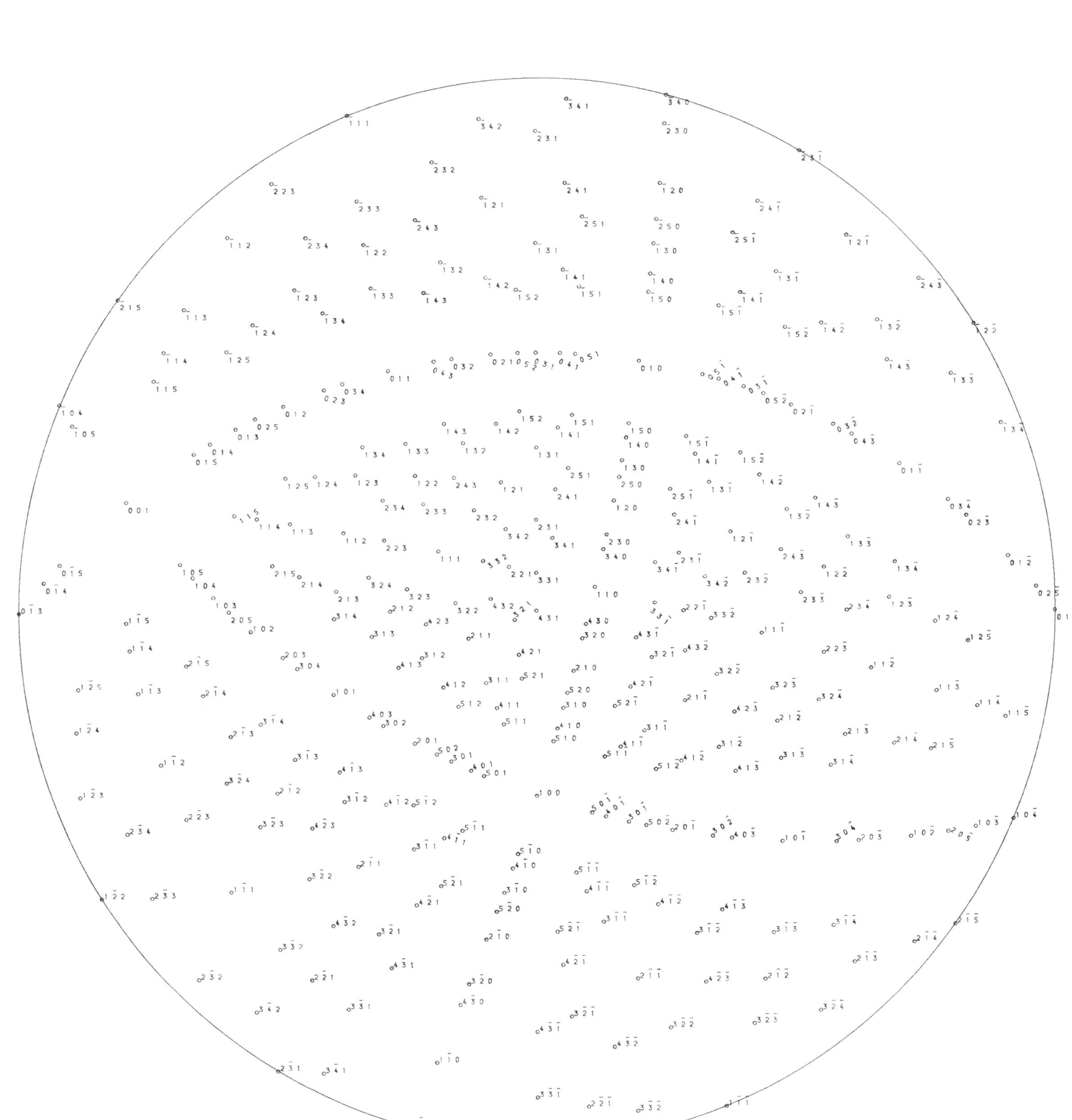

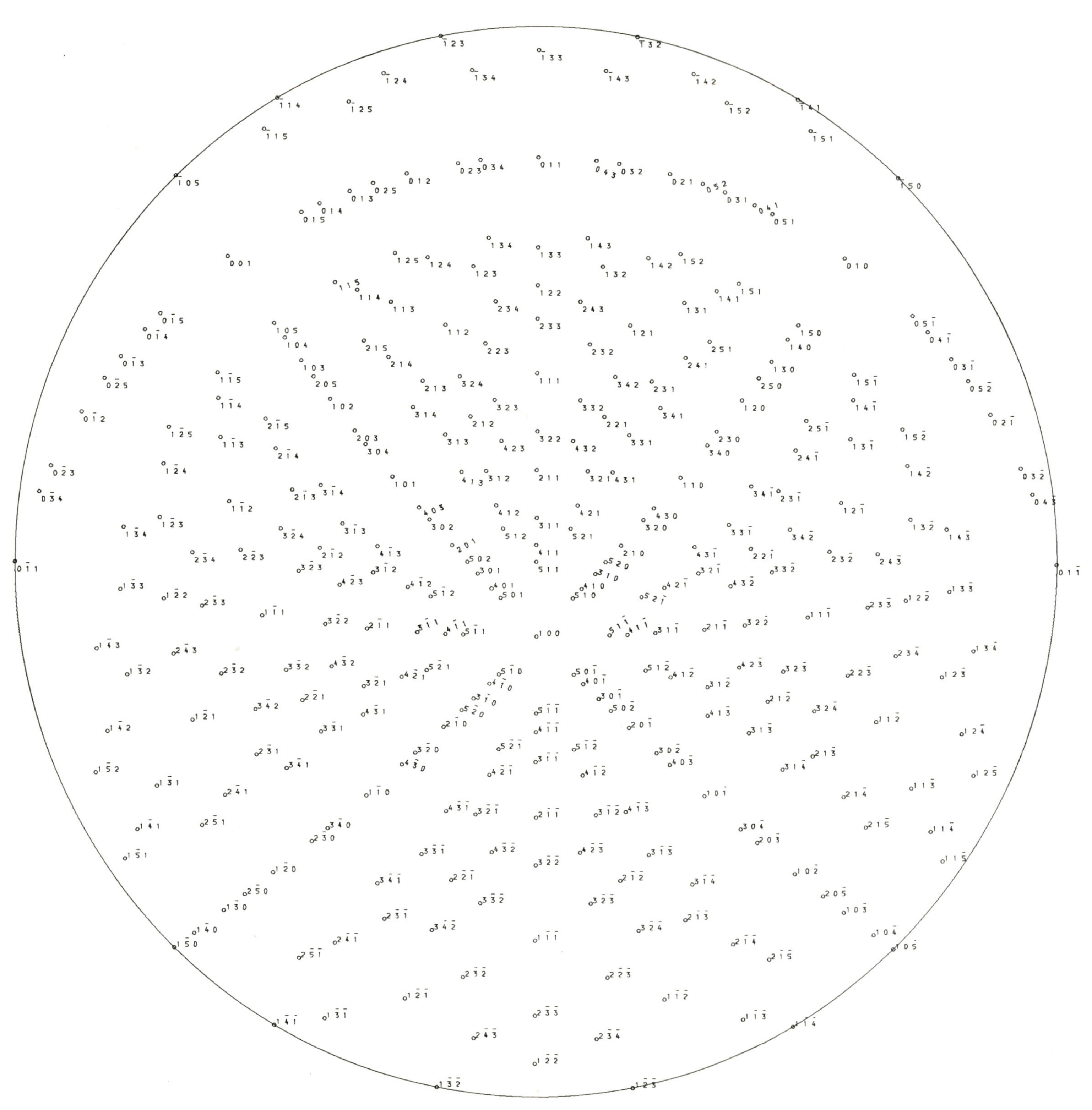

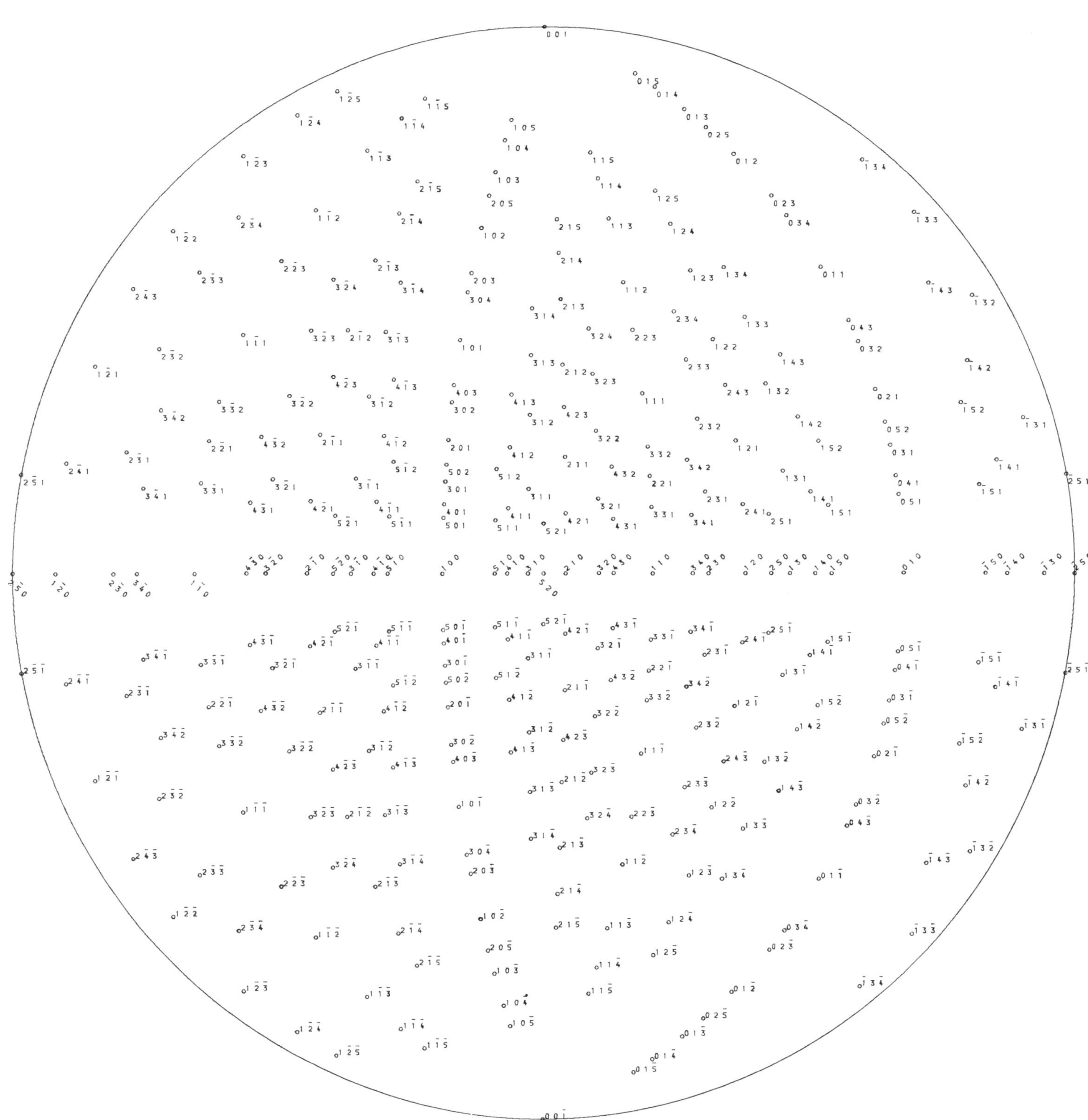

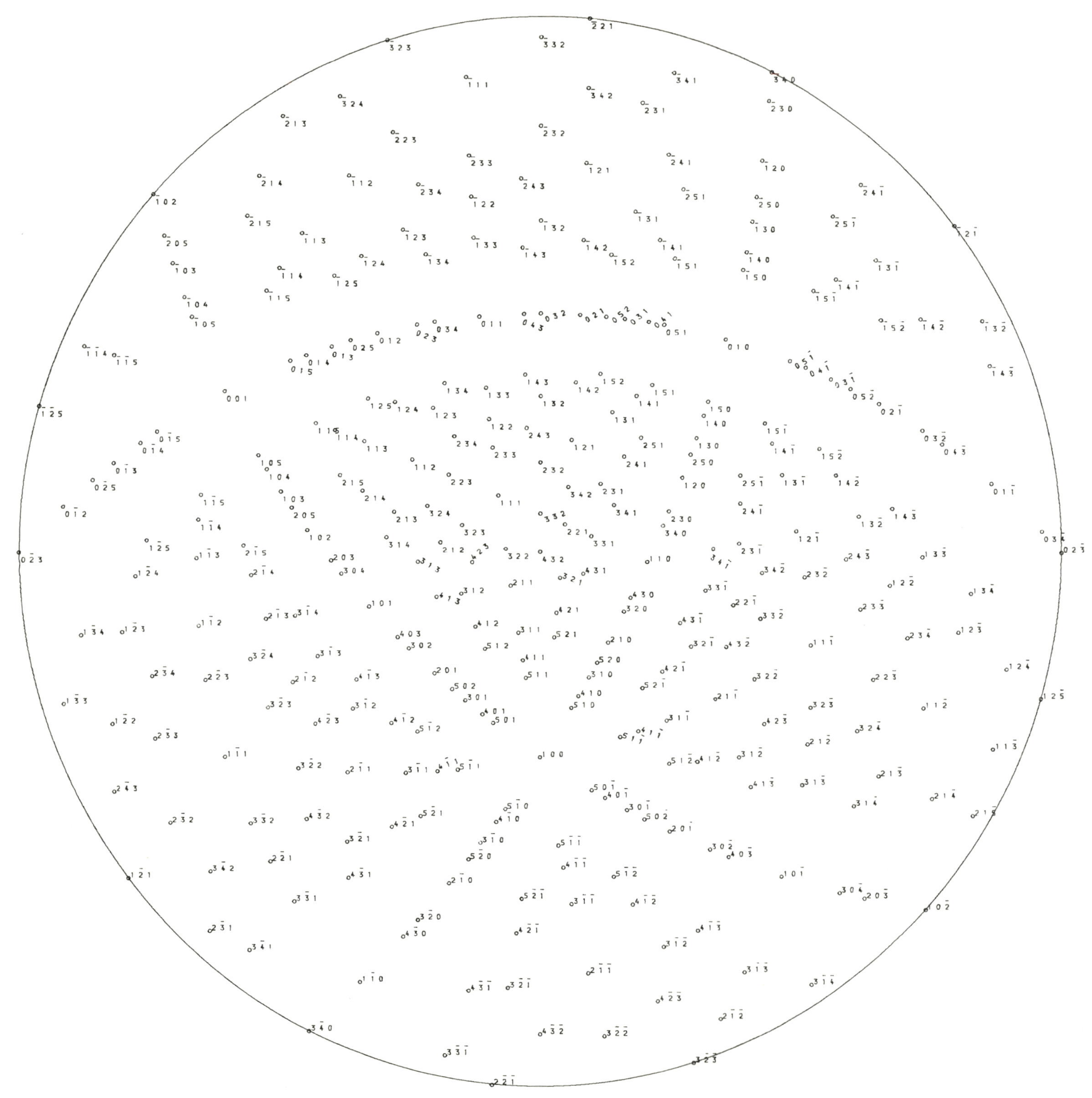

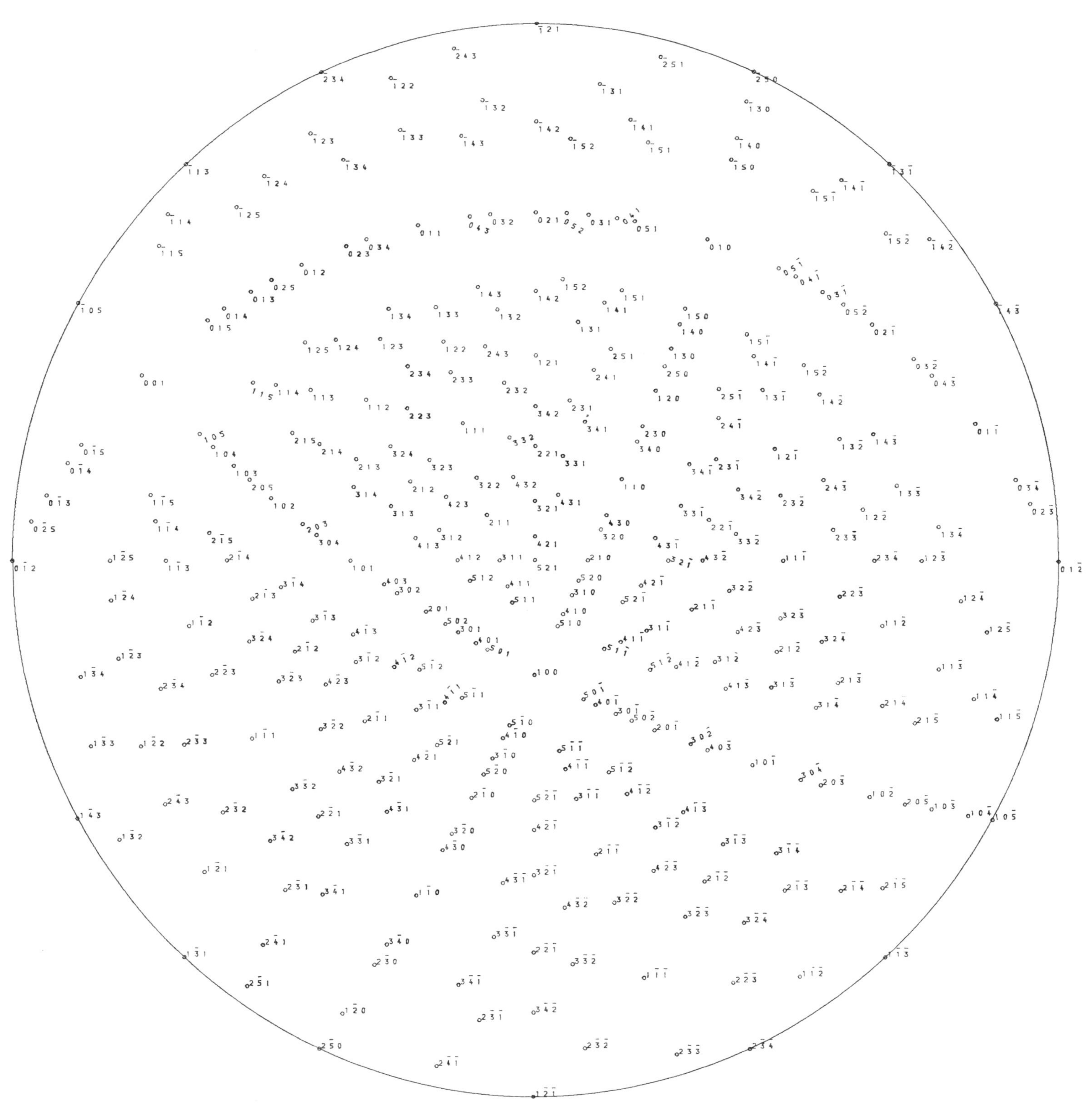

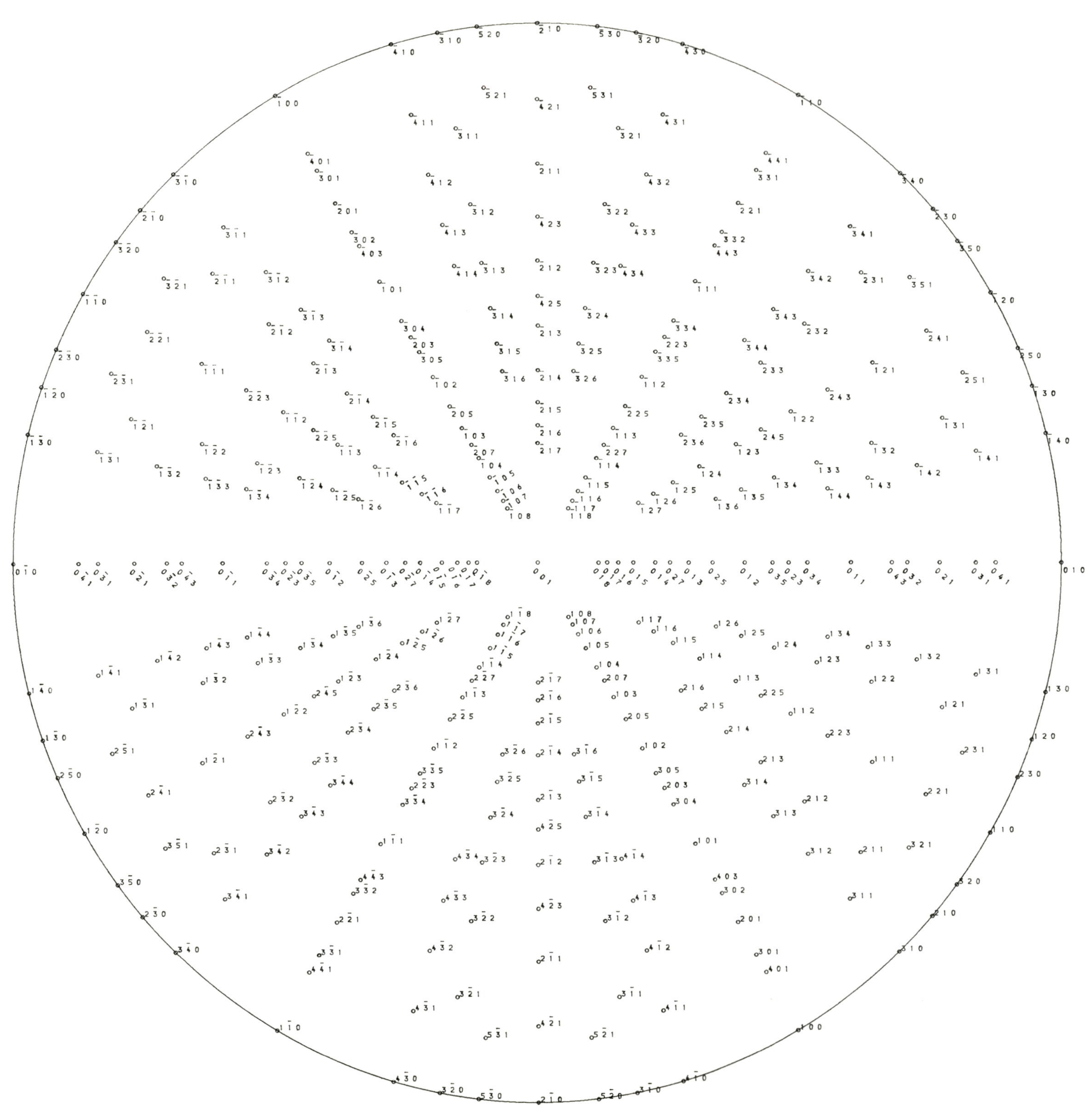

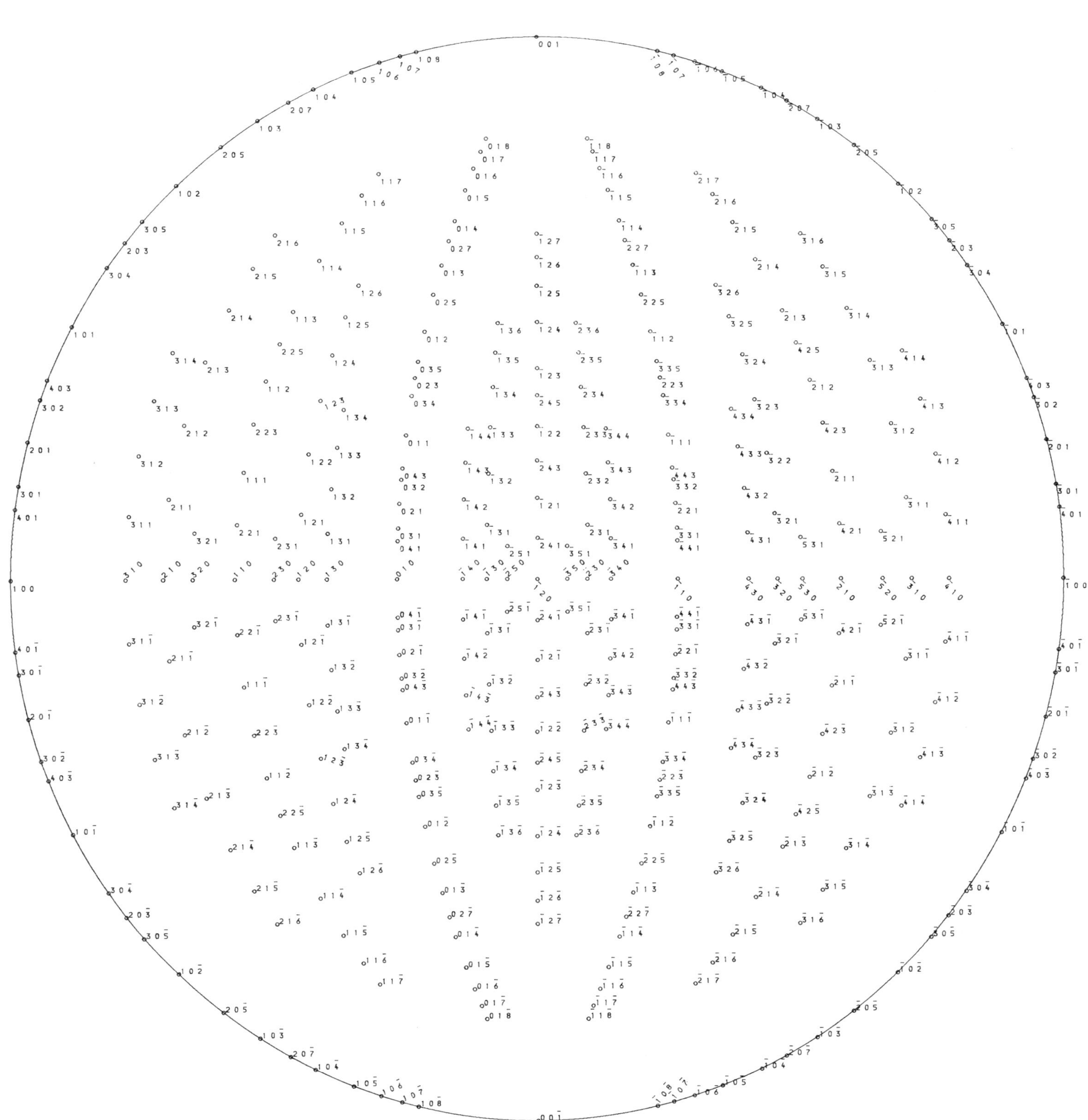

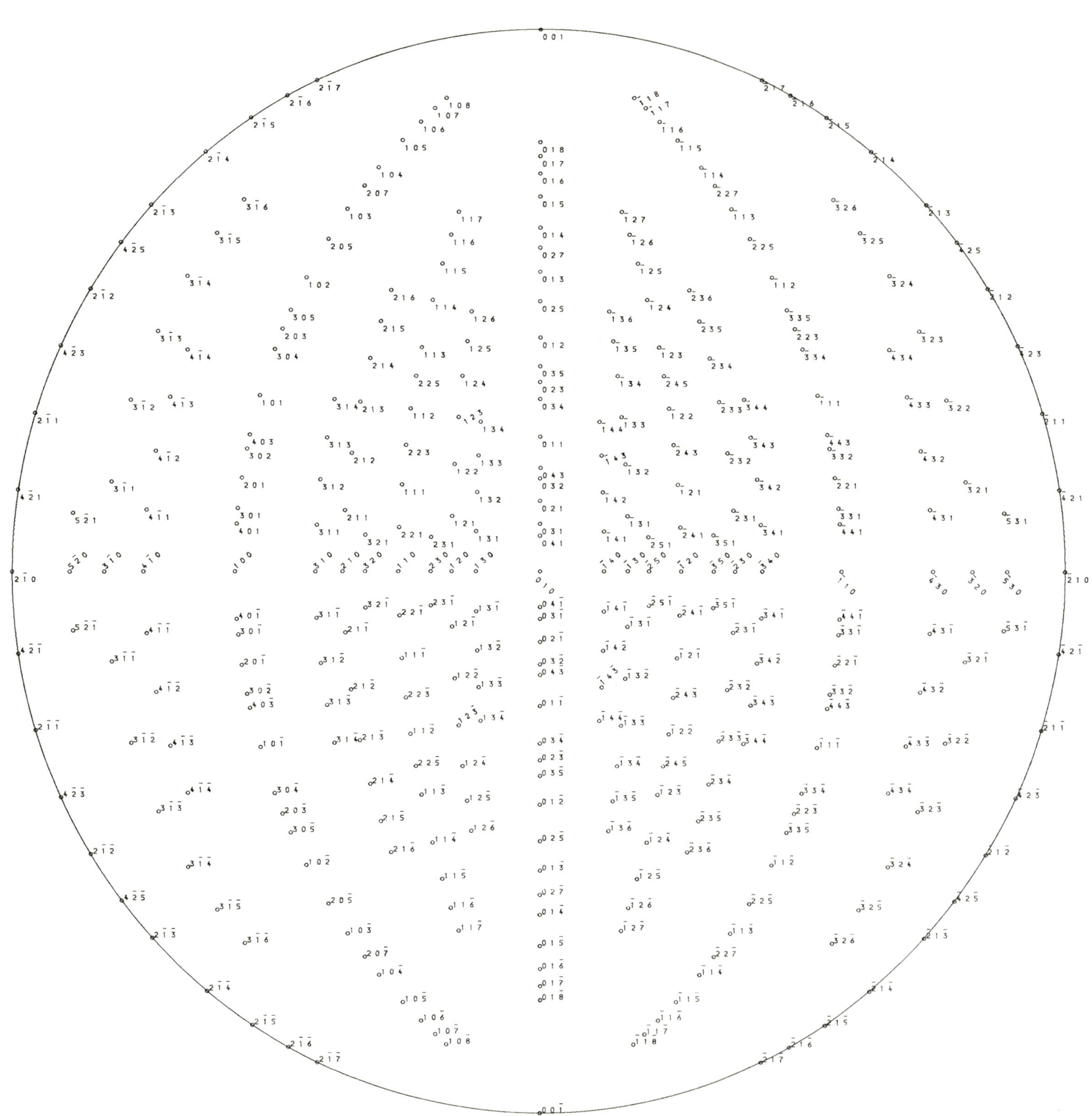

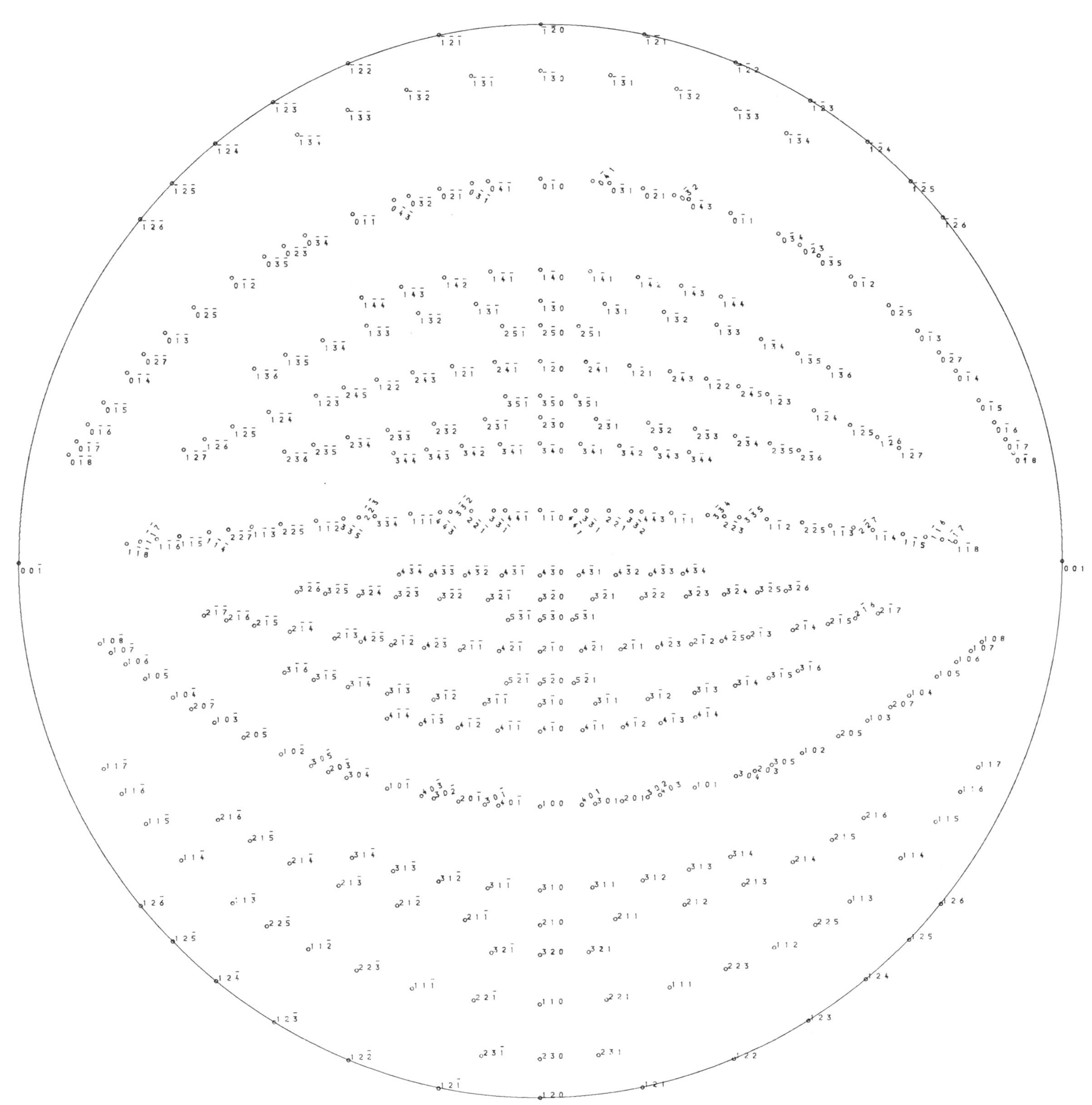

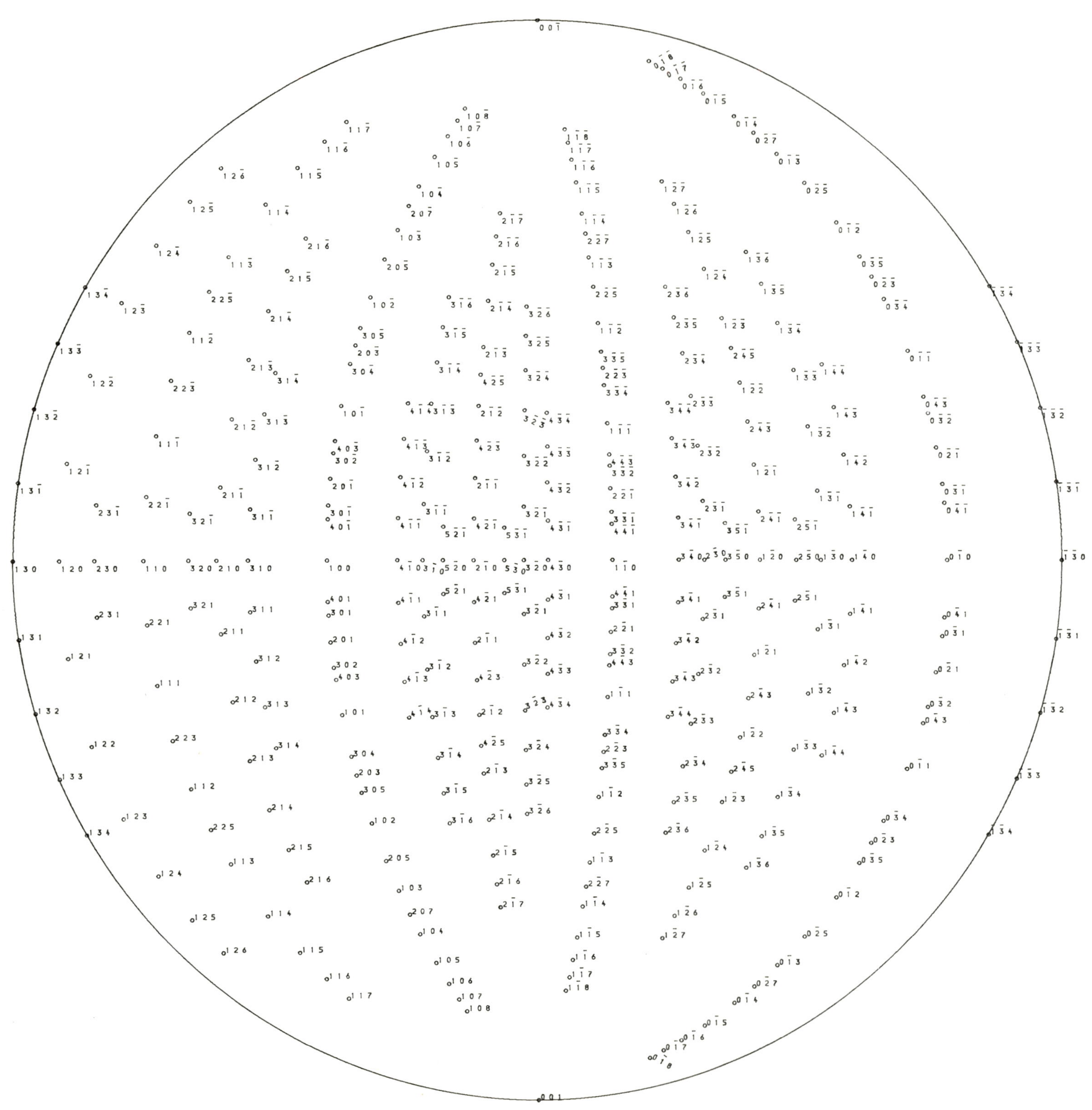

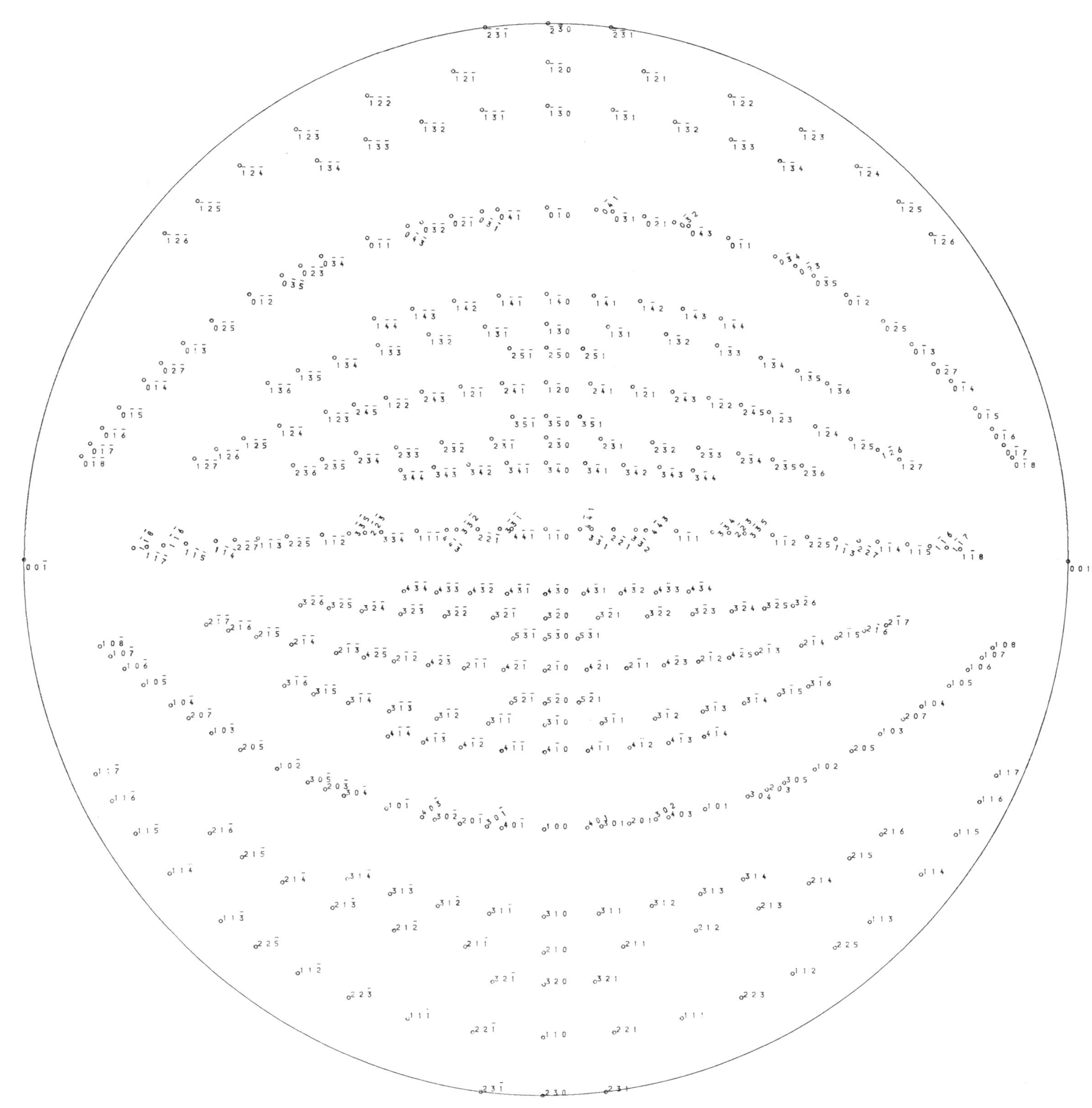

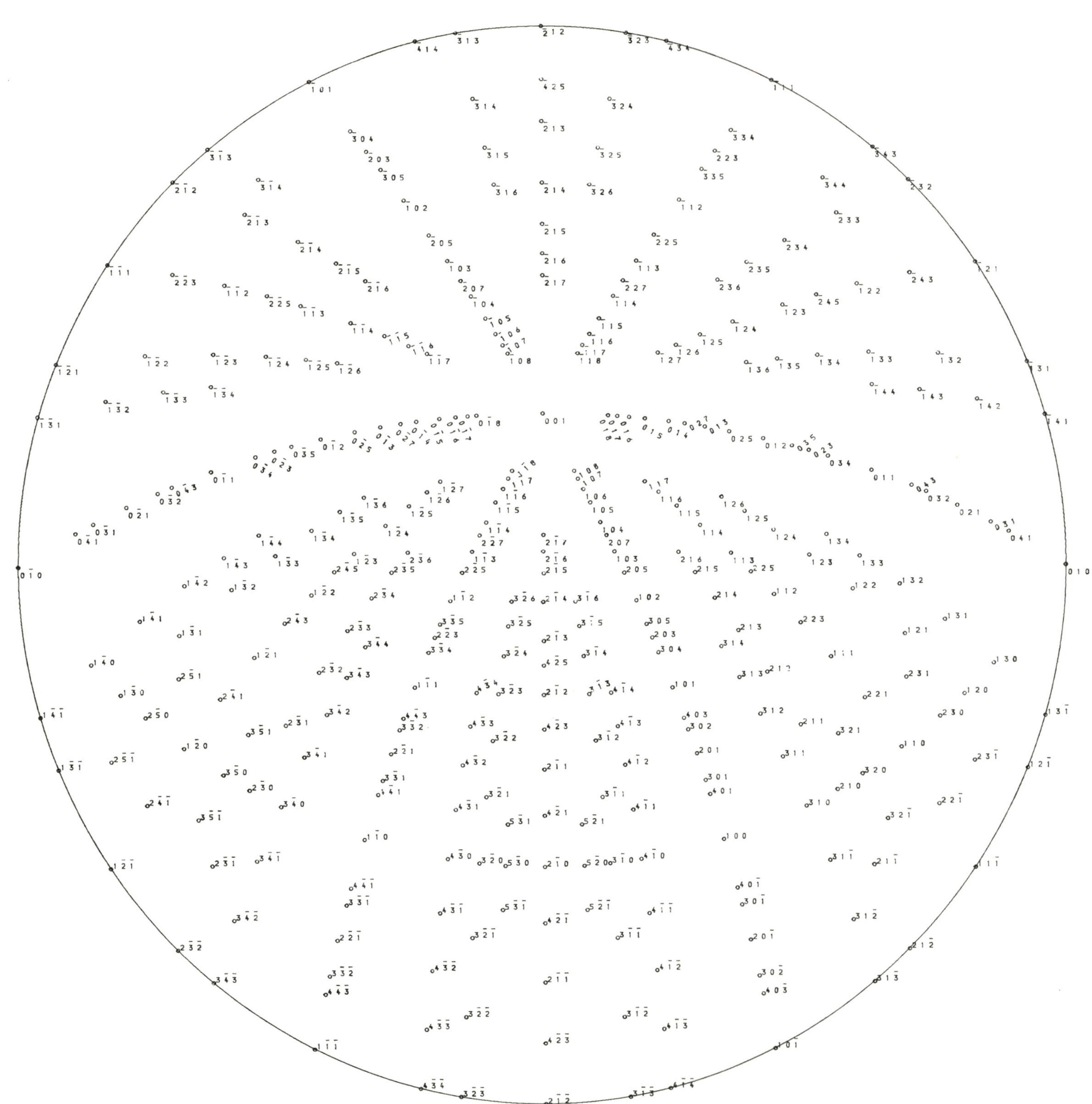

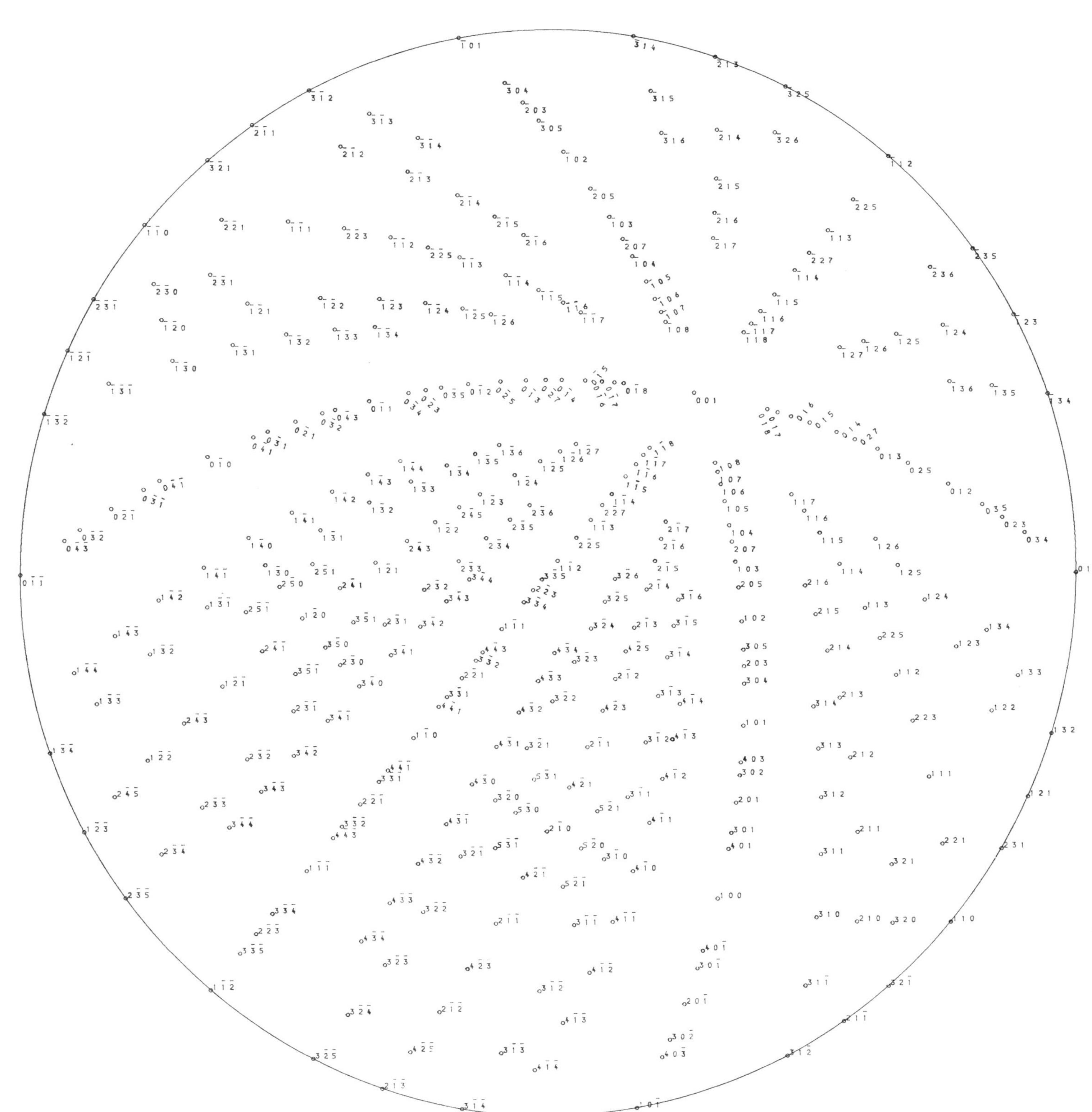

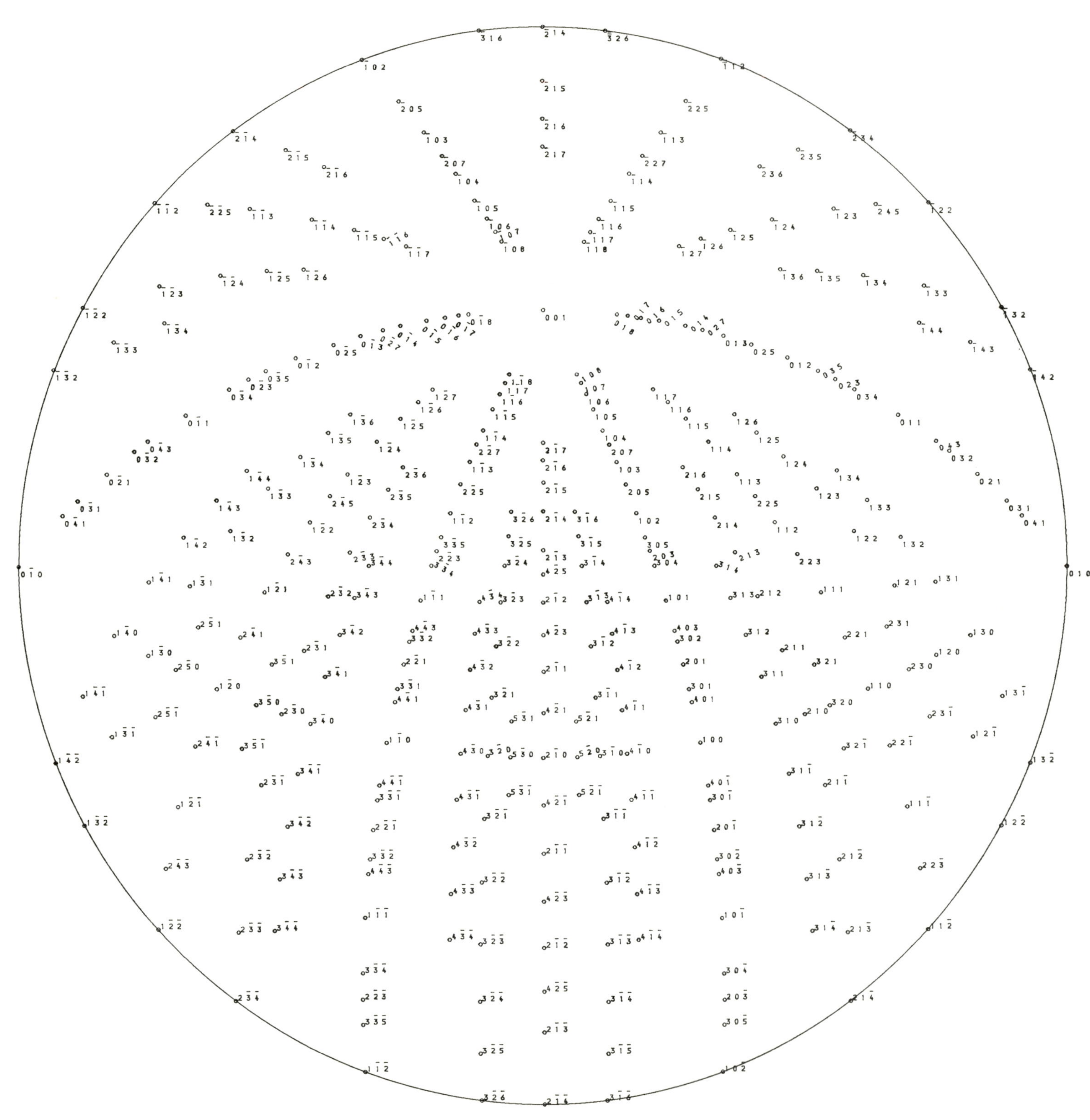

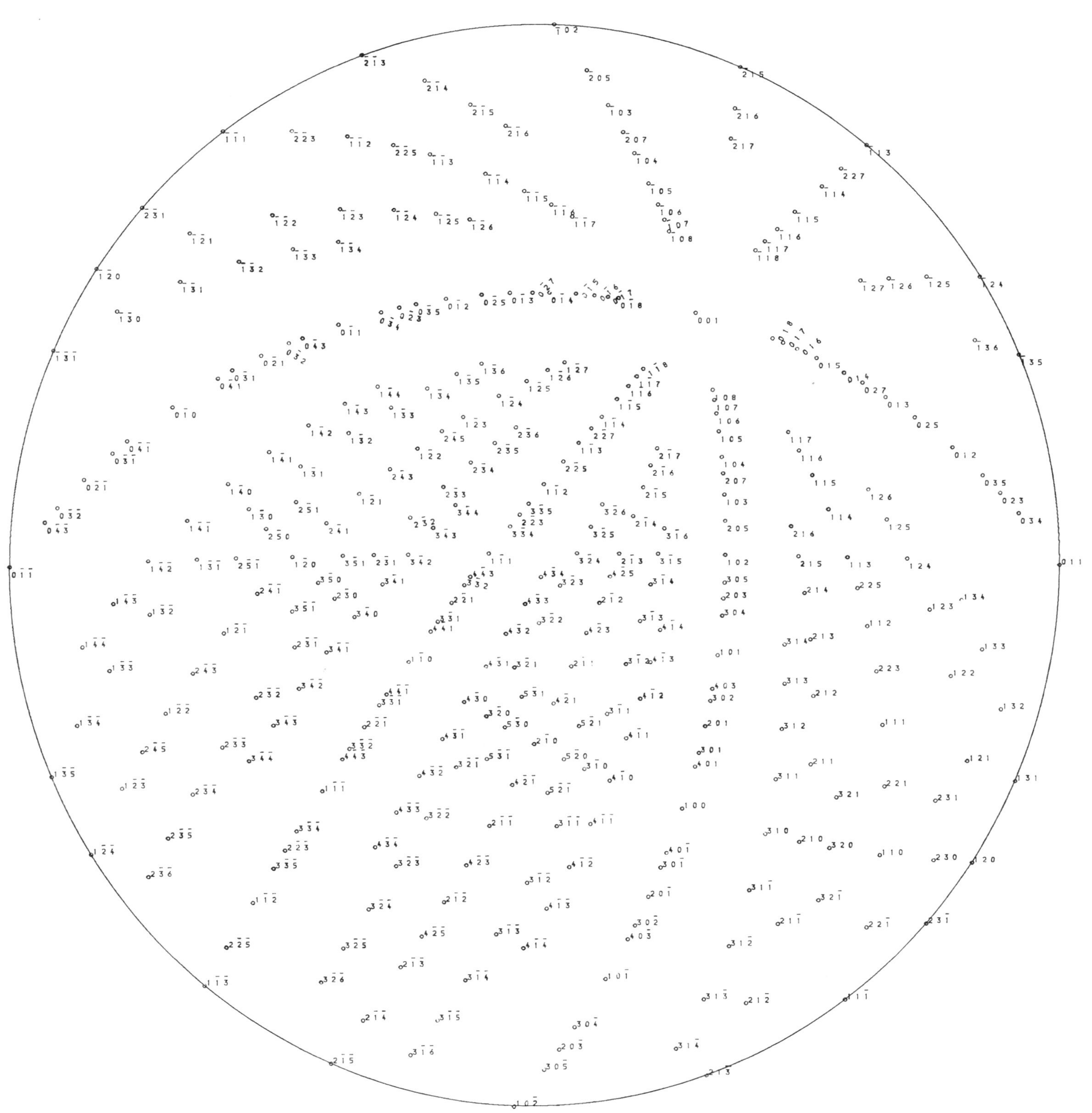

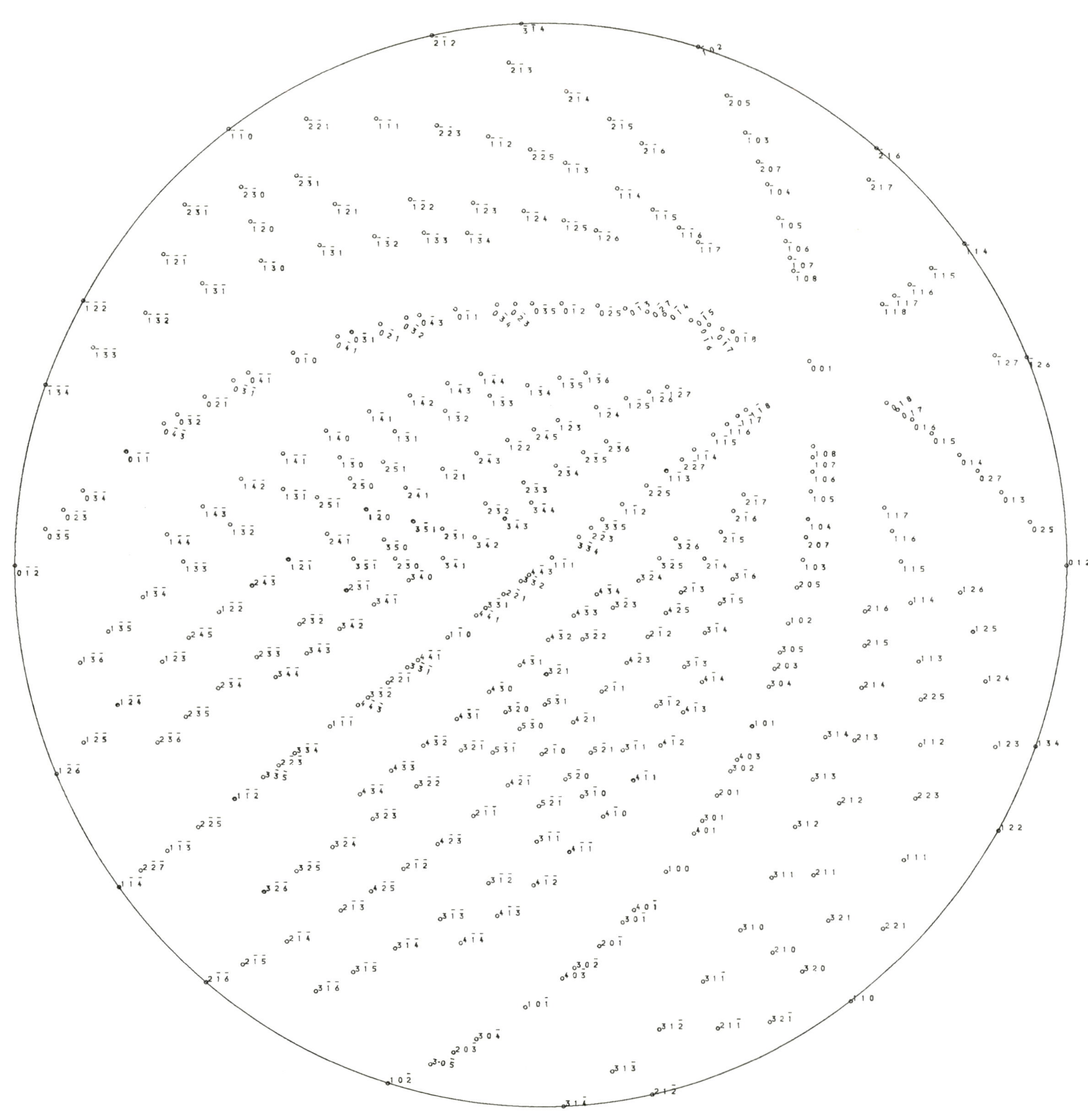

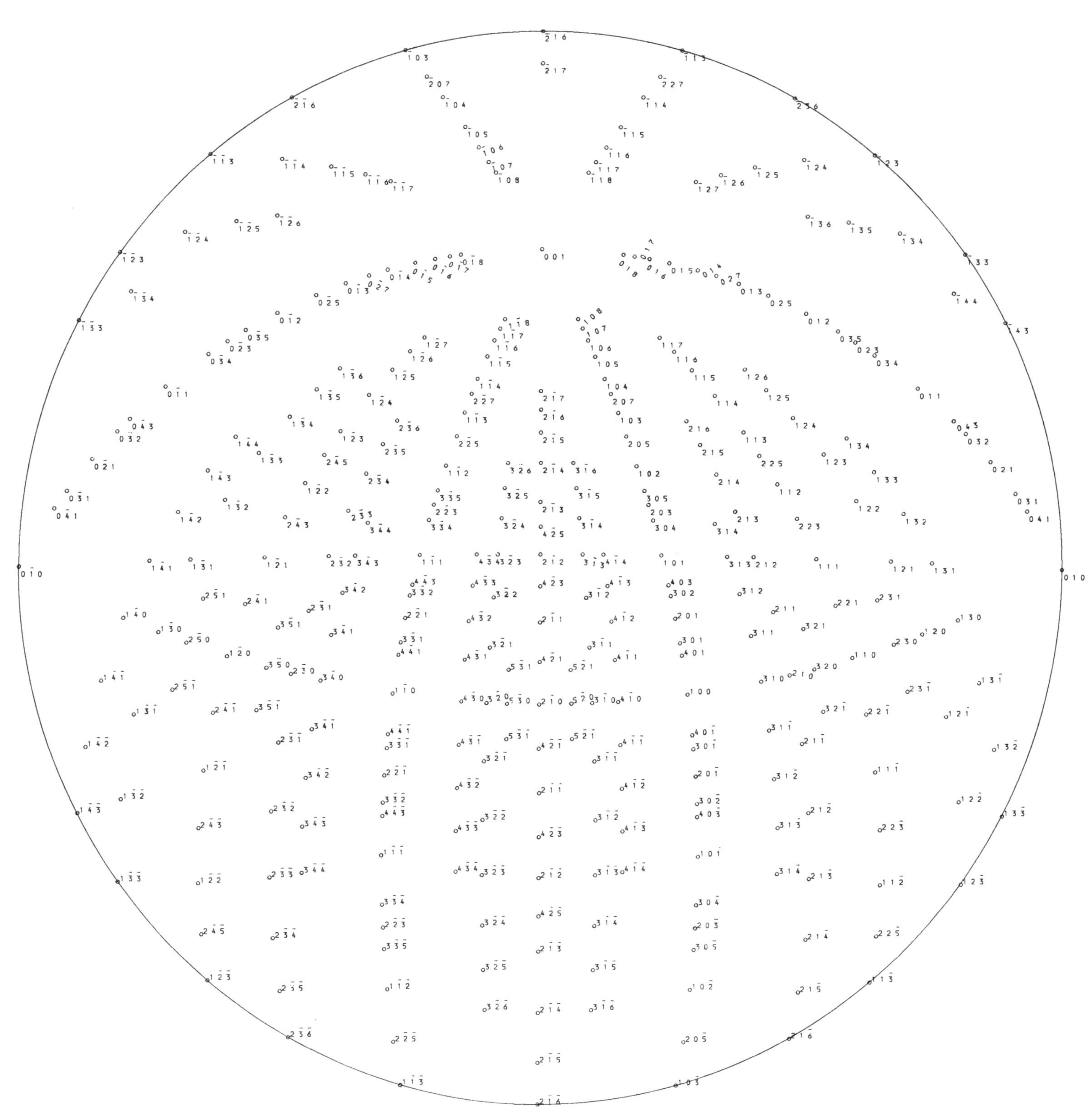

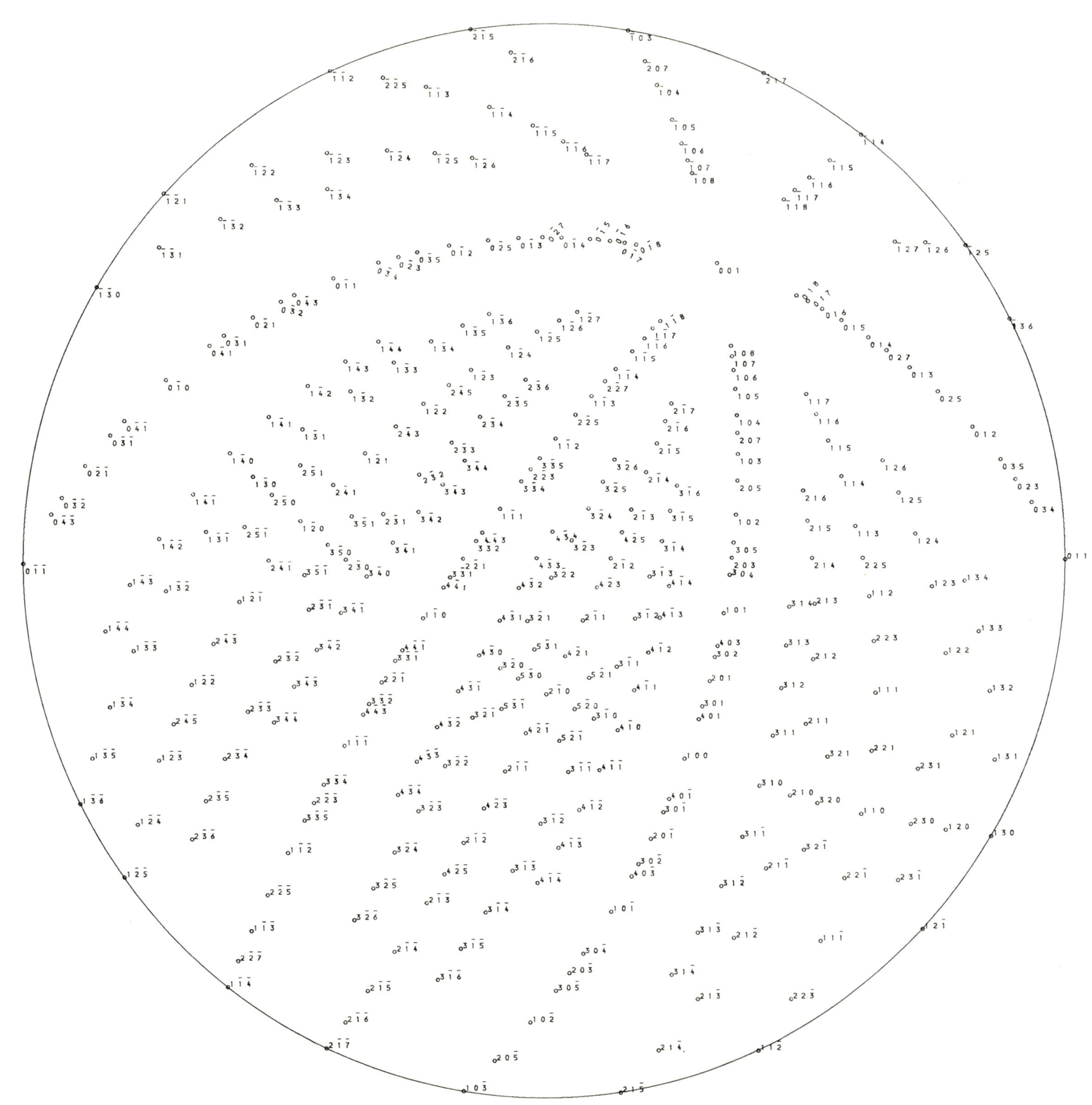

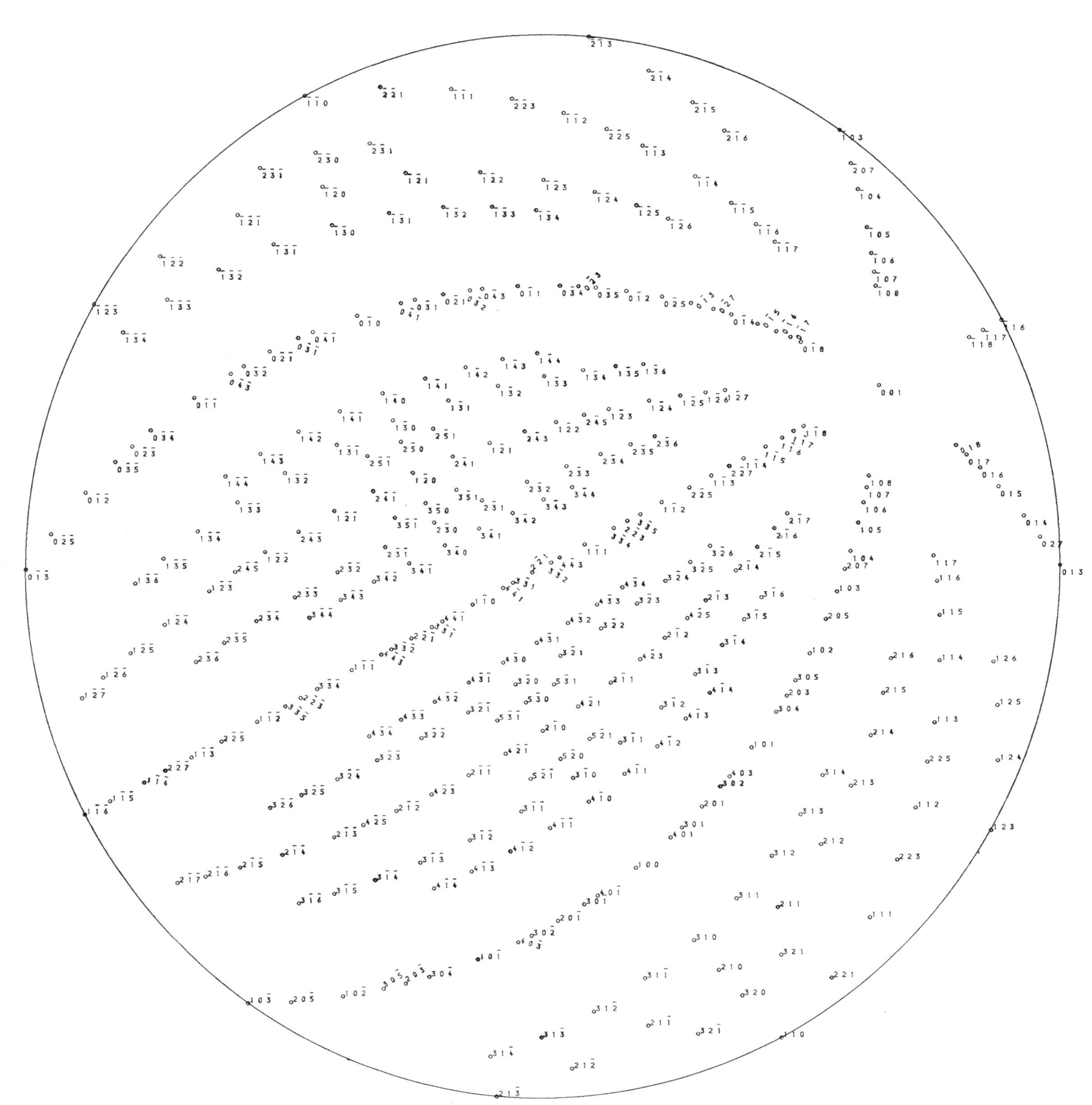

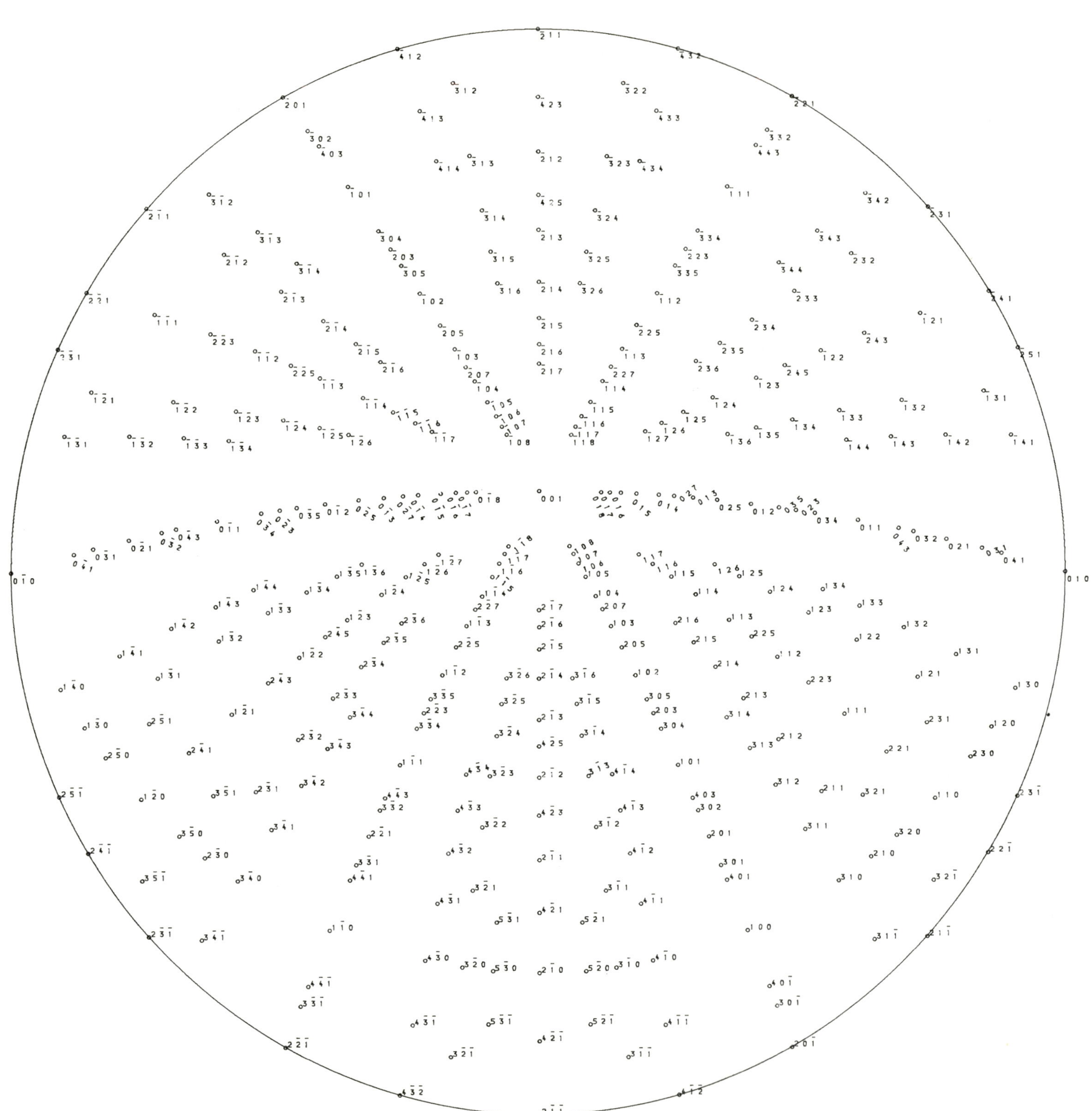

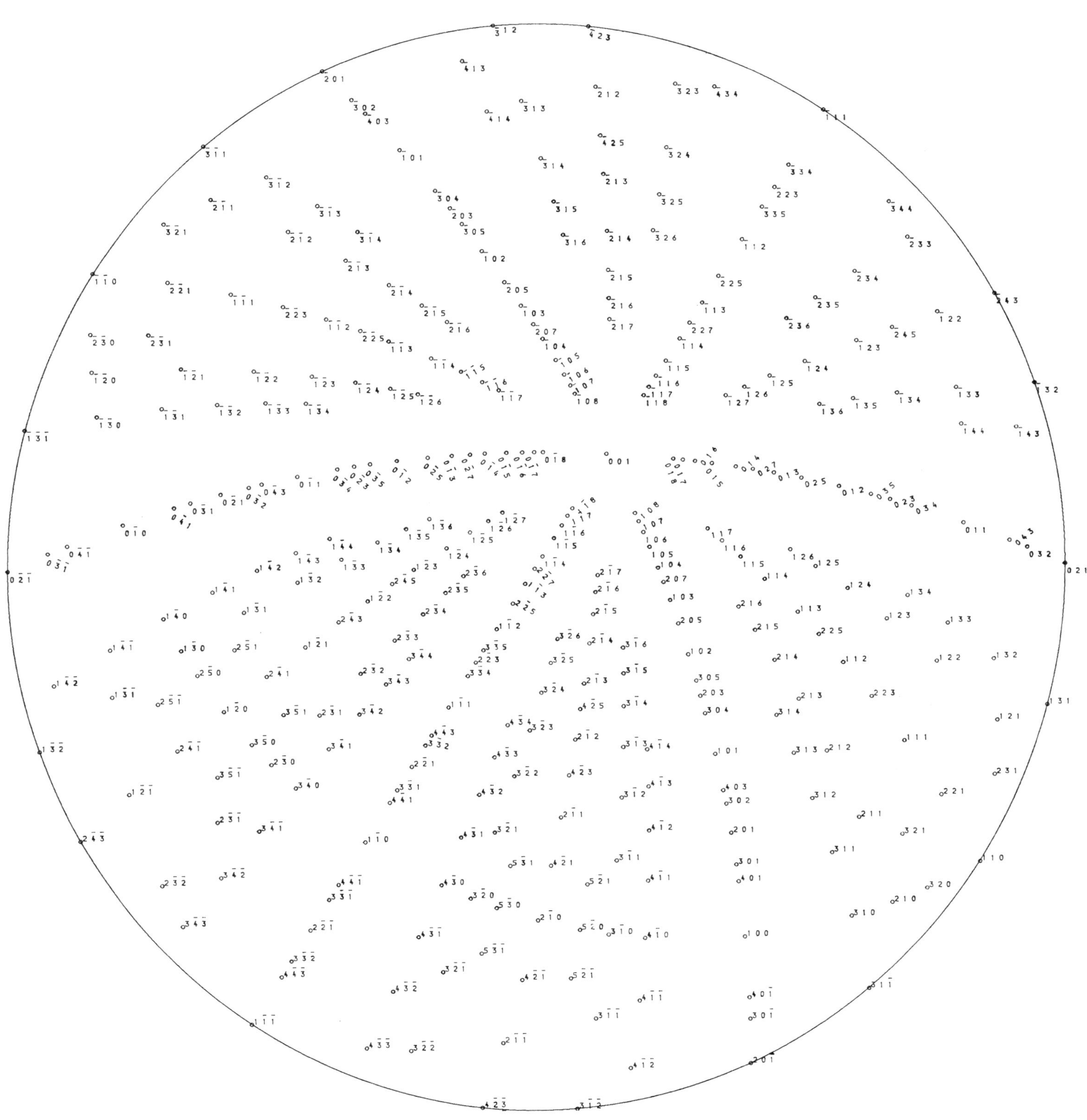

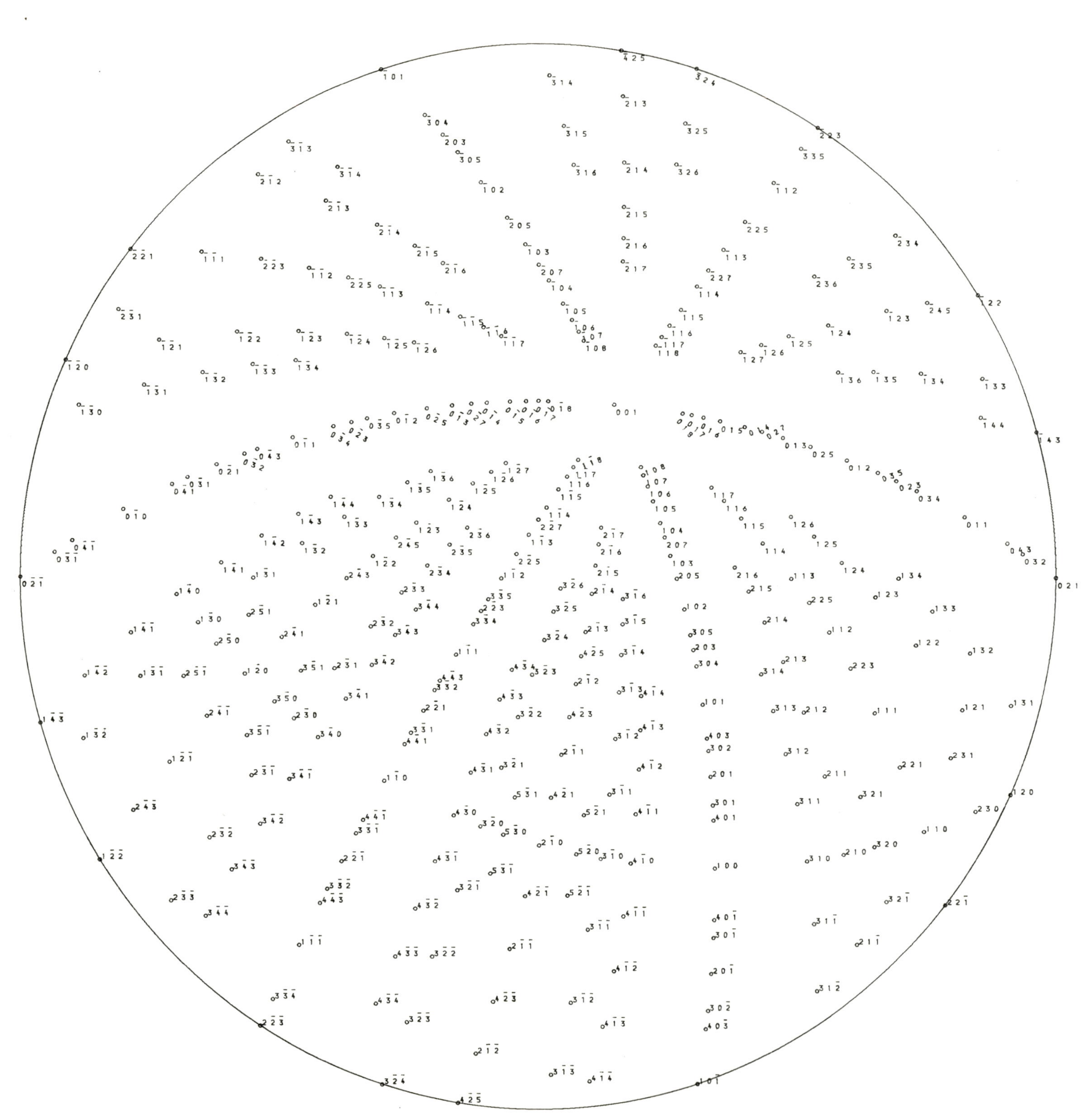

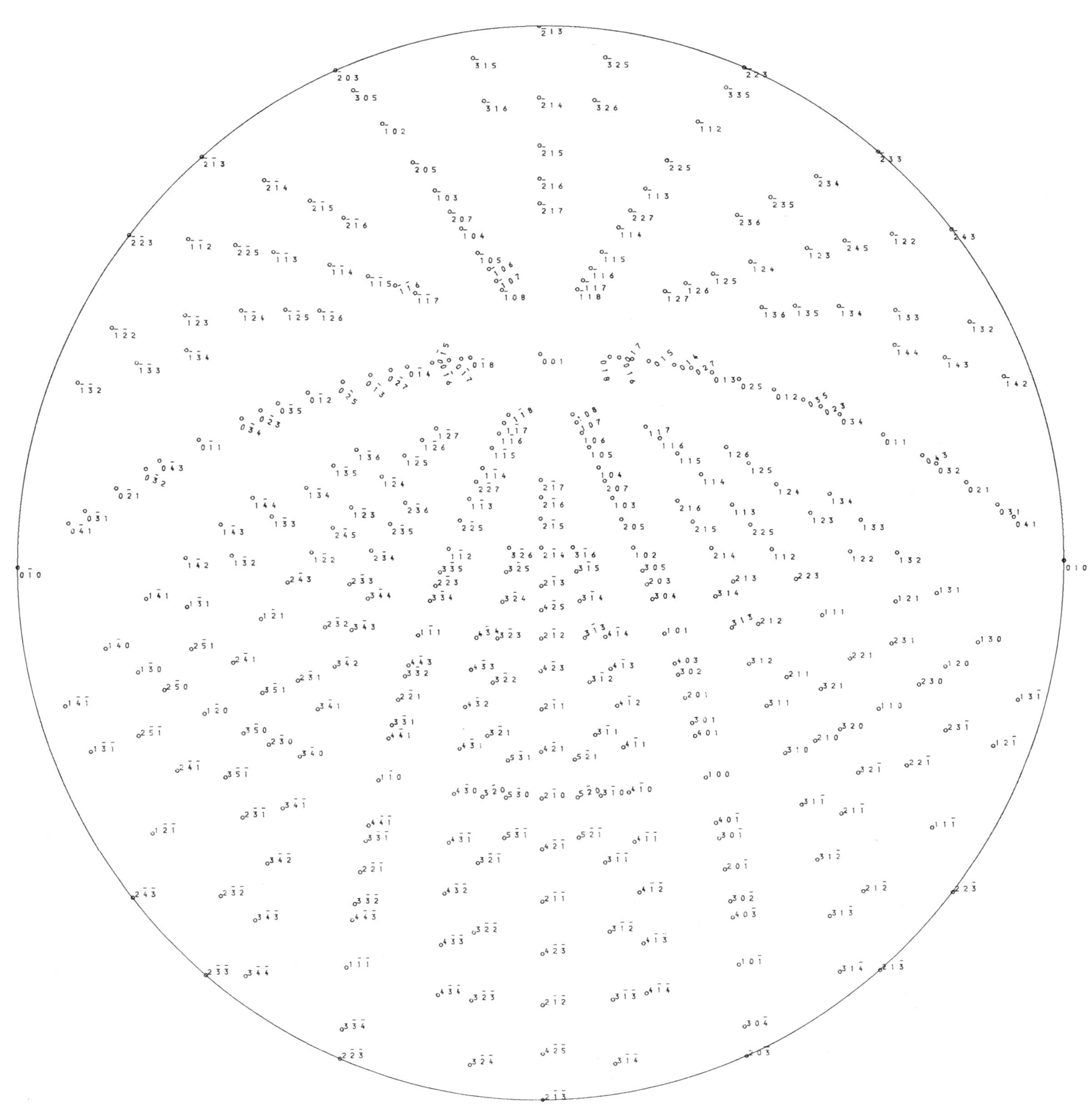

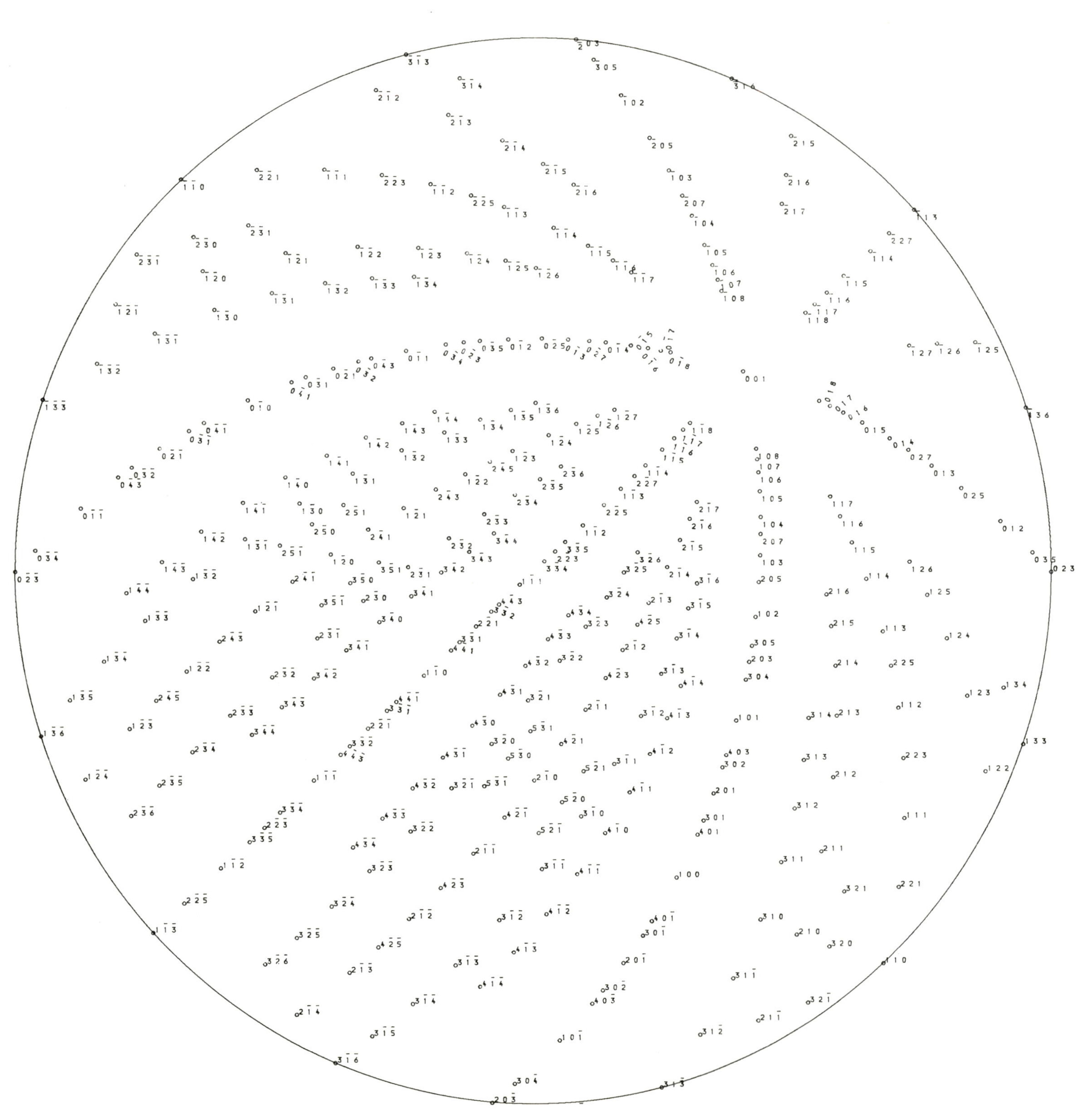

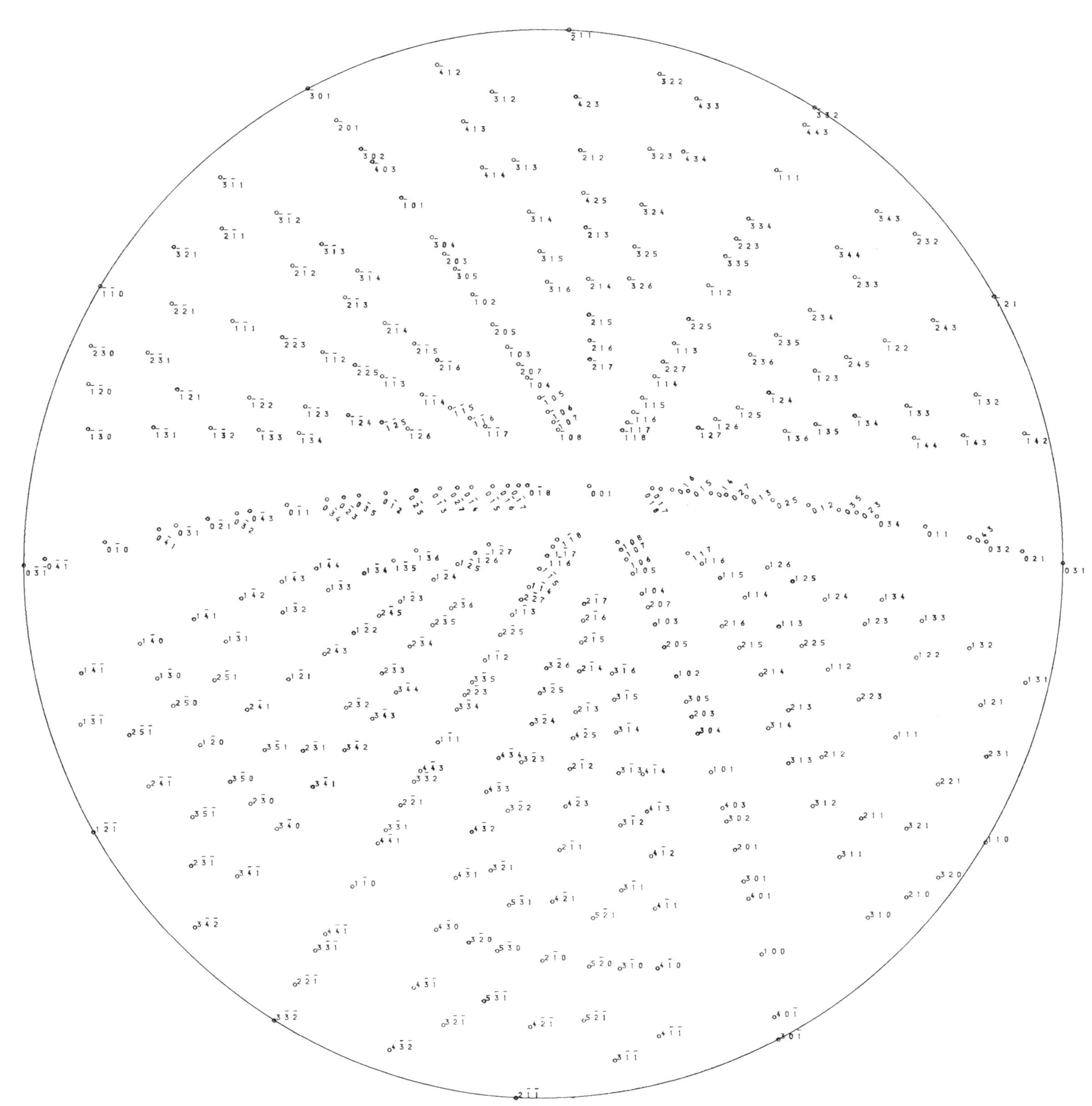

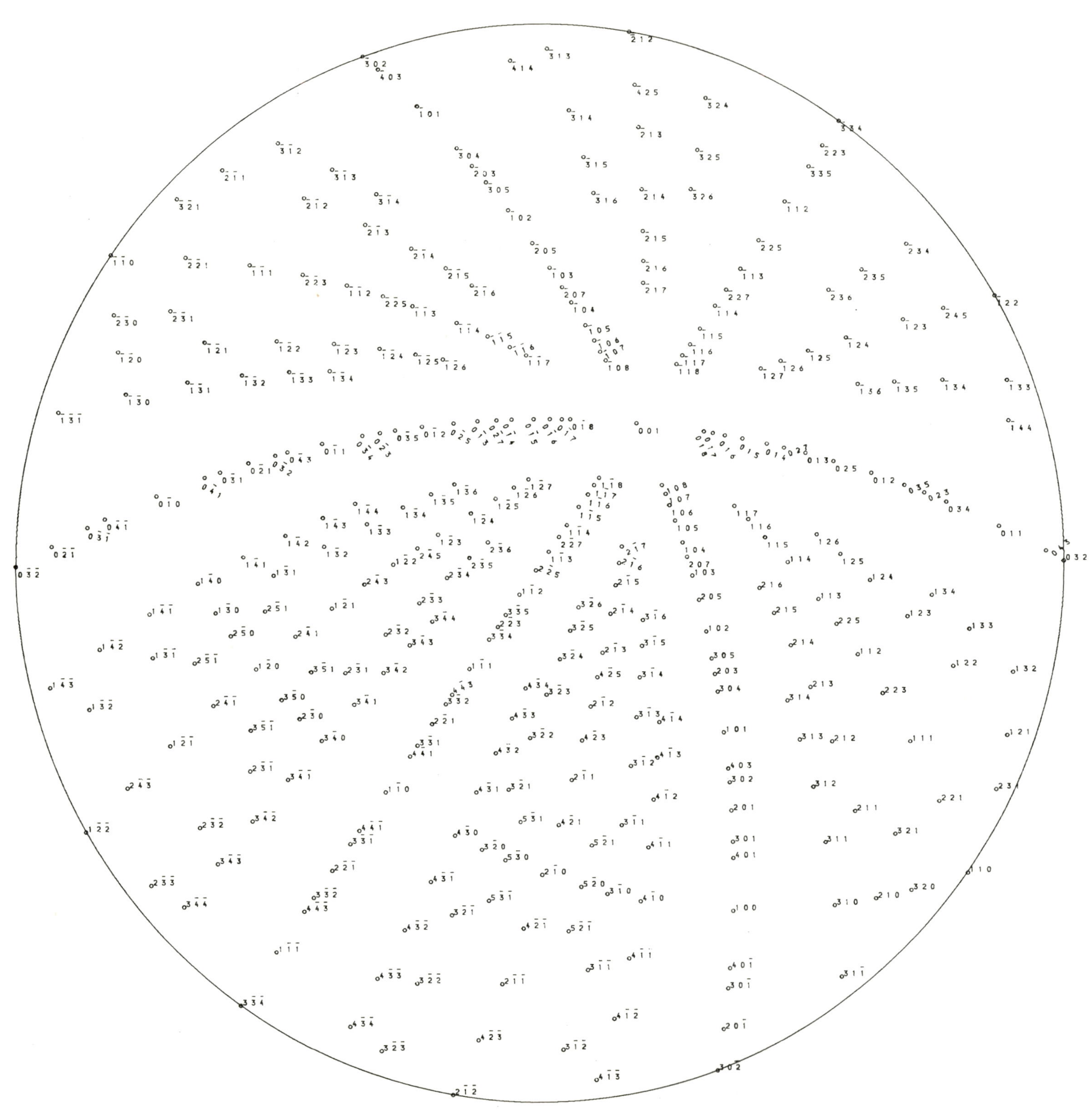

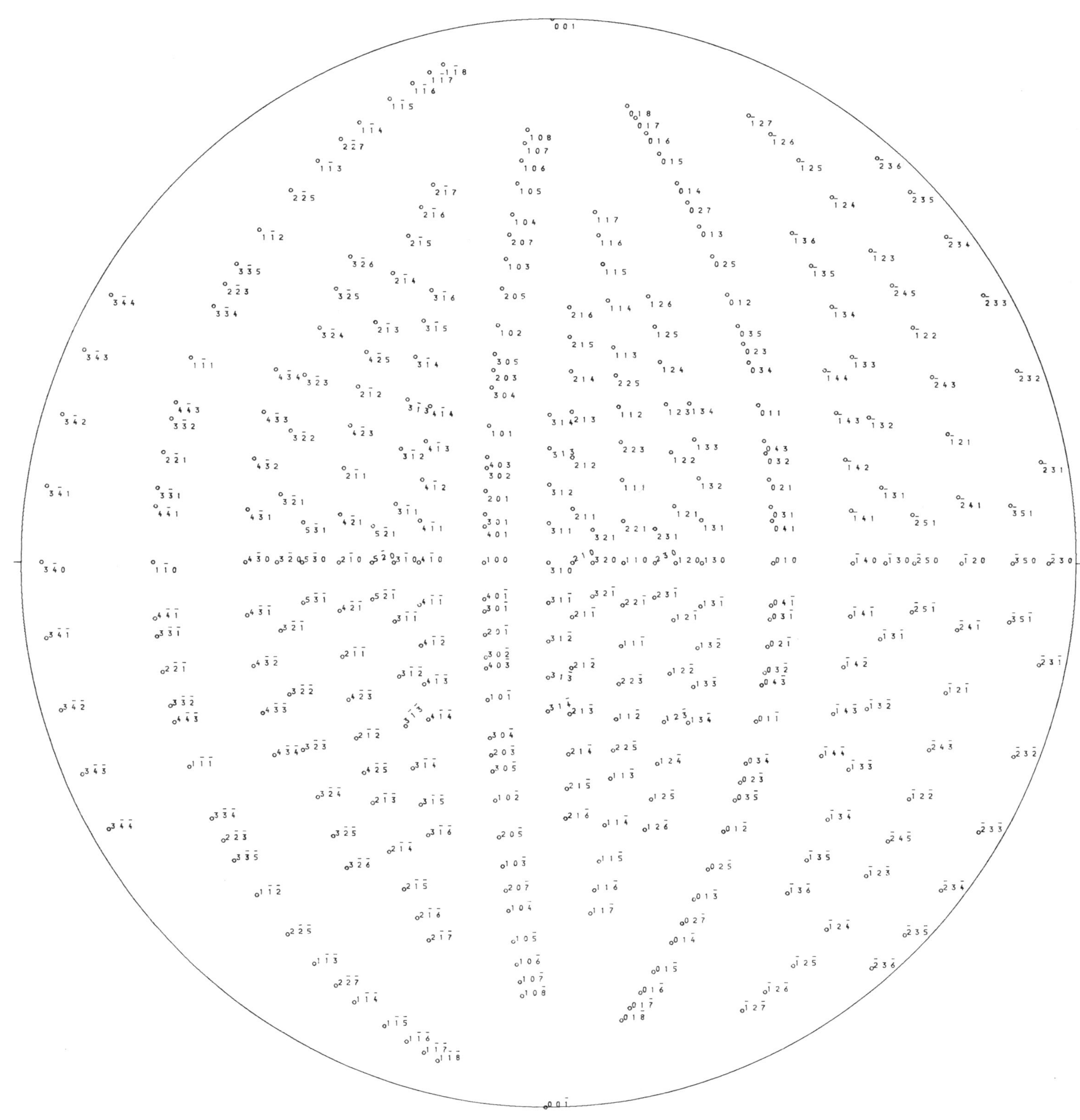

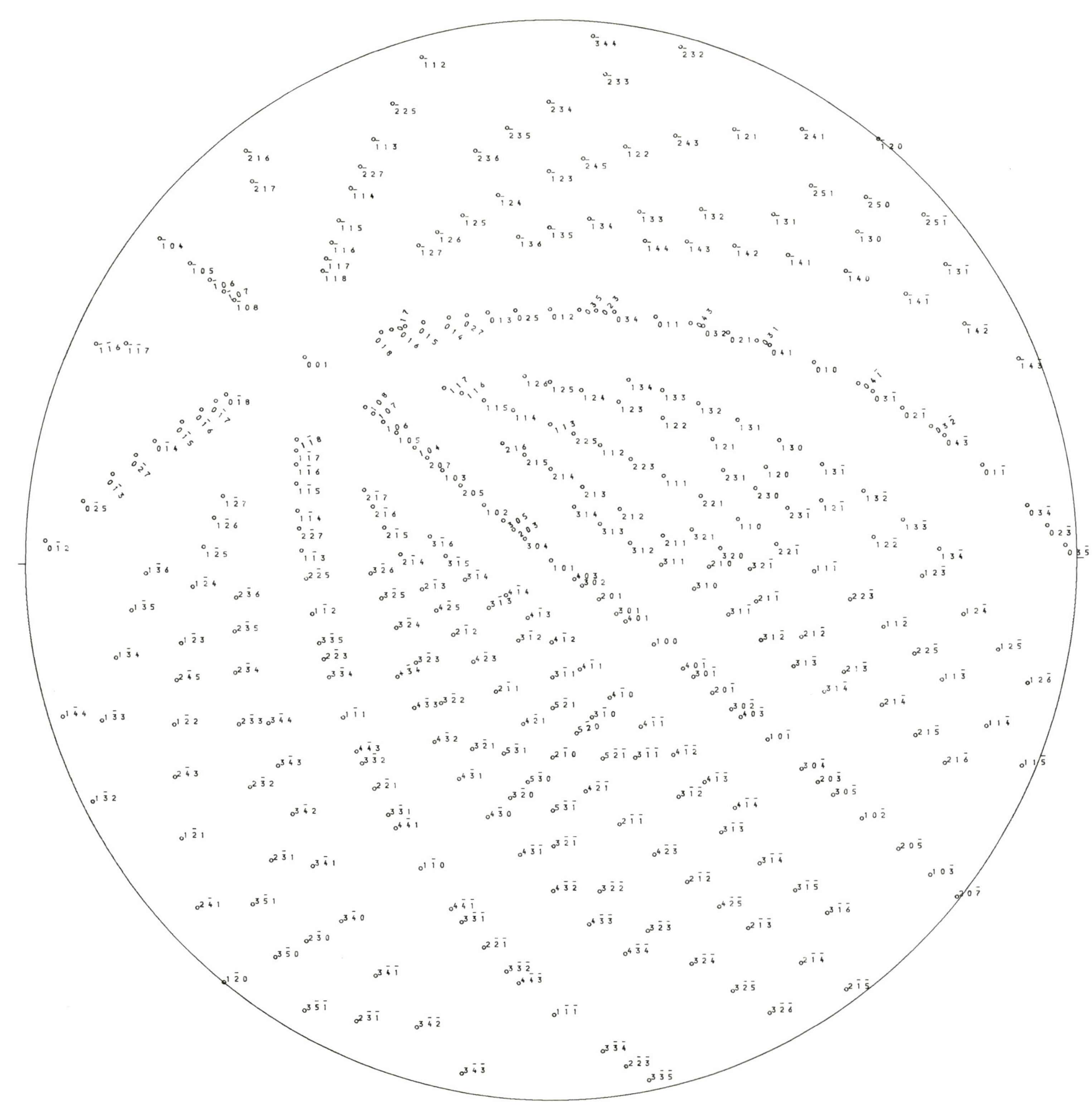

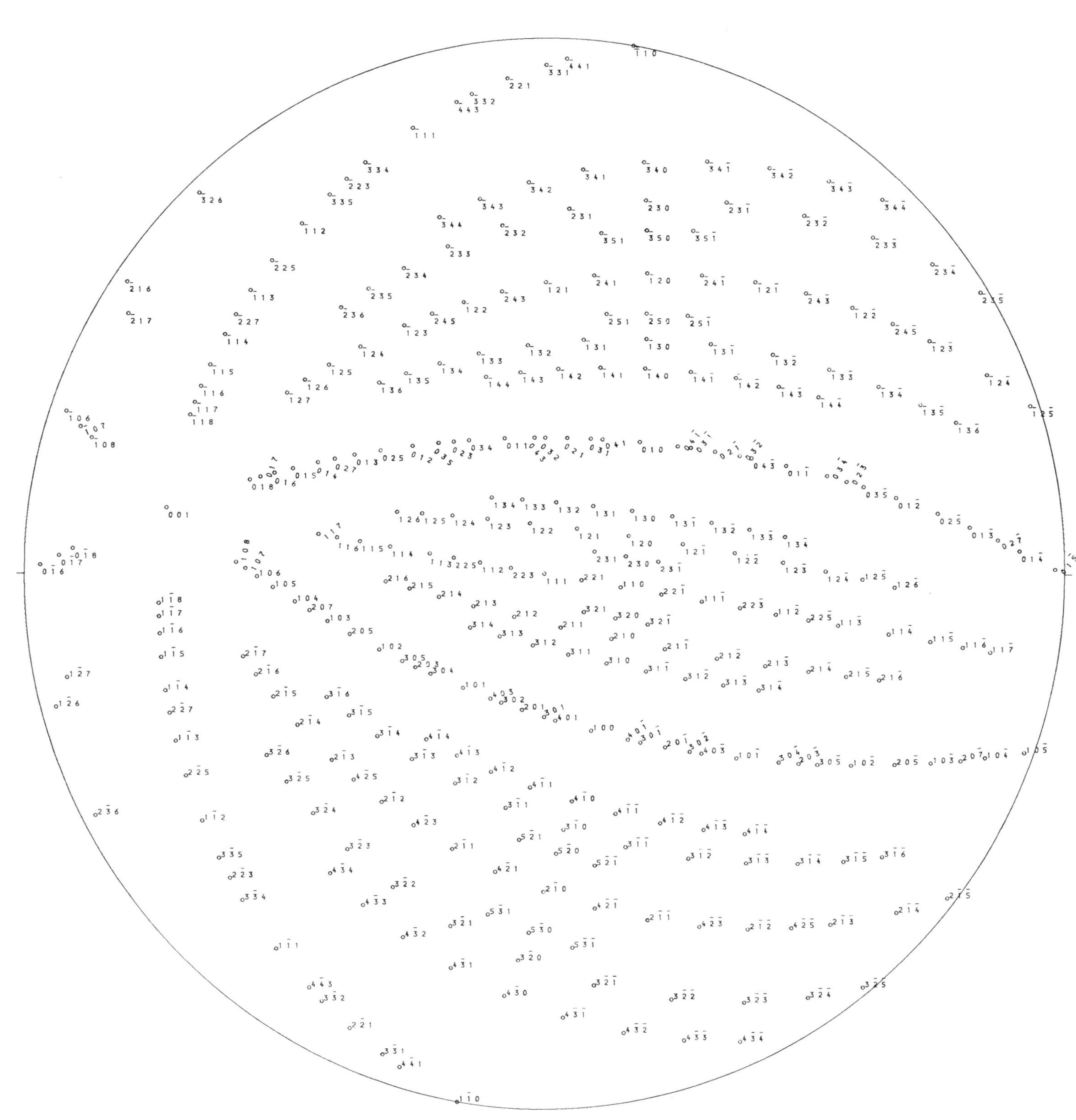

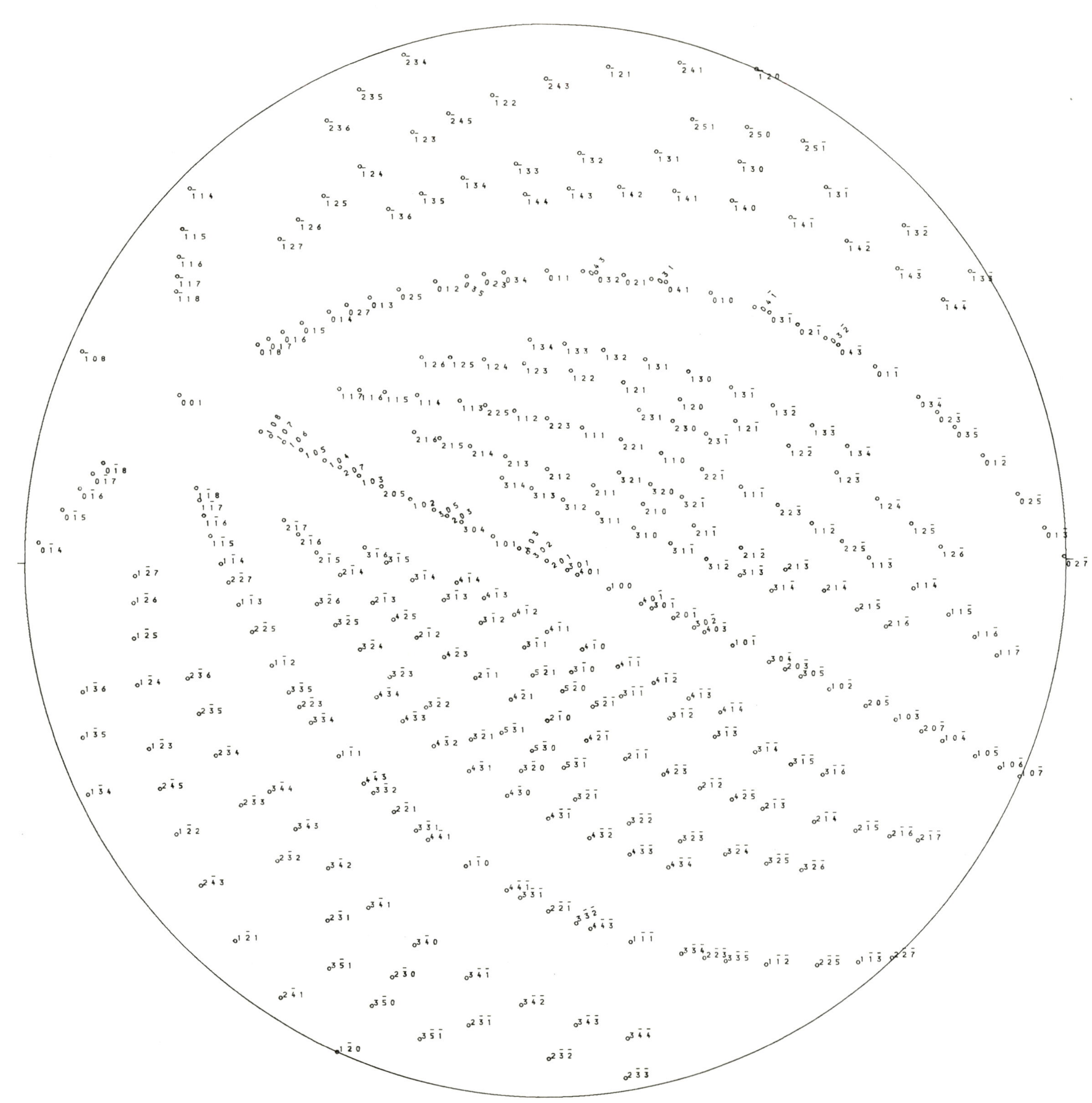

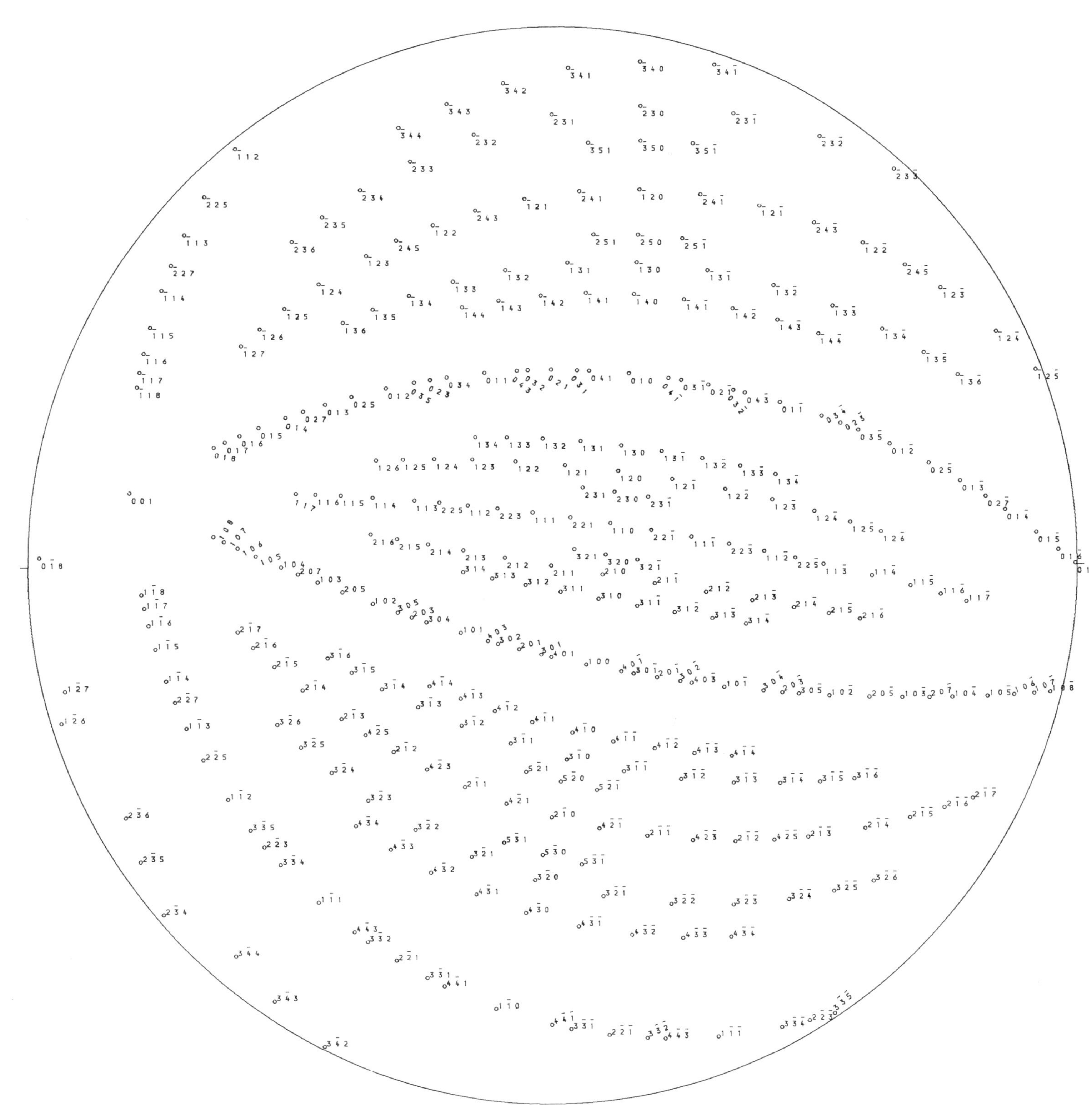

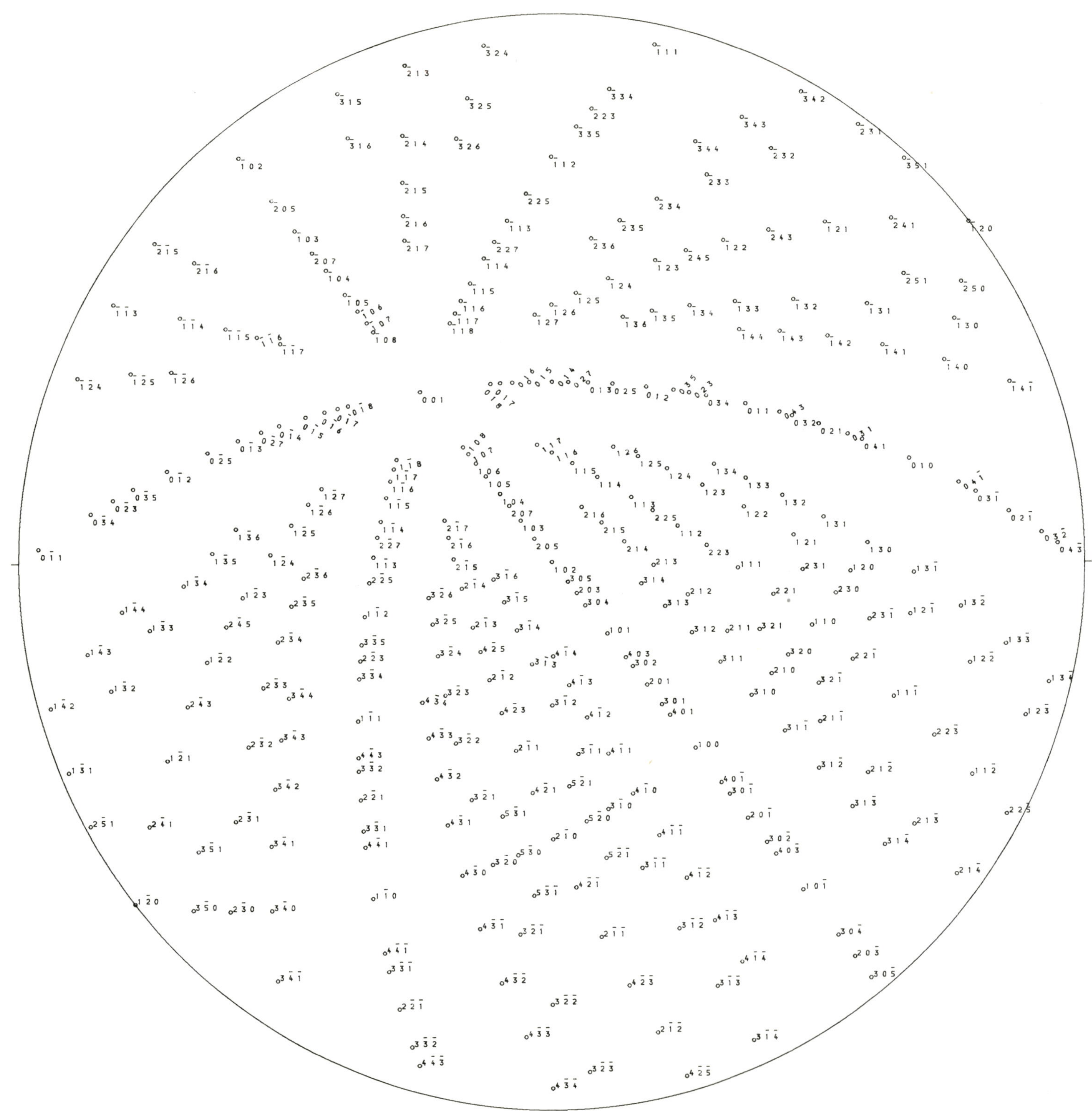

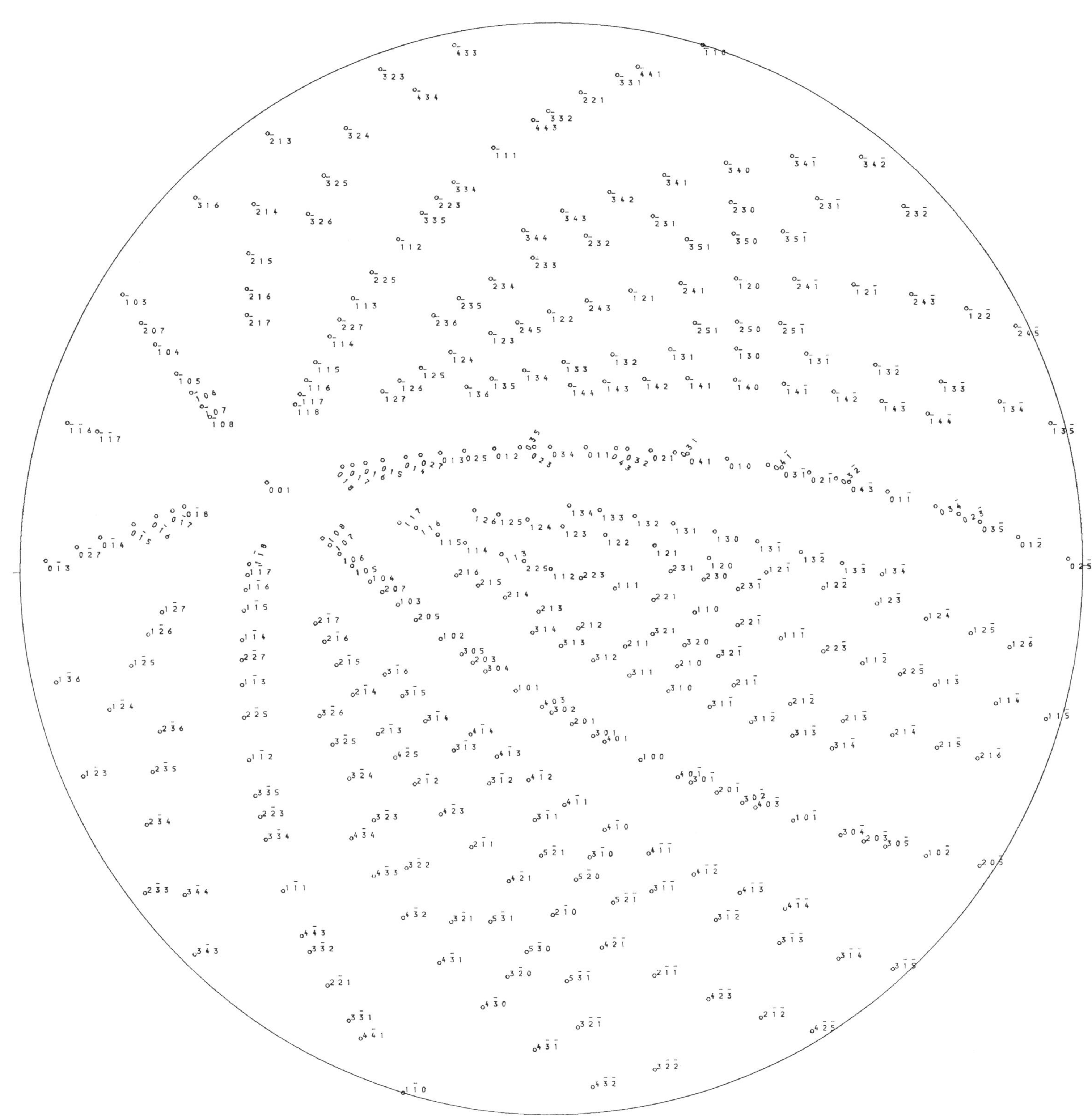

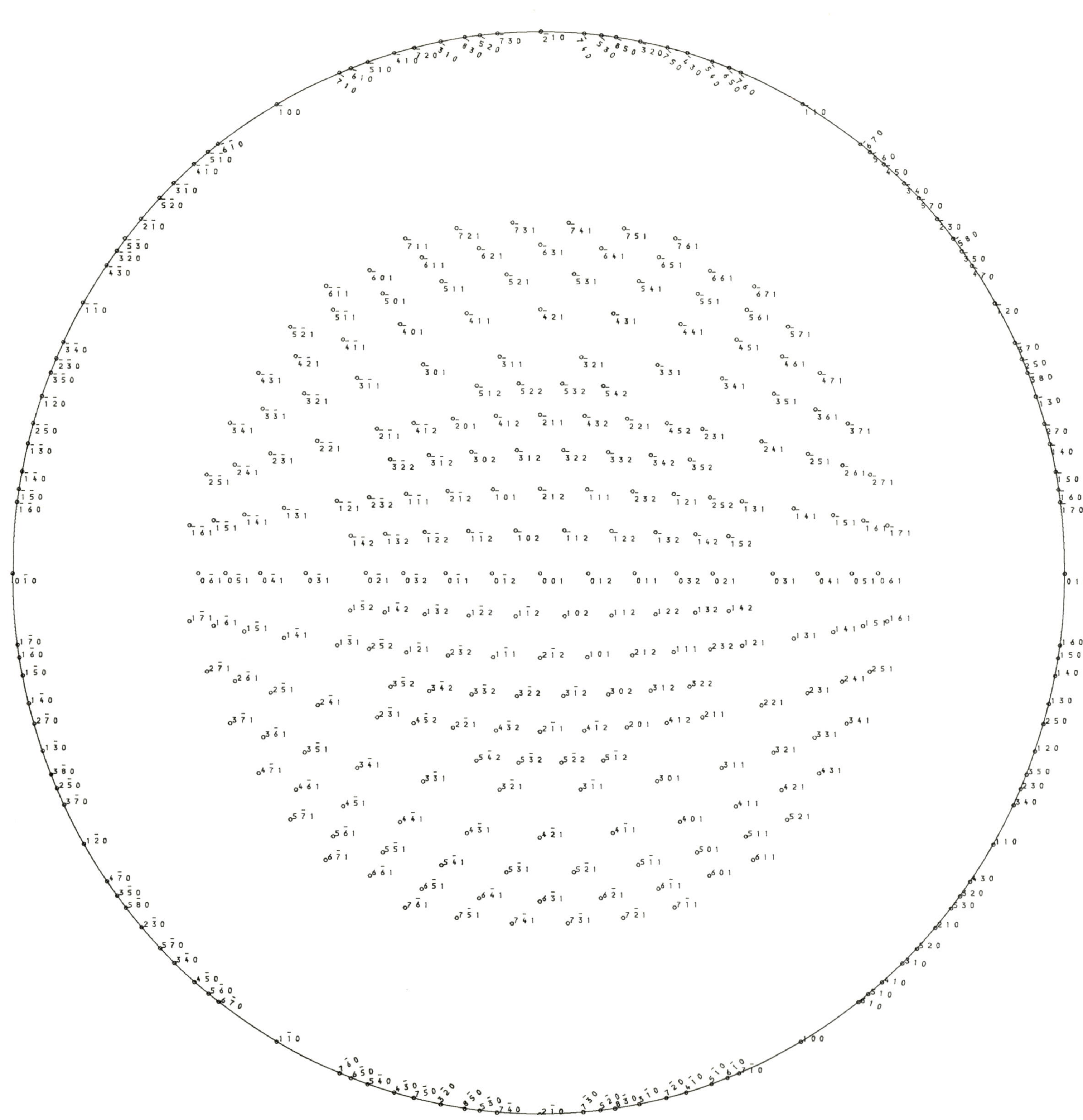

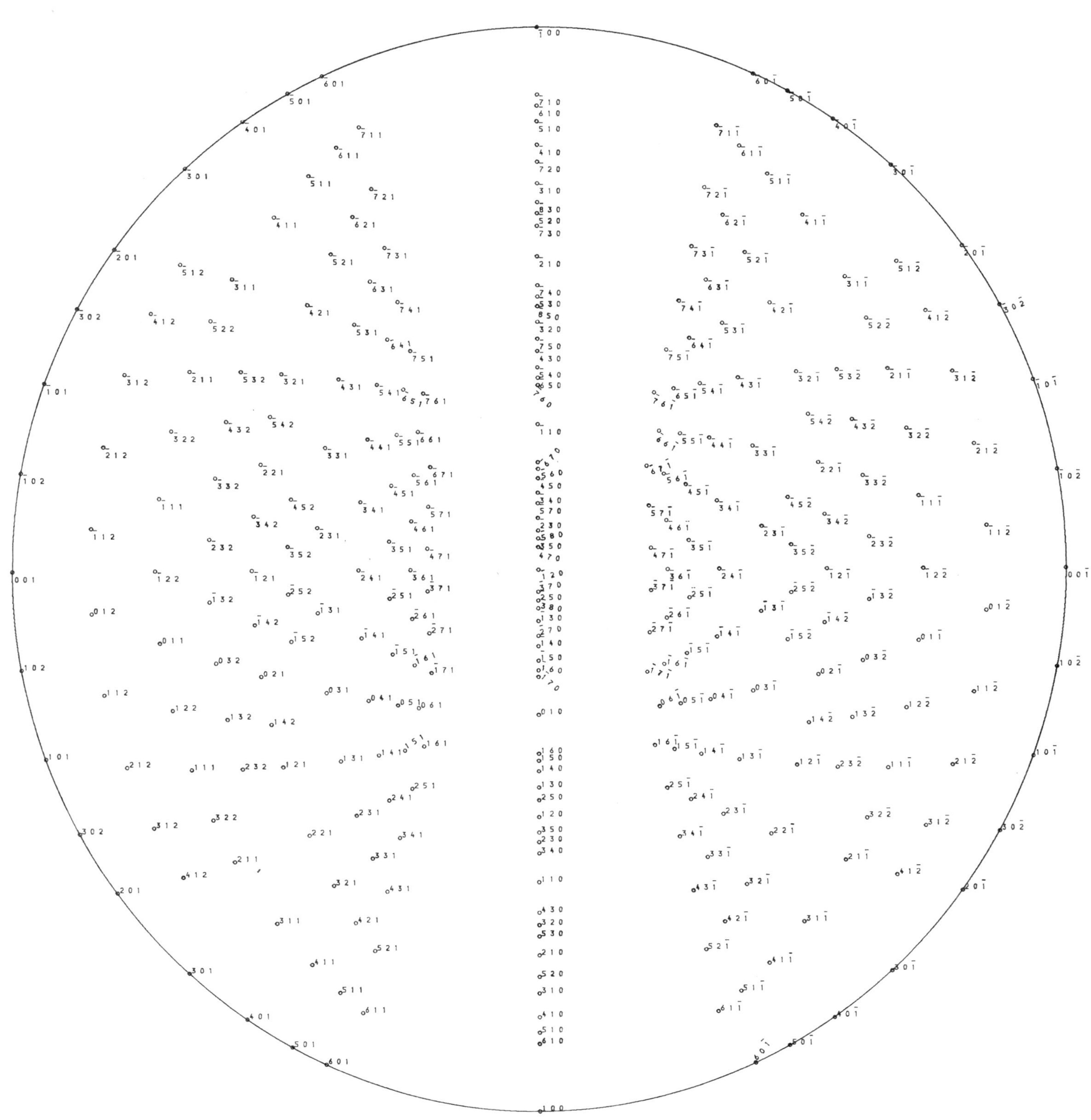

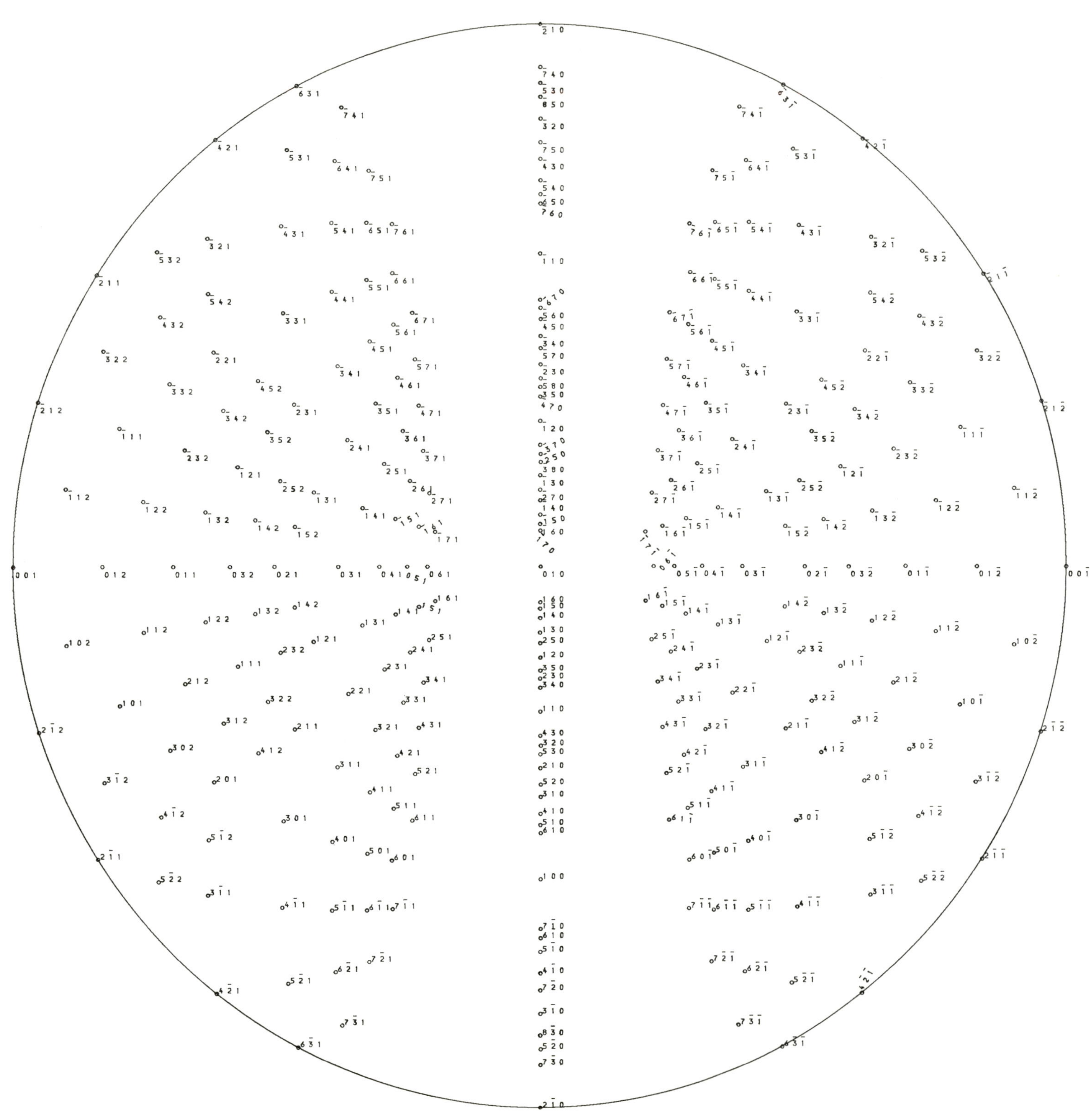

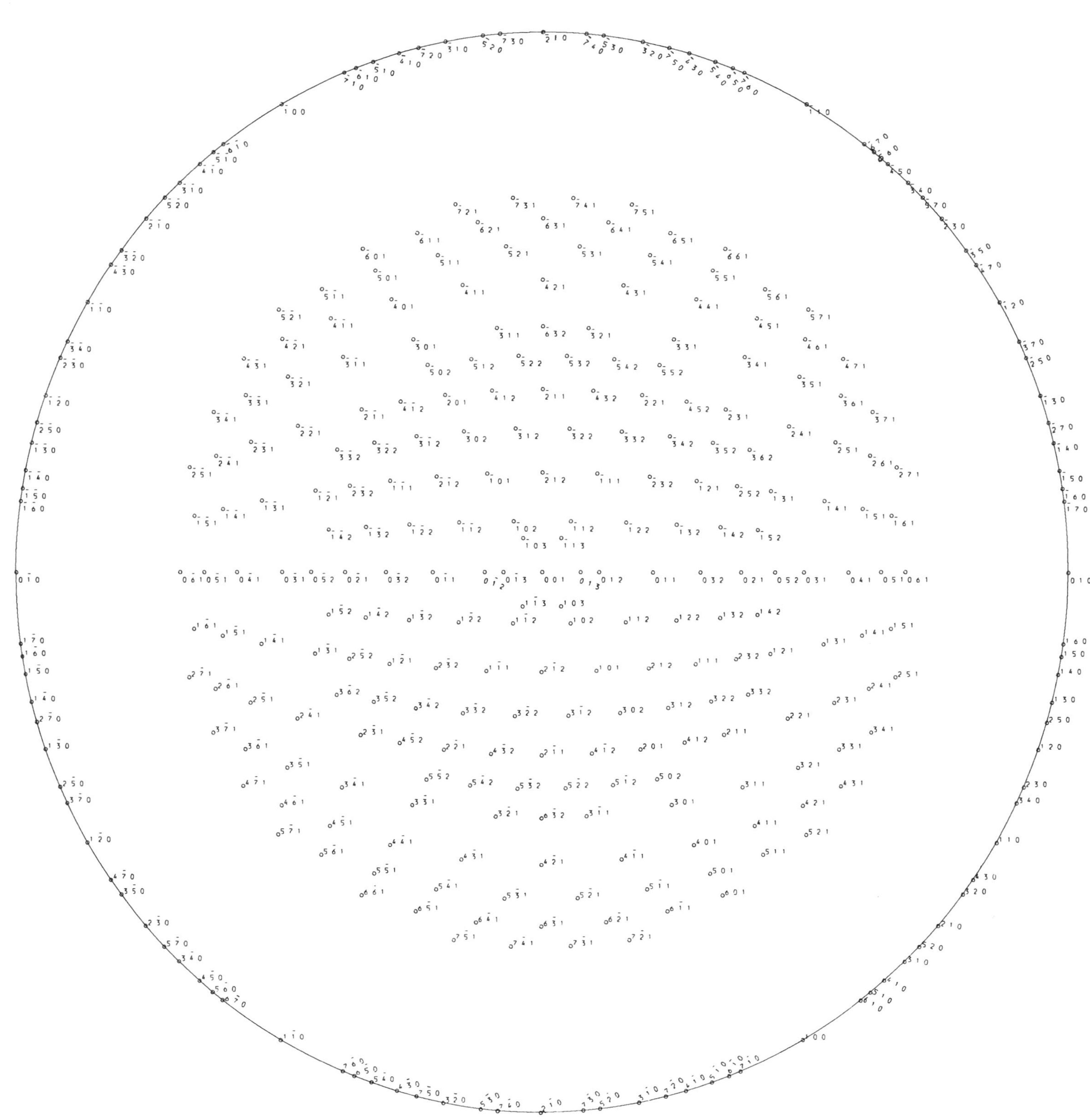

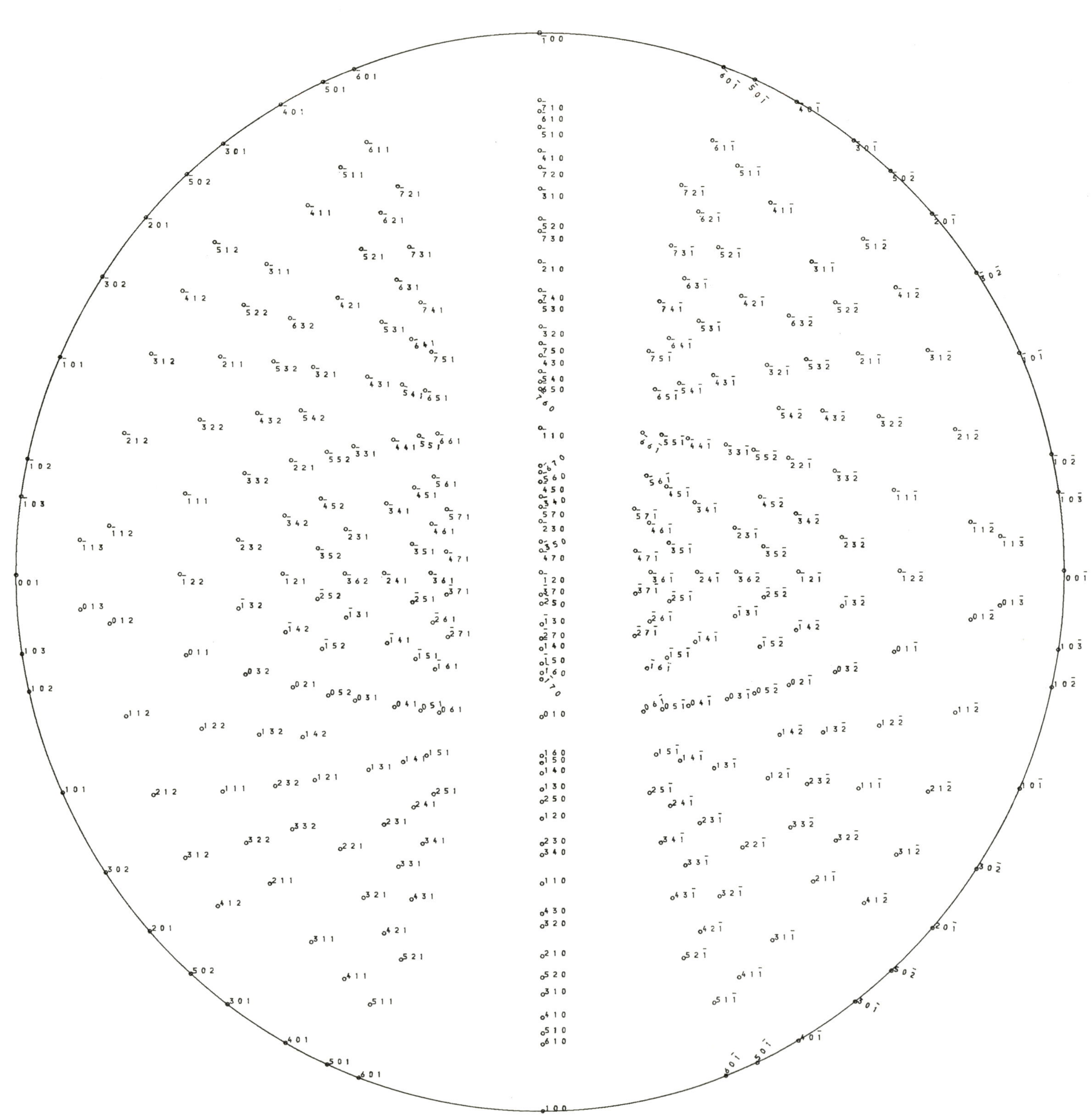

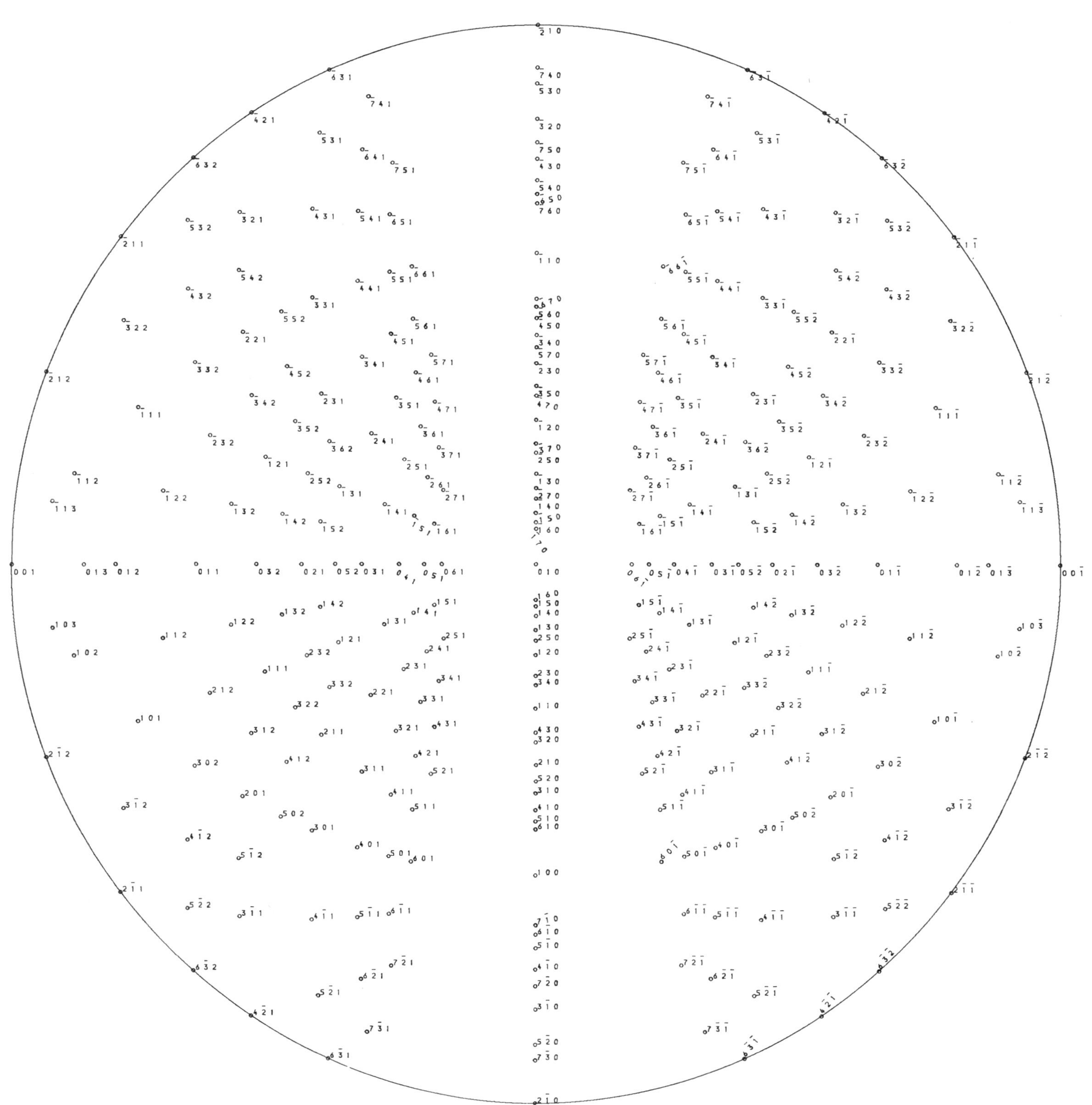

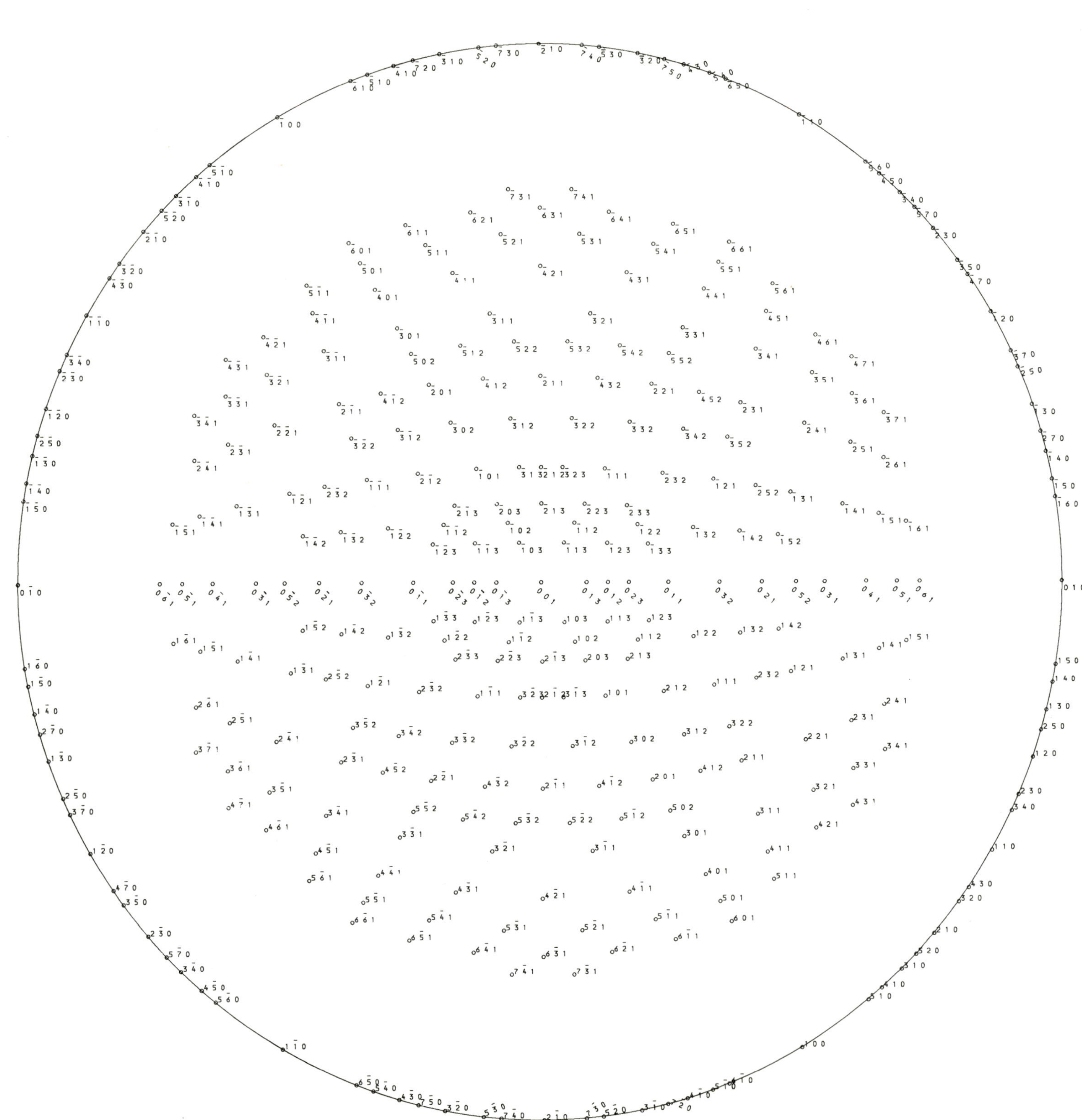

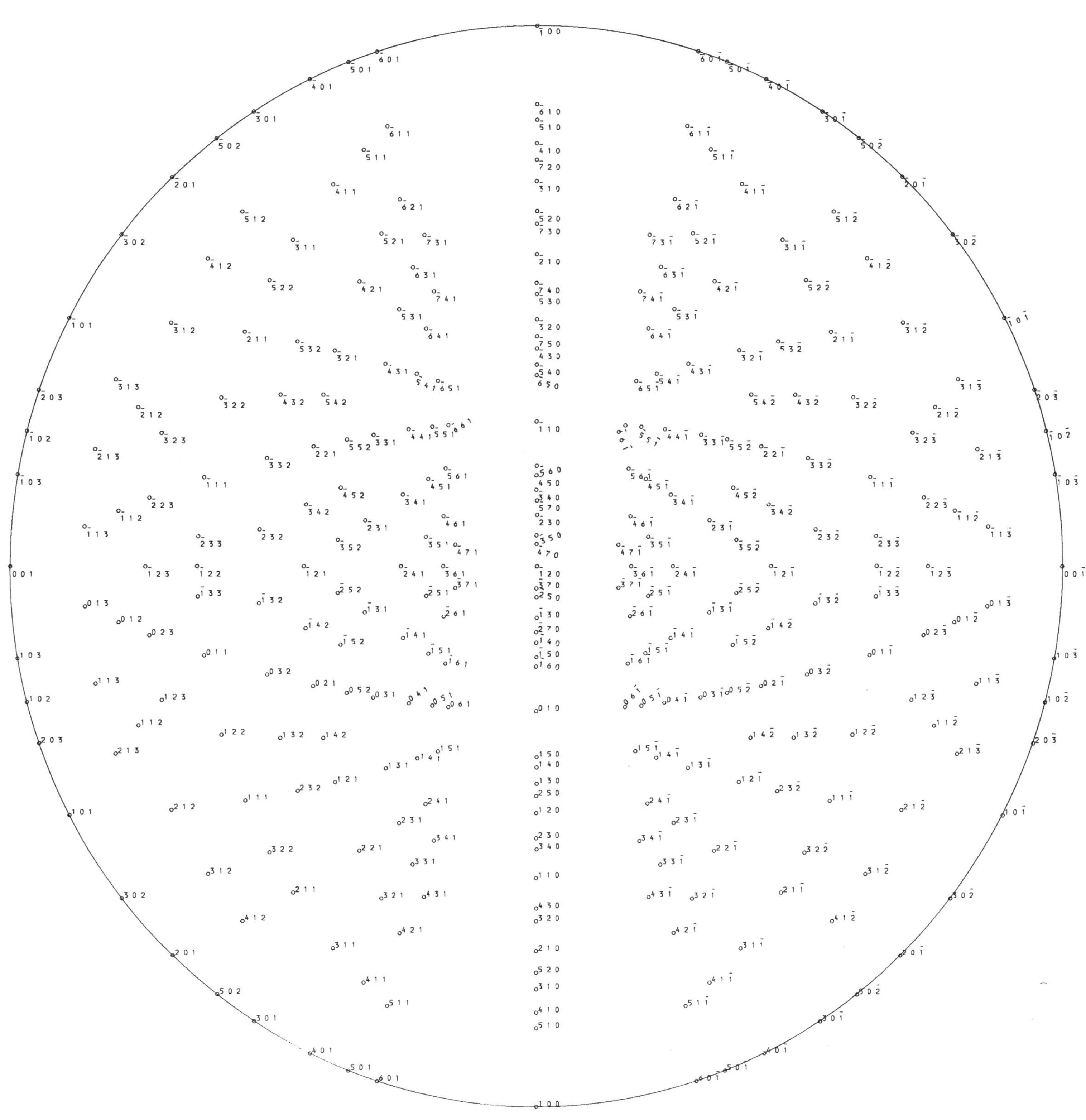

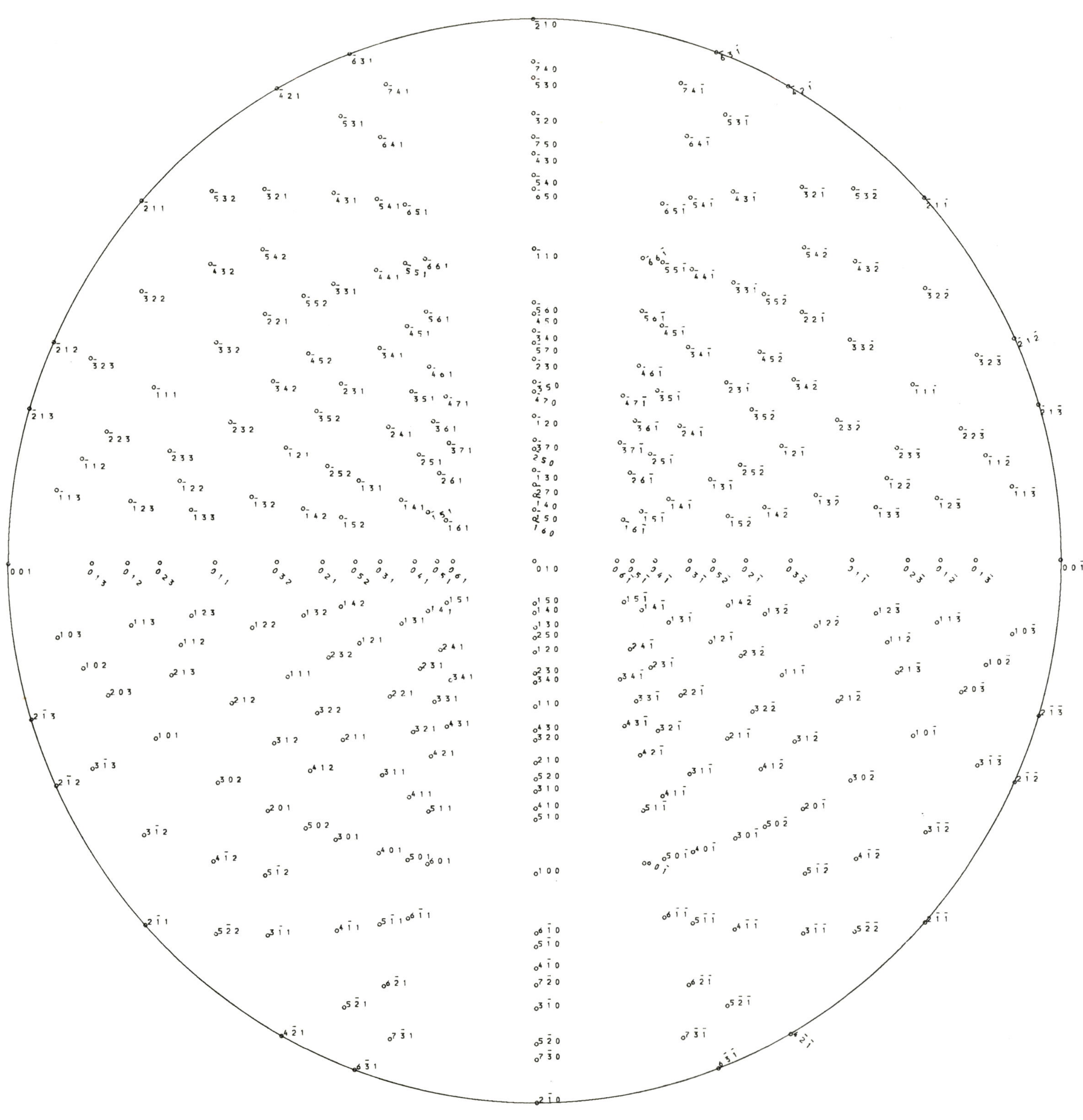

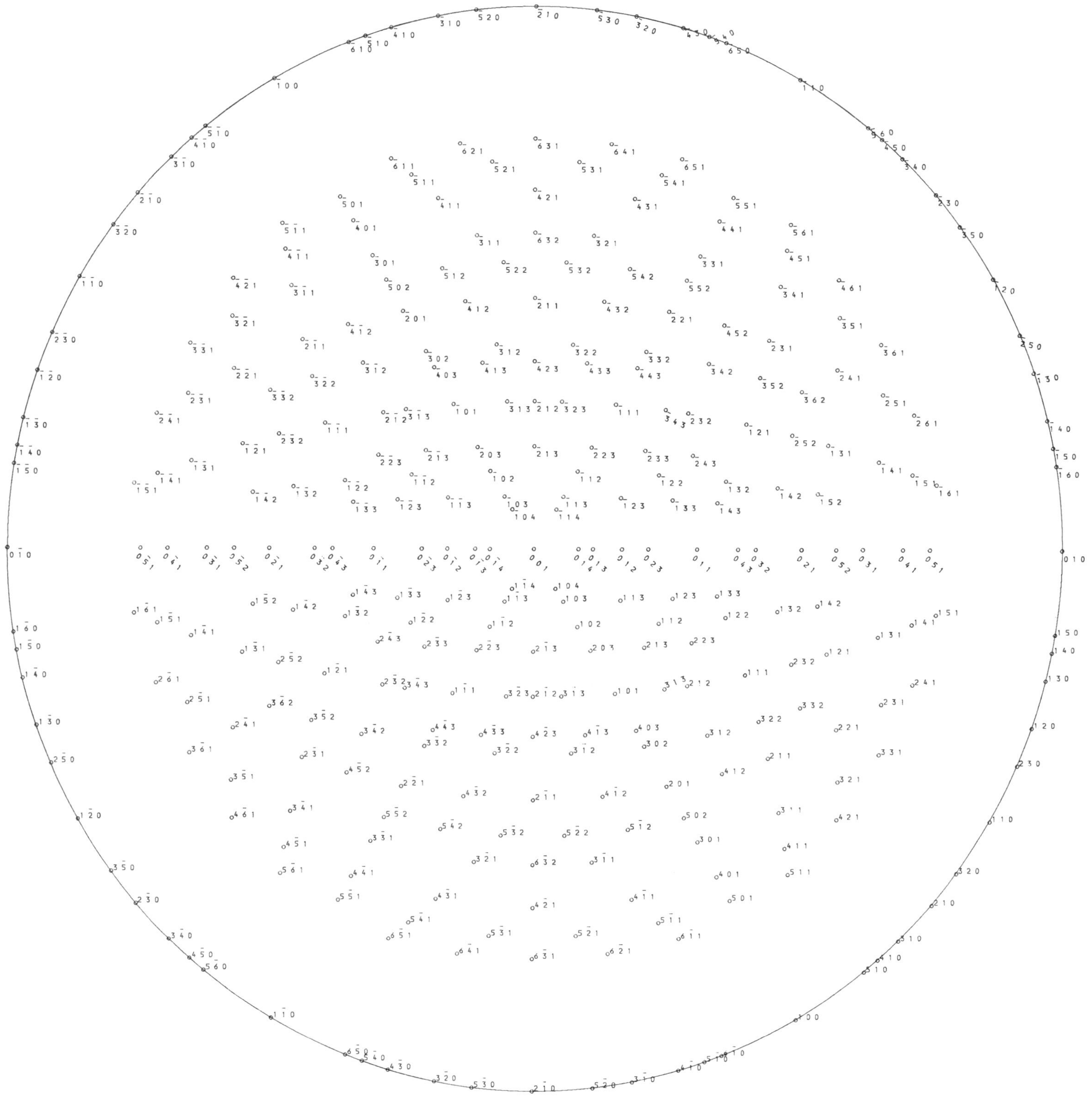

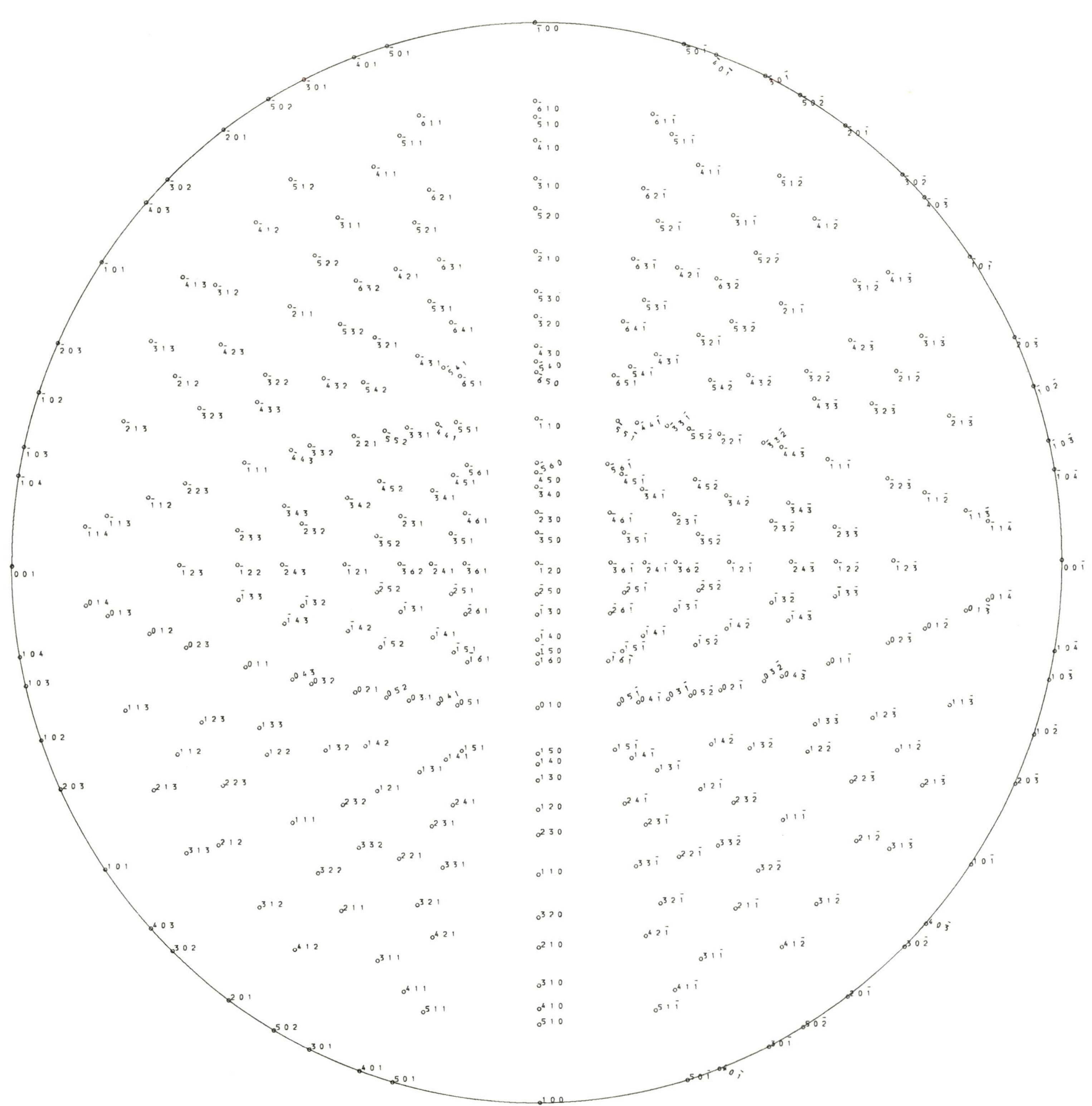

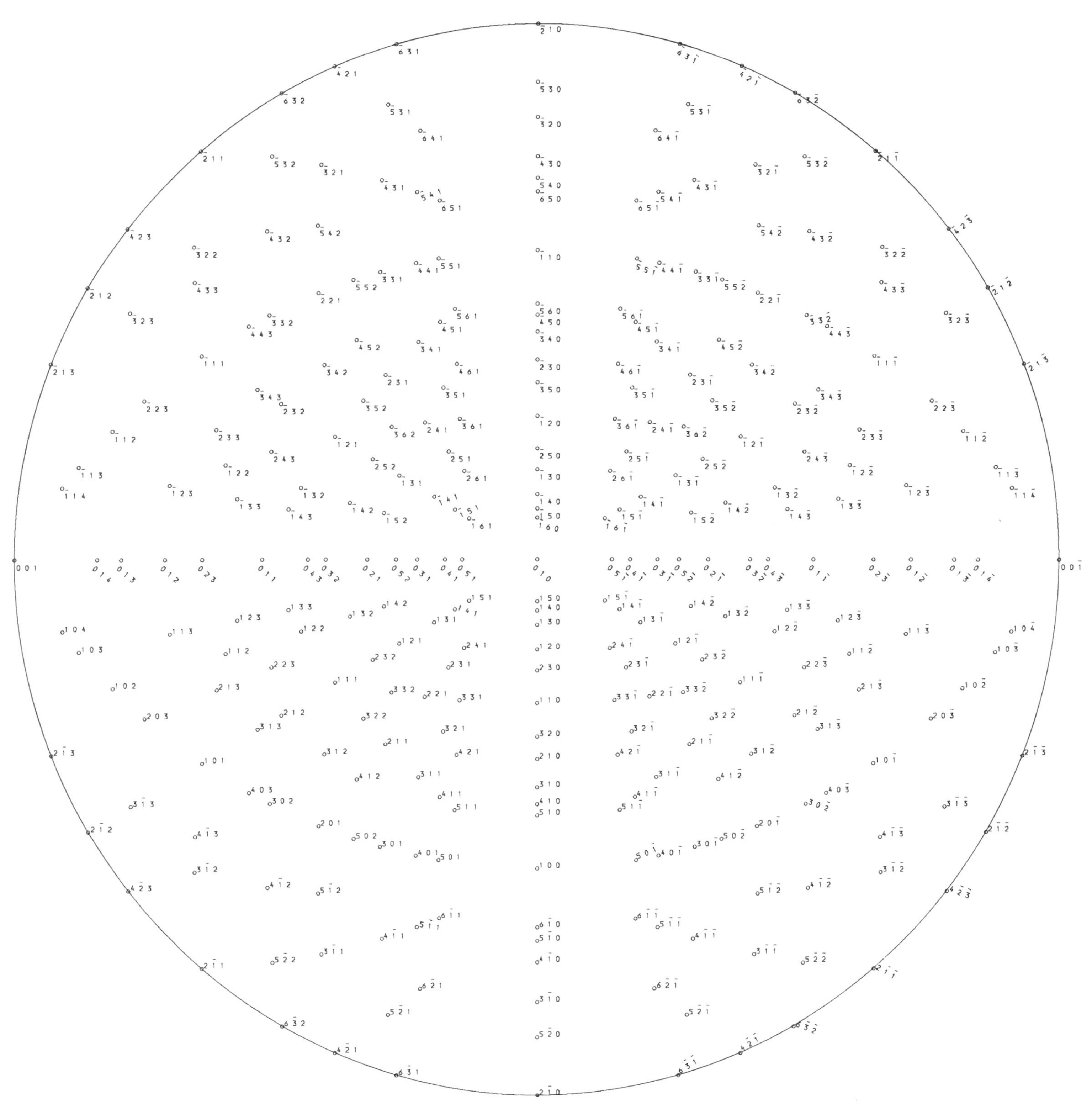

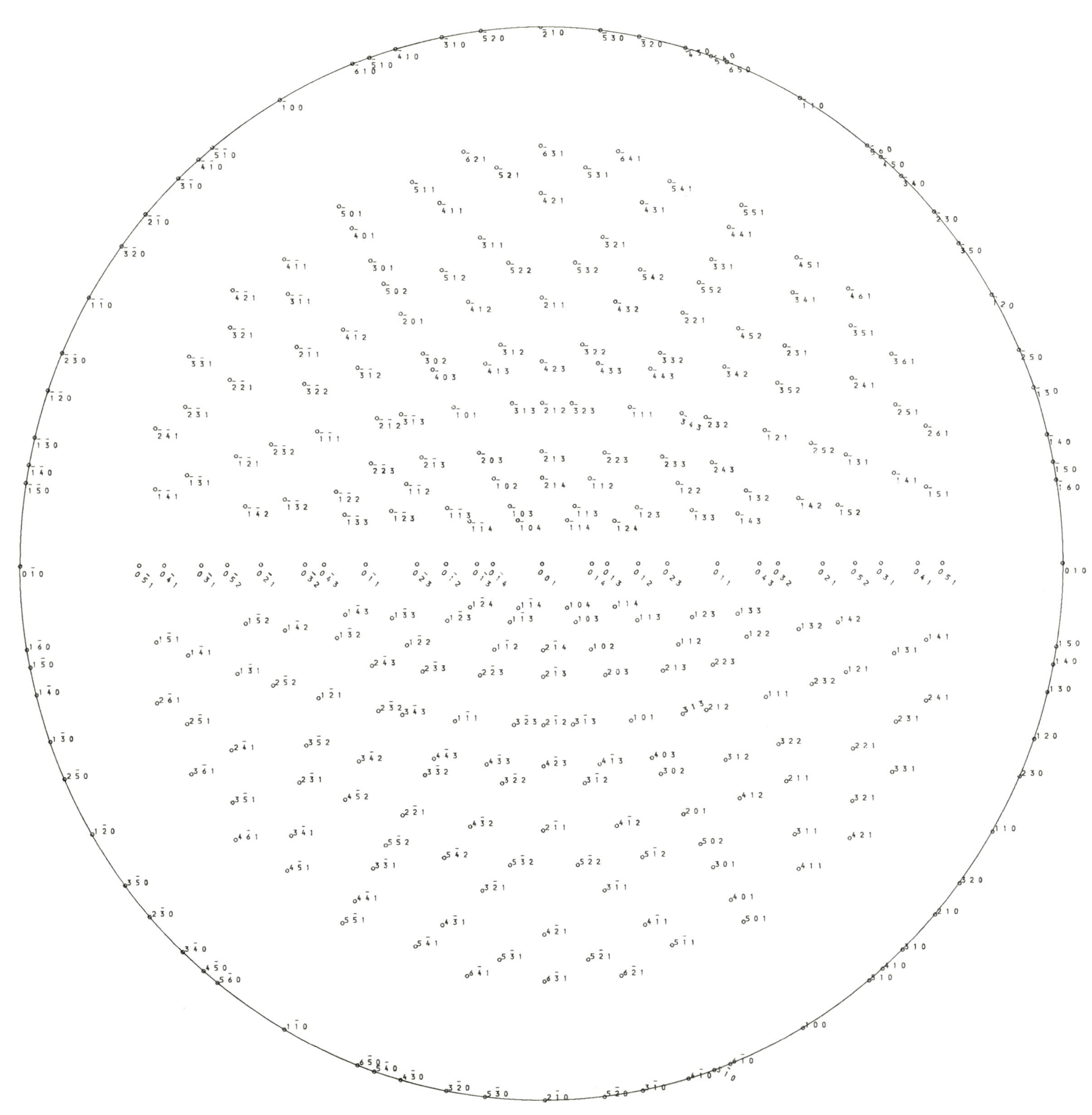

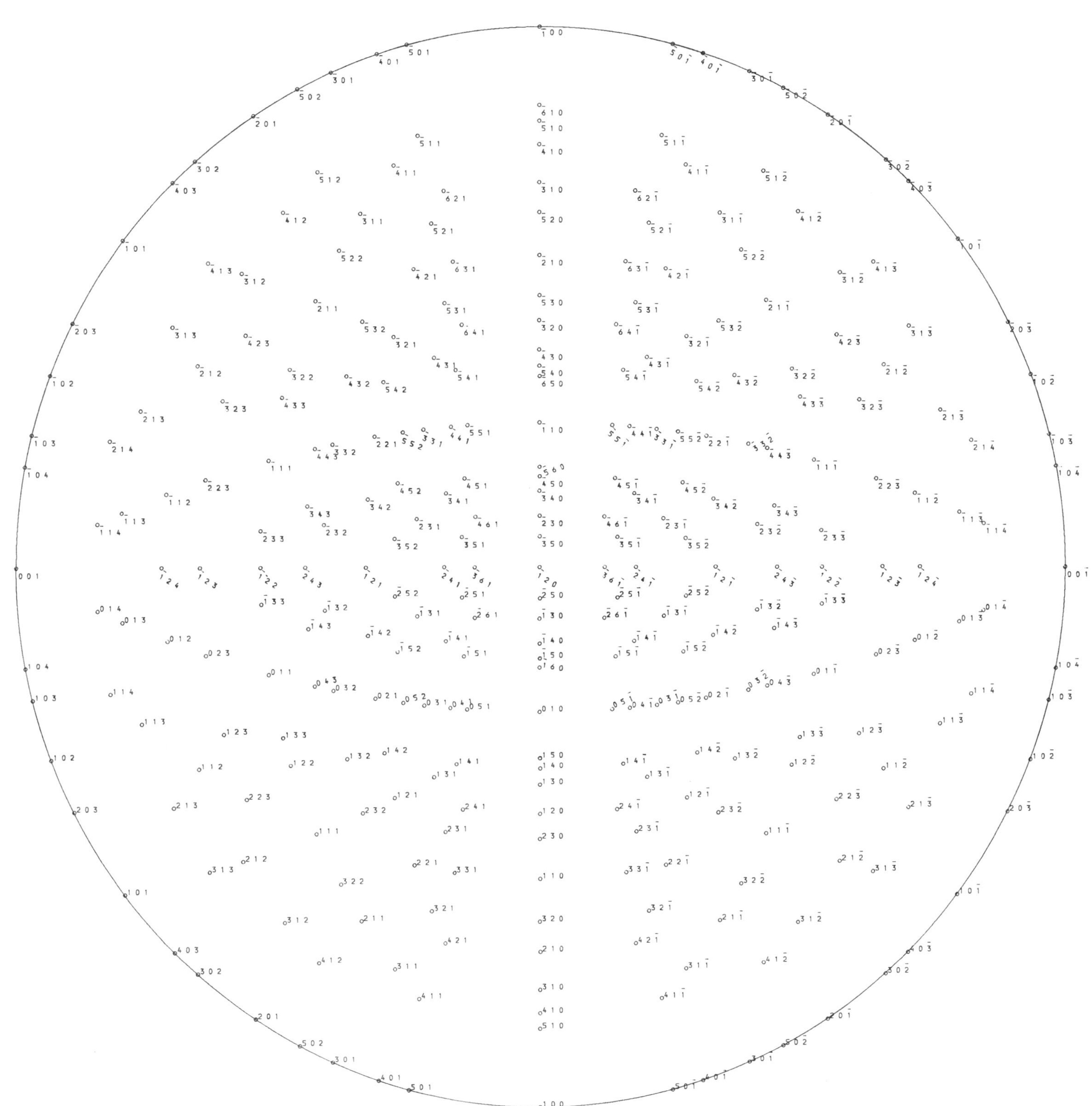

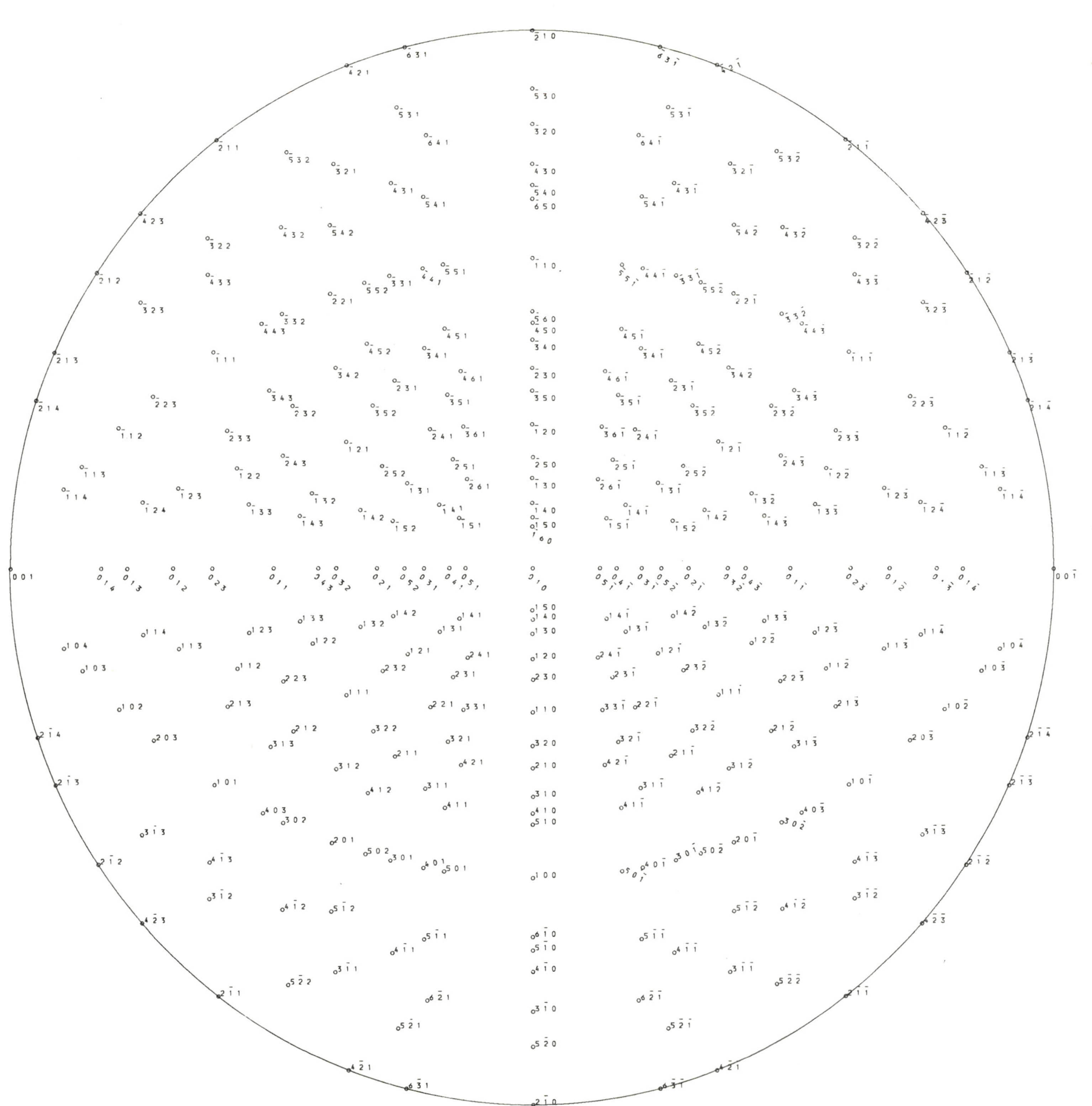

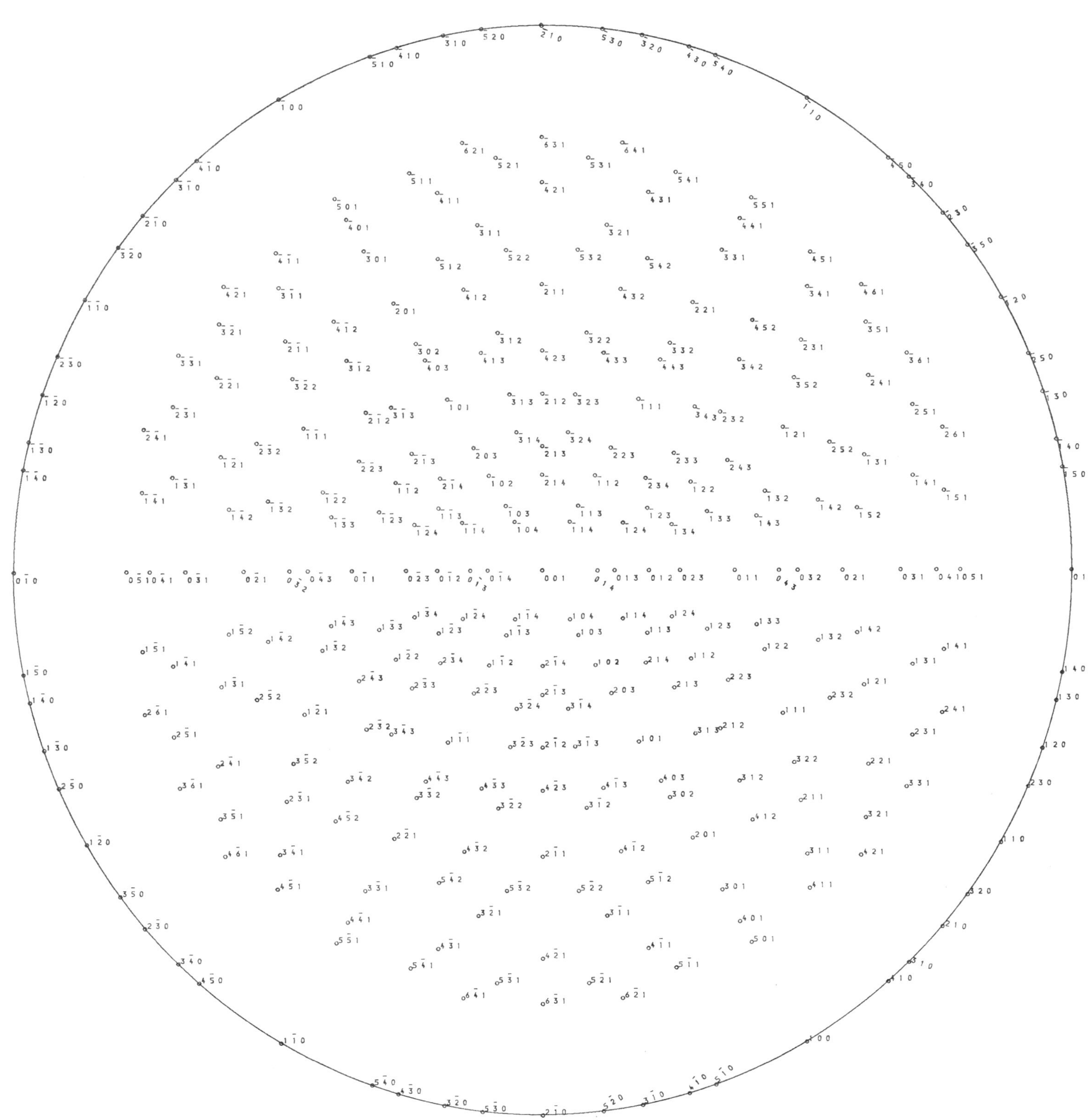

68 Alpha silicon nitride Si_3N_4 Axial ratio 0.7250 010 otherwise $\bar{1}2\bar{1}0$ axis

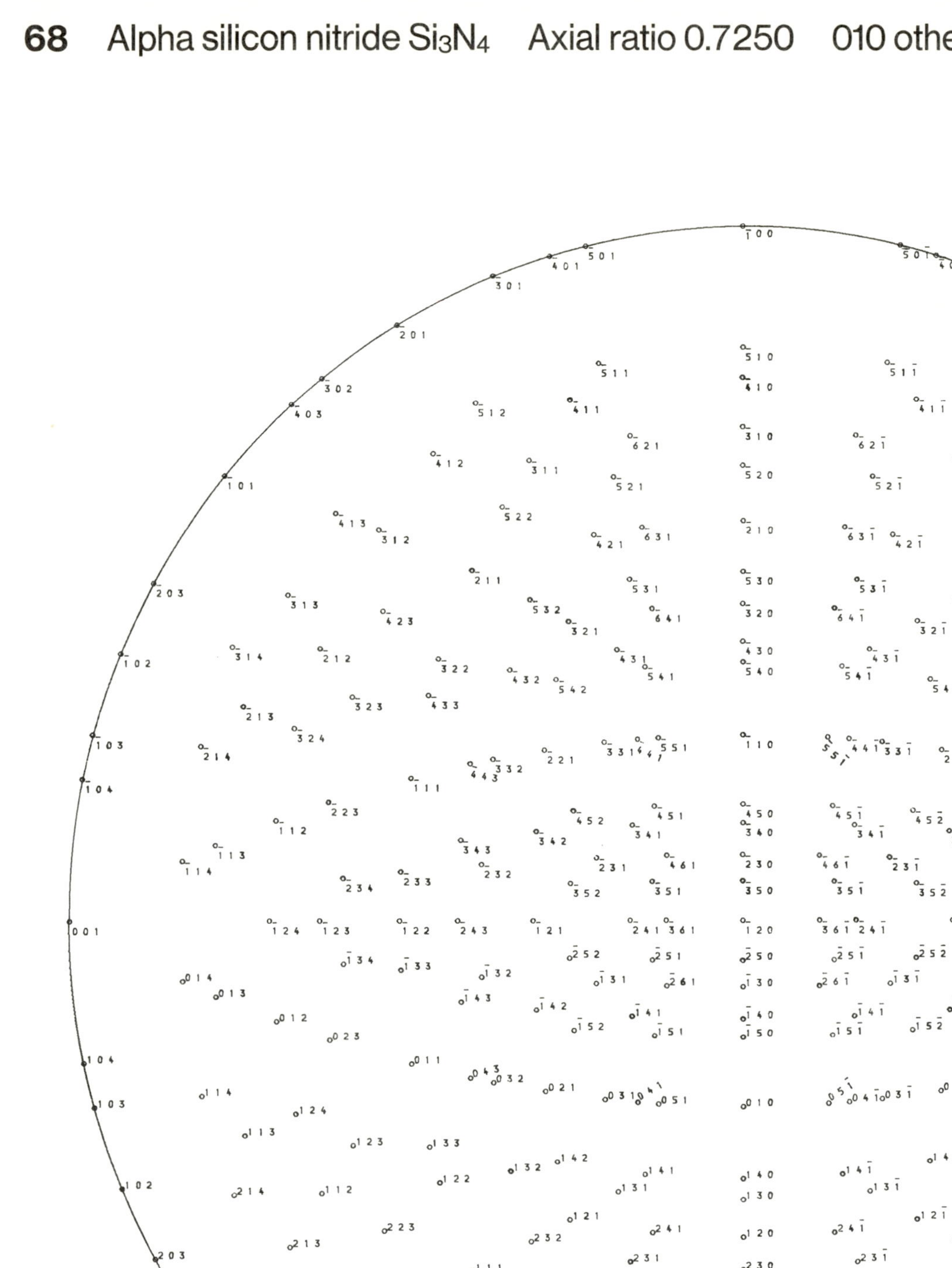

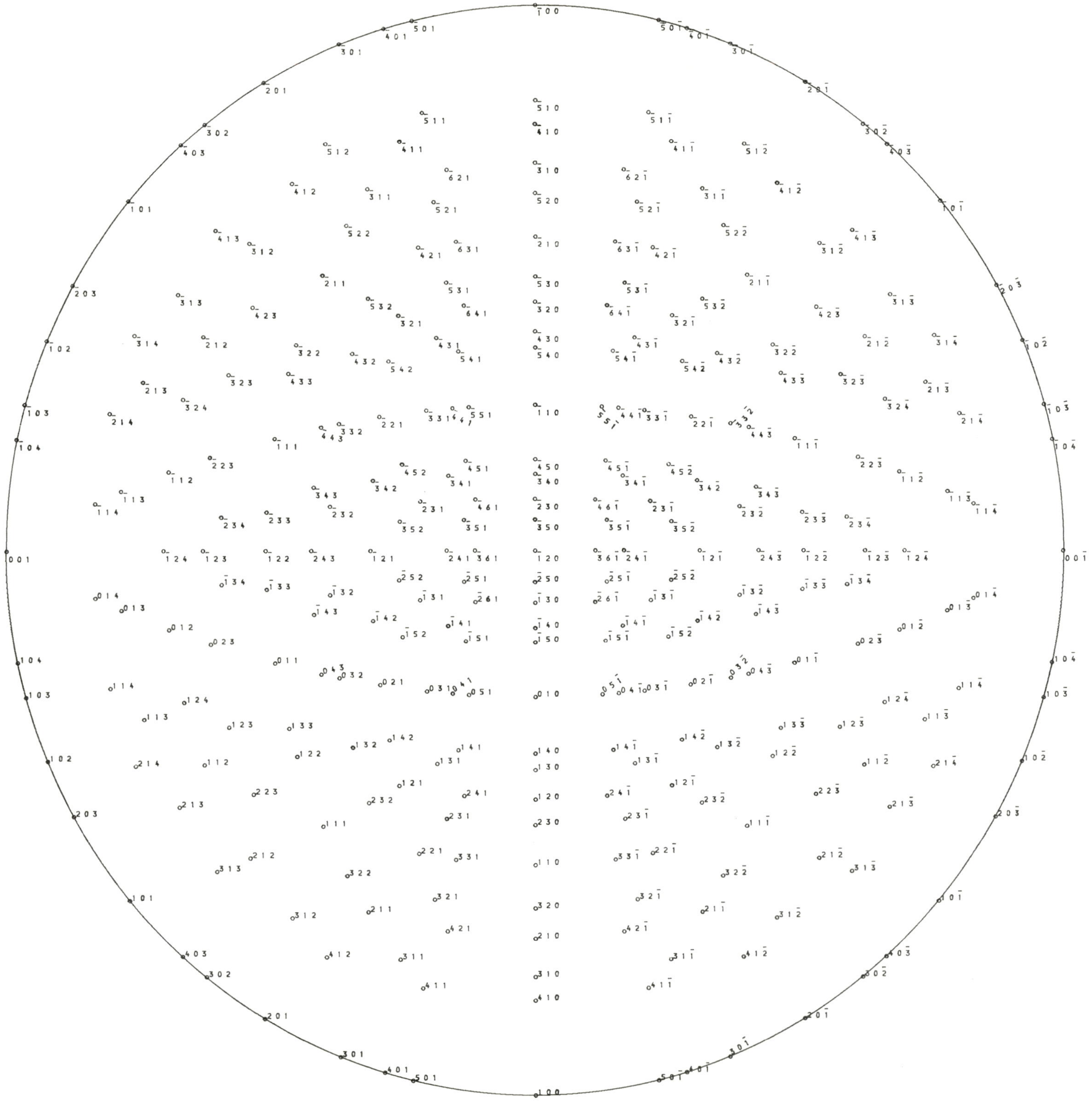

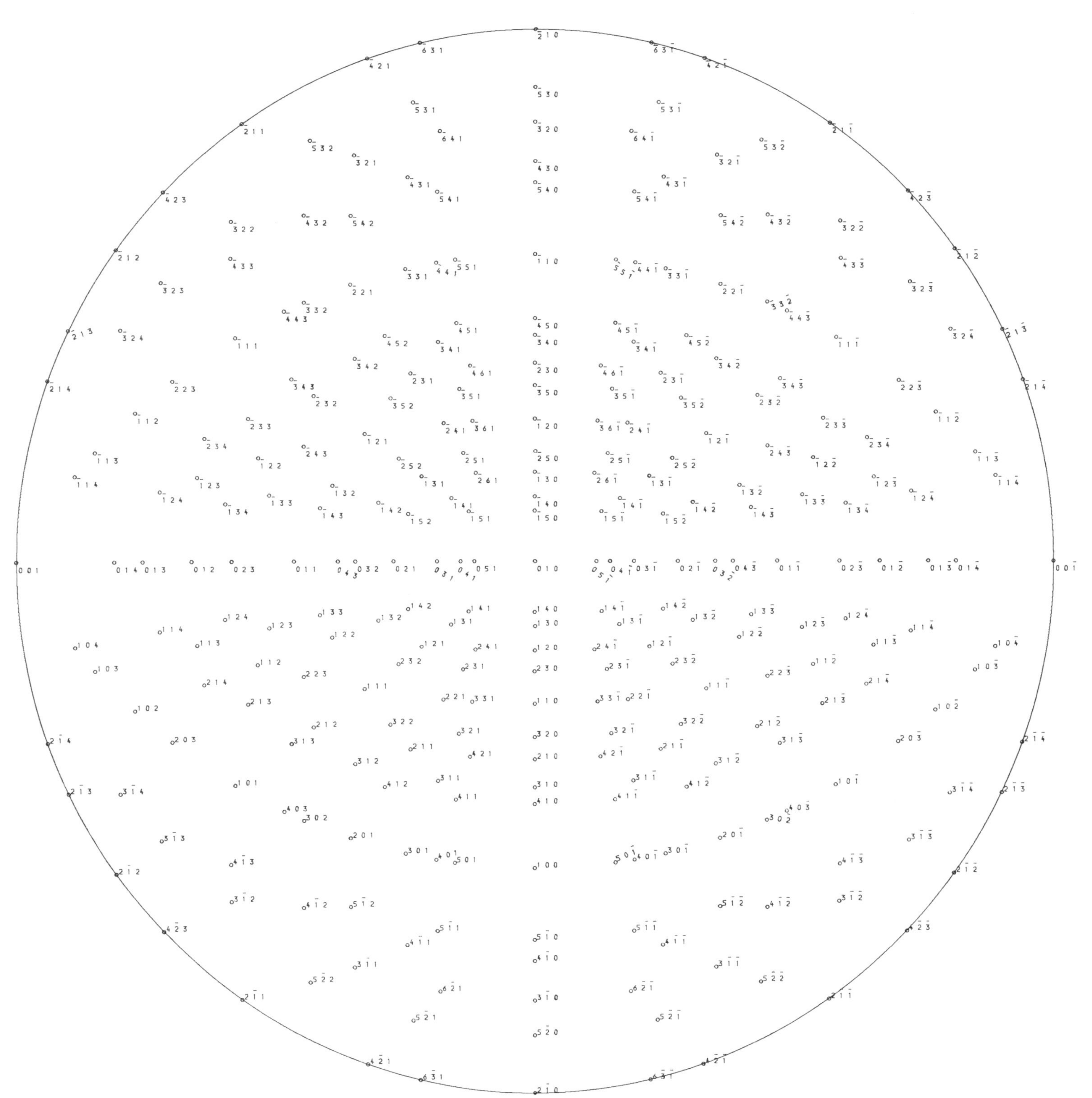

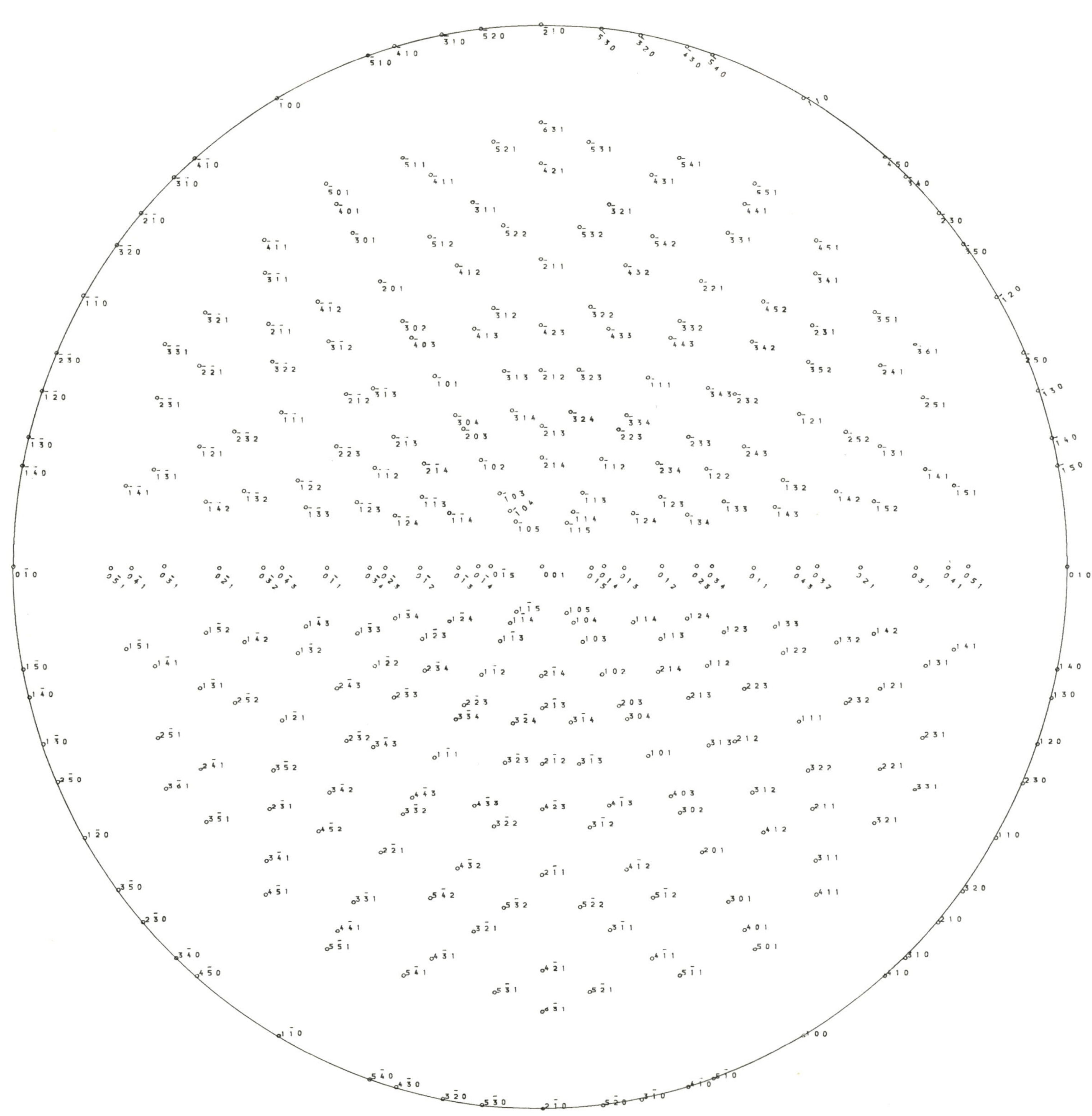

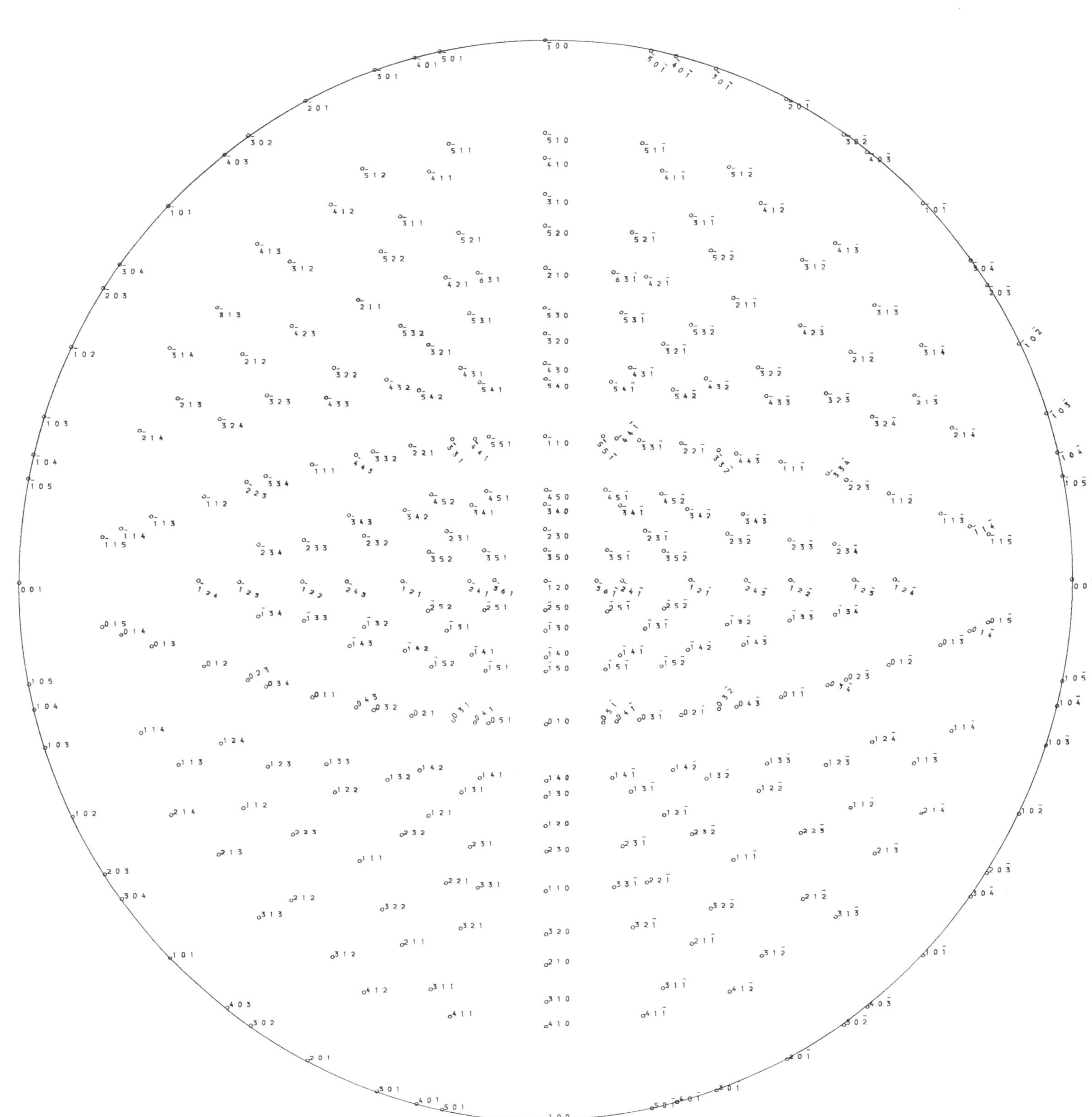

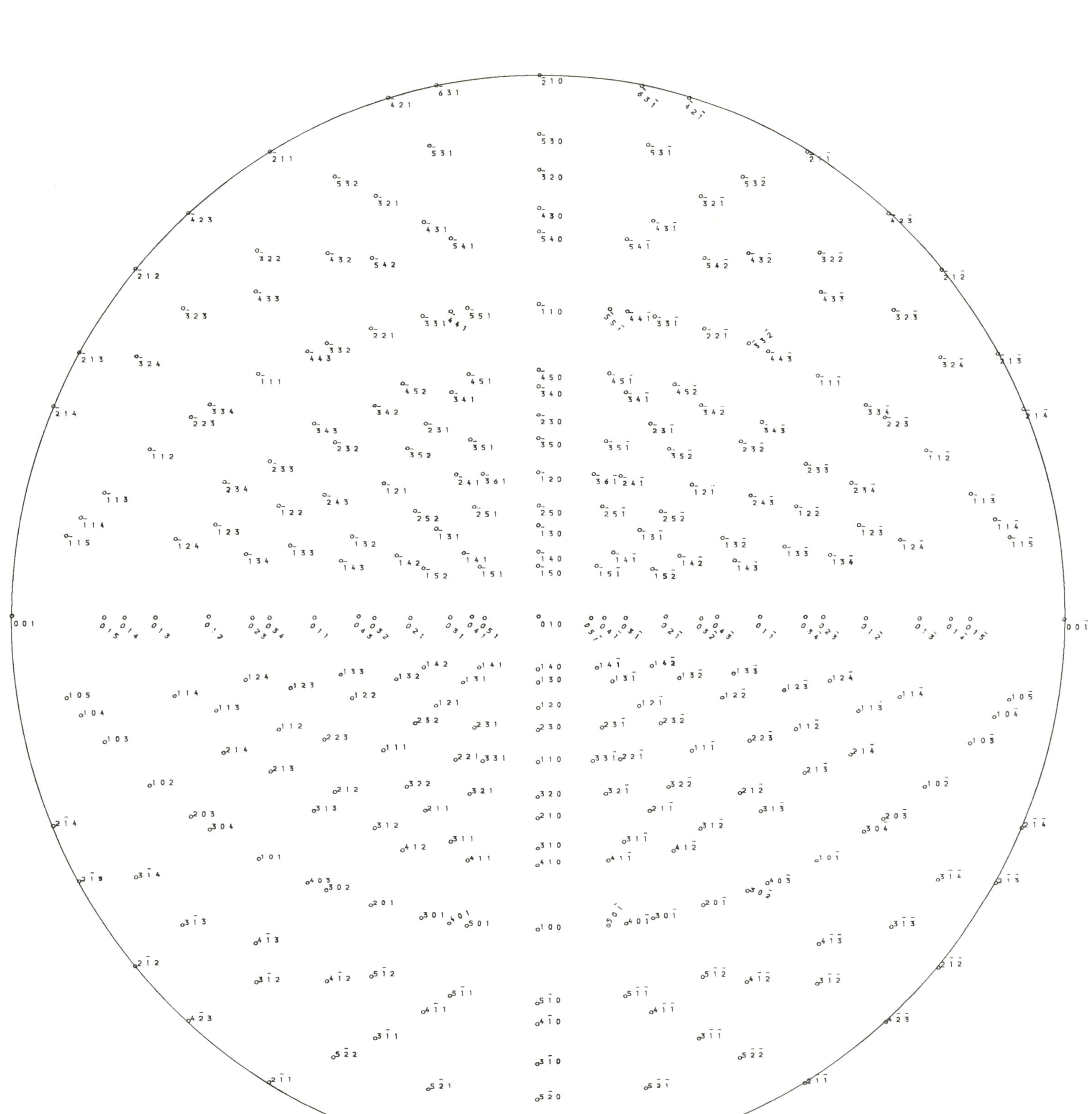

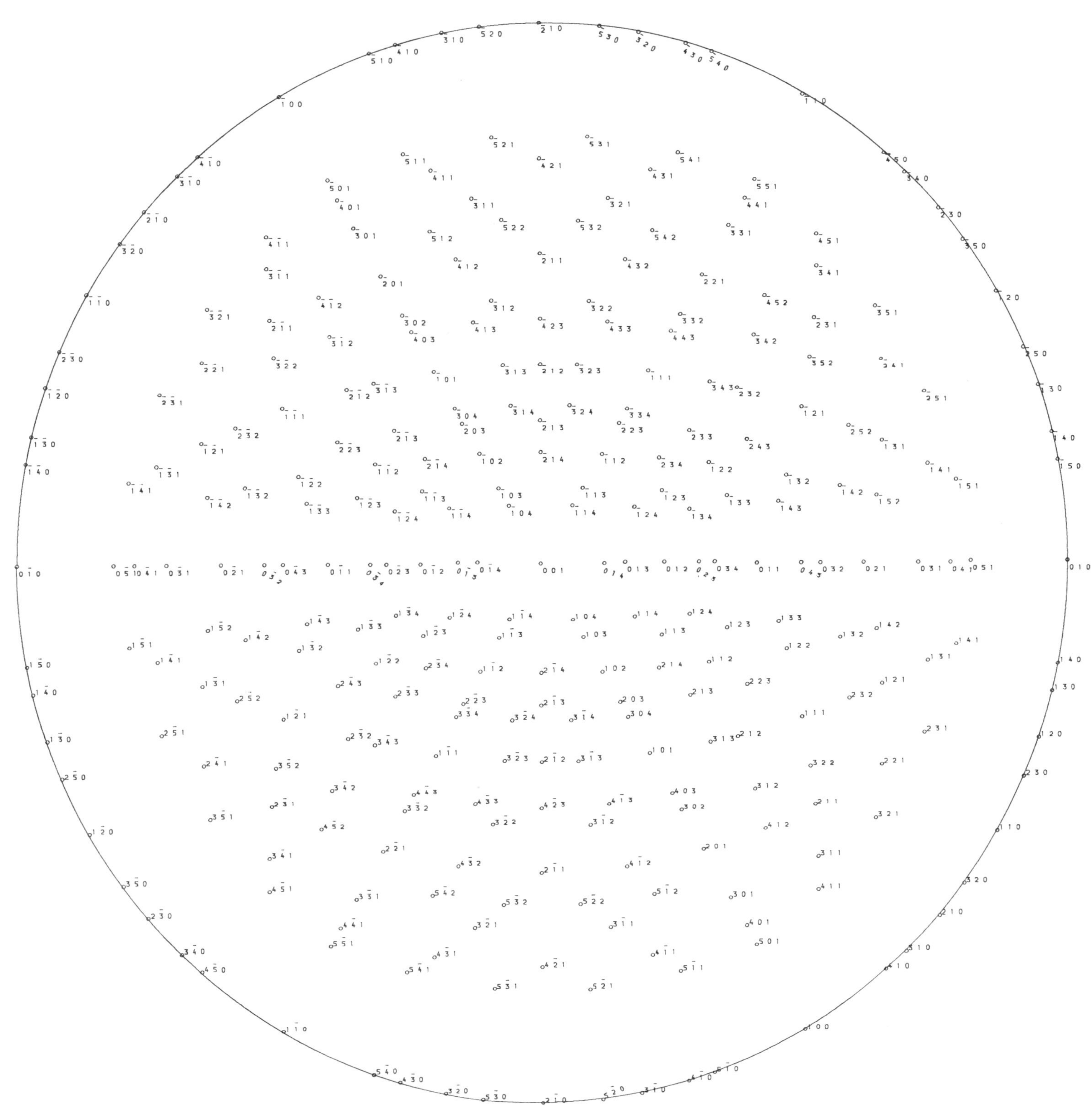

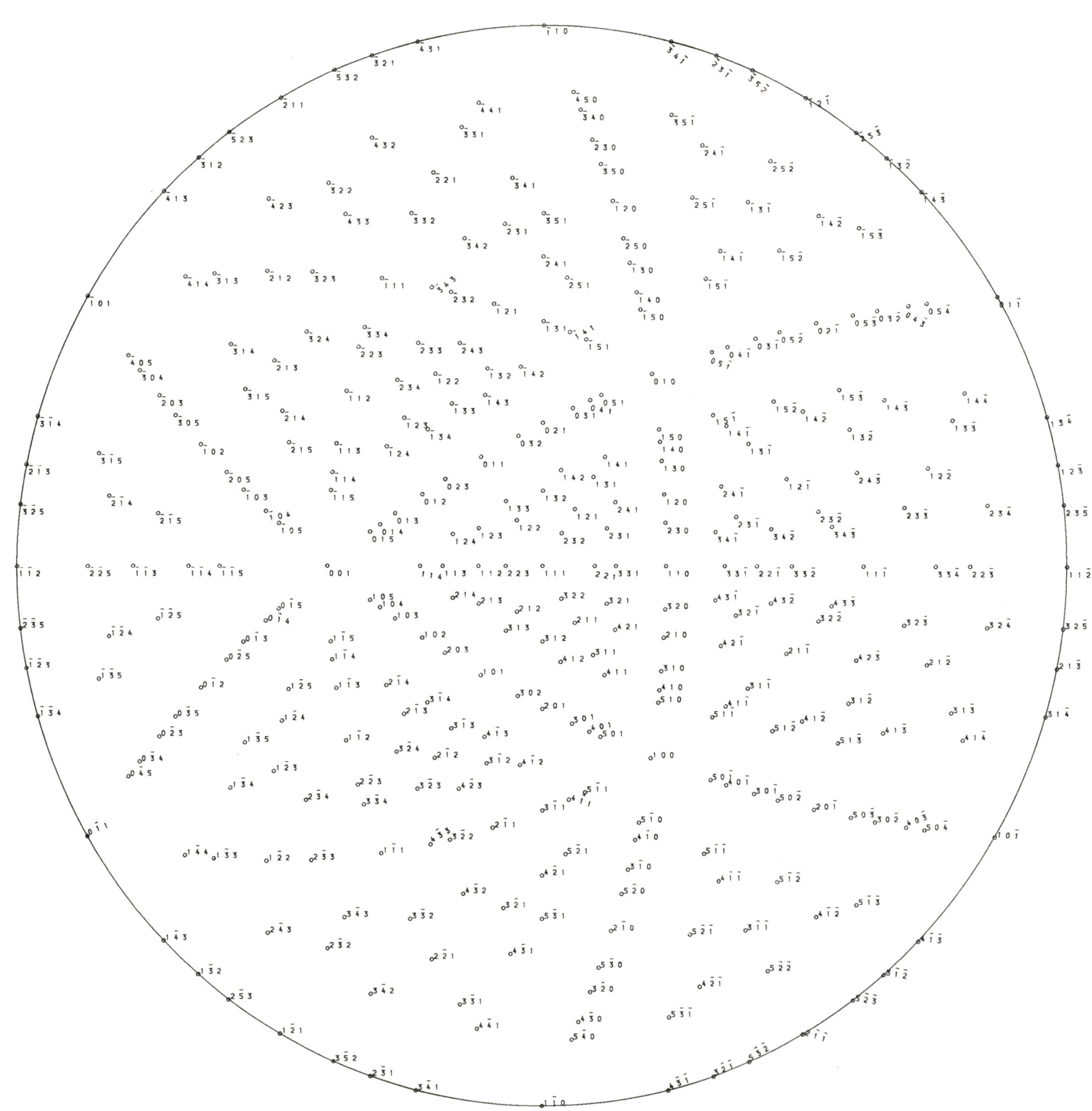

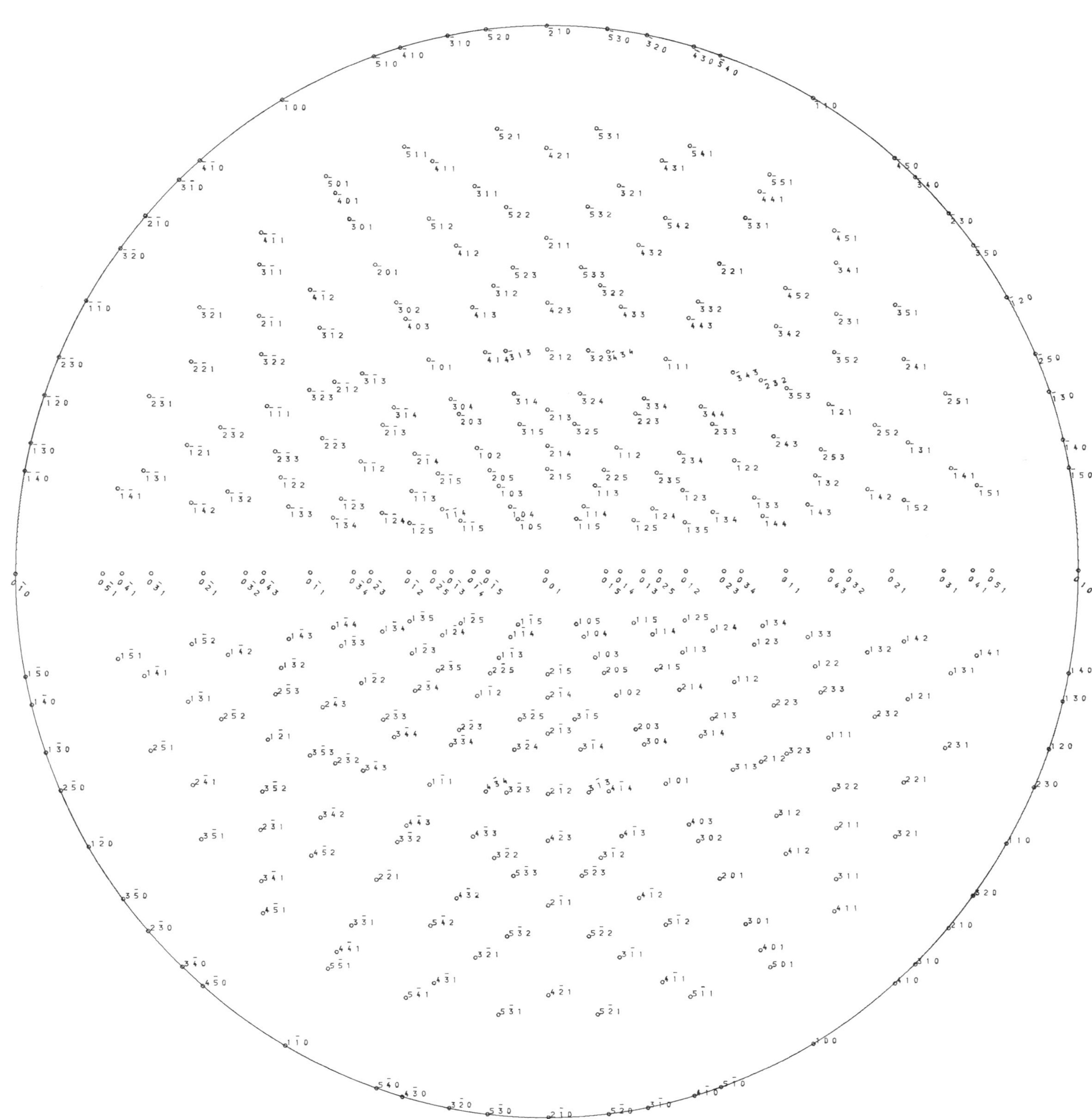

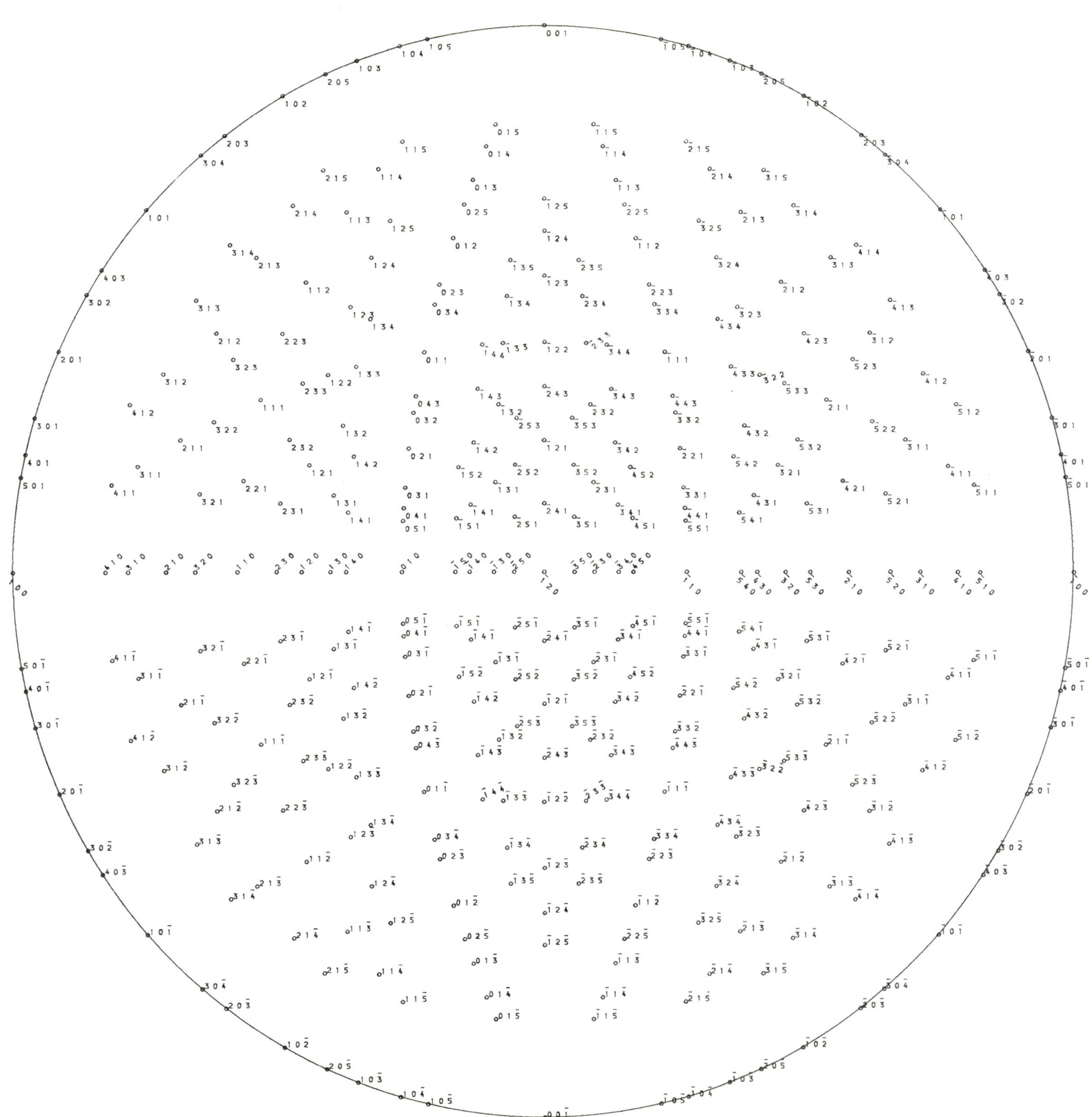

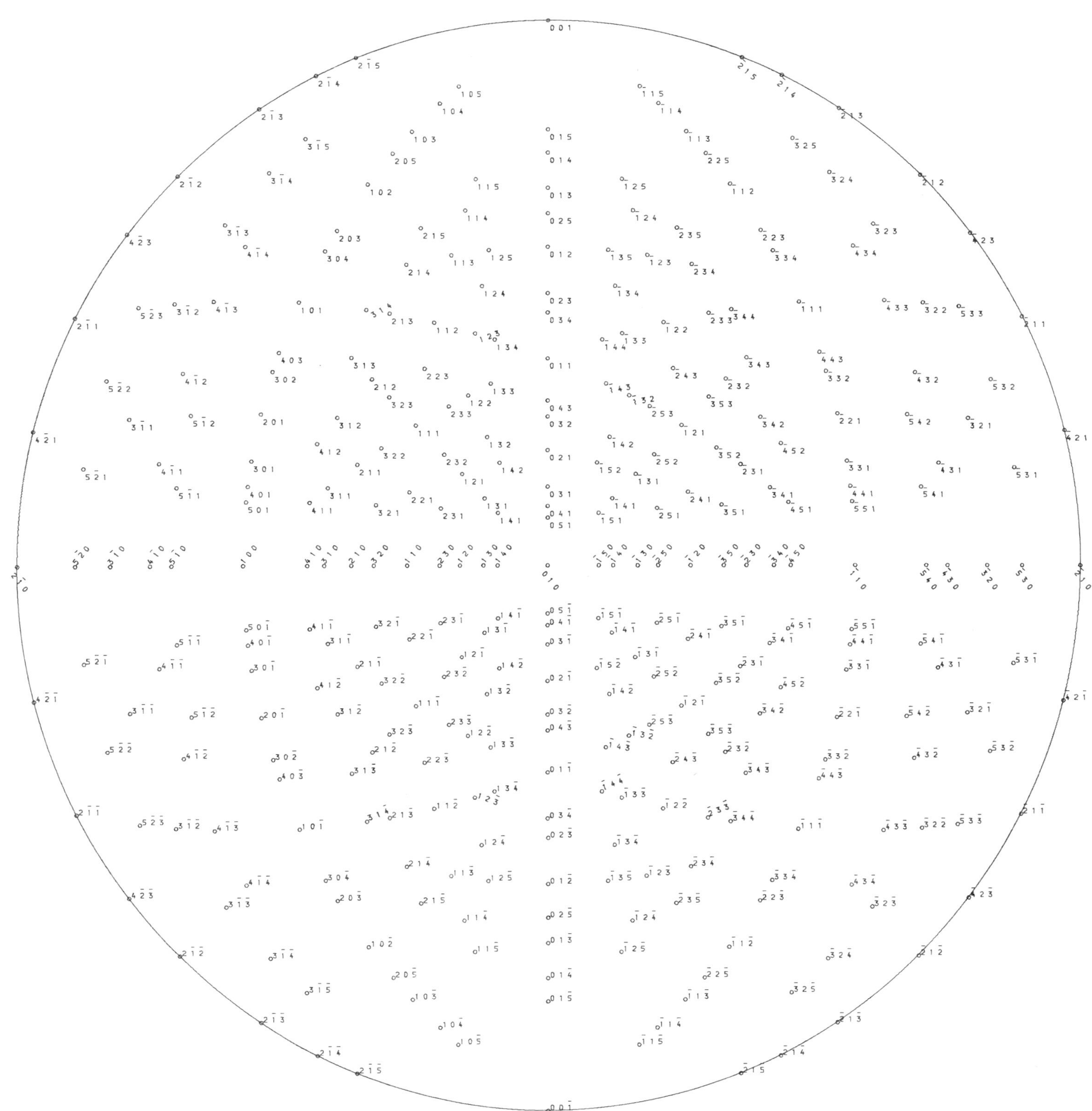

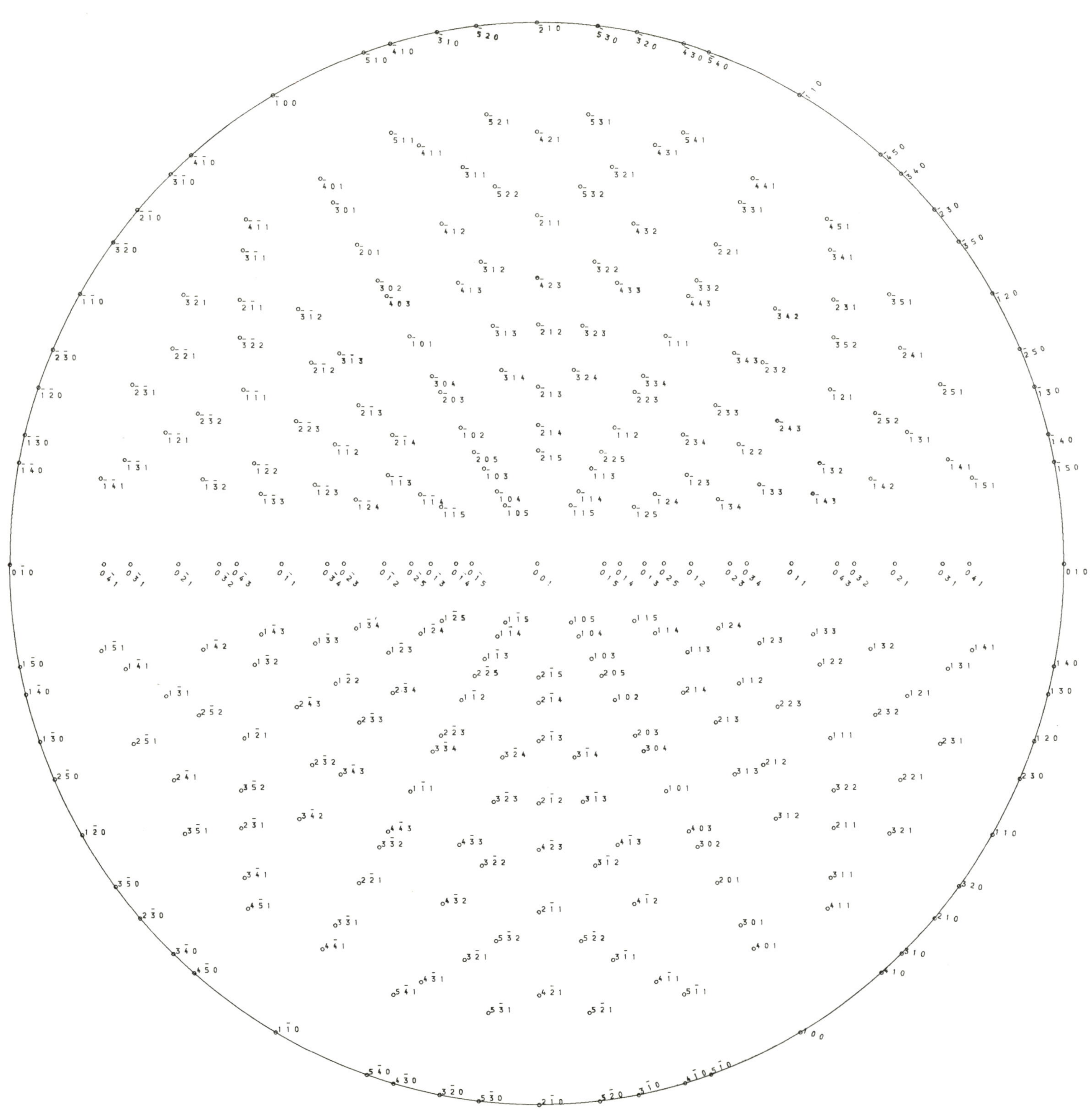

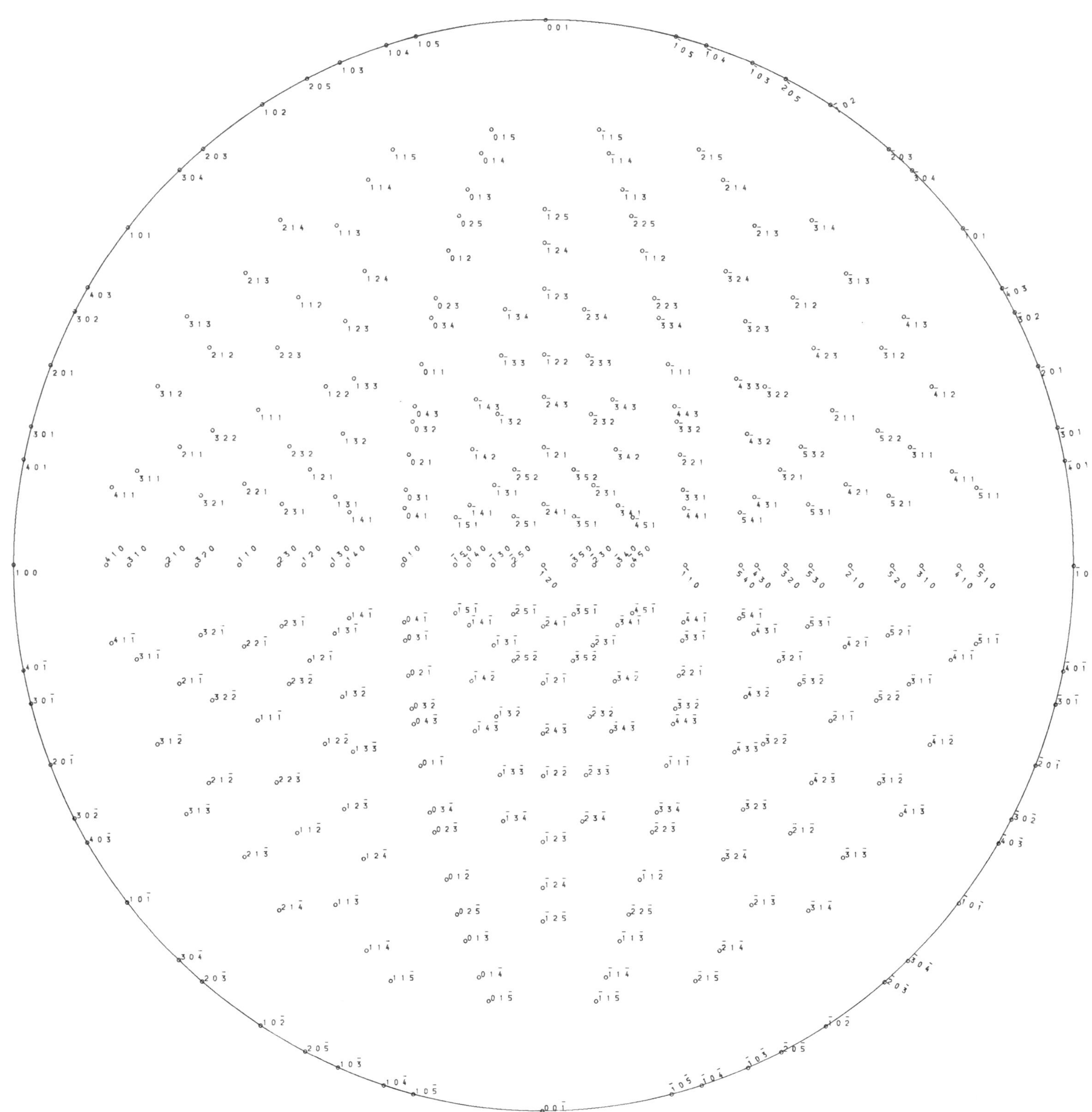

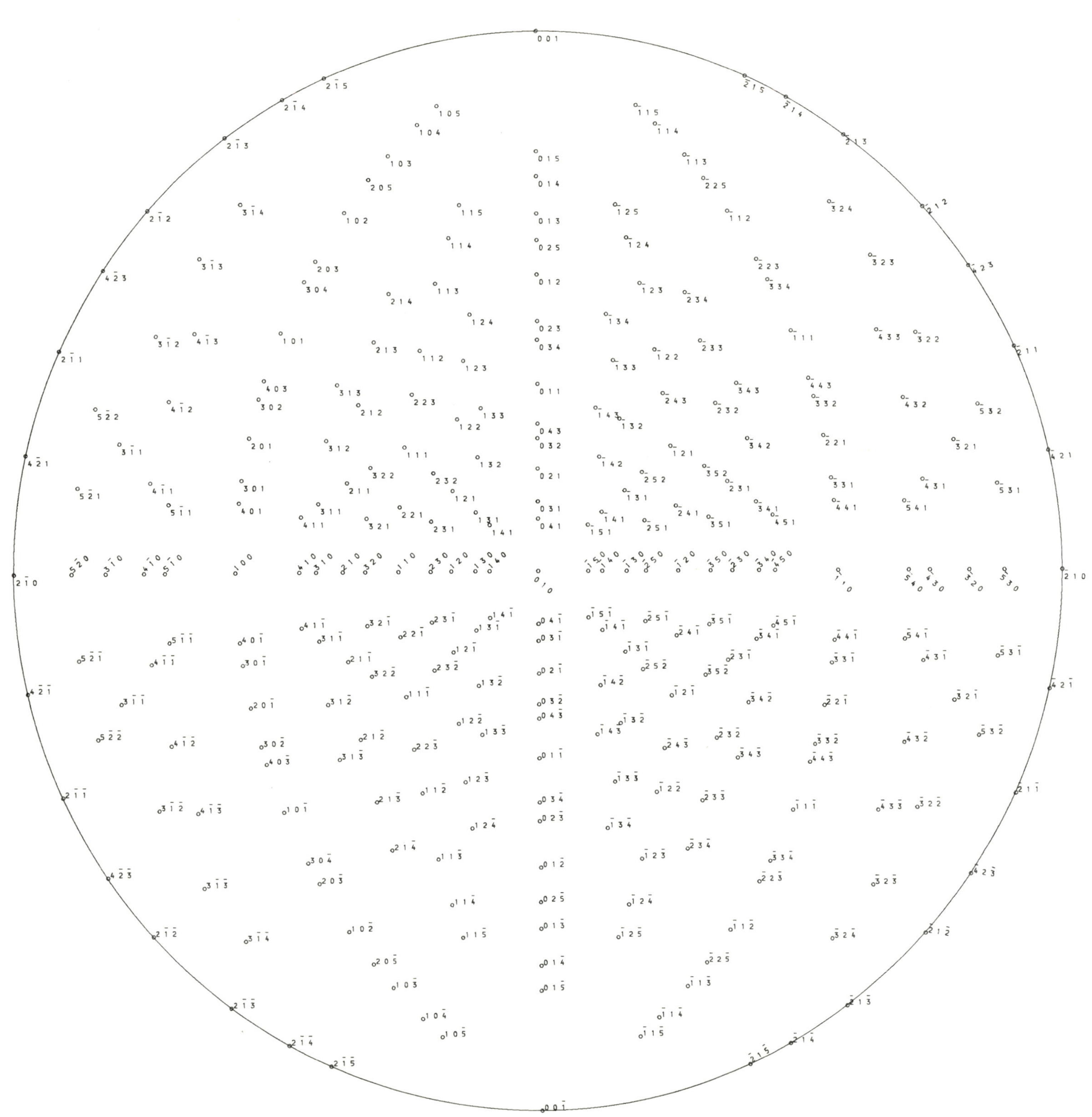

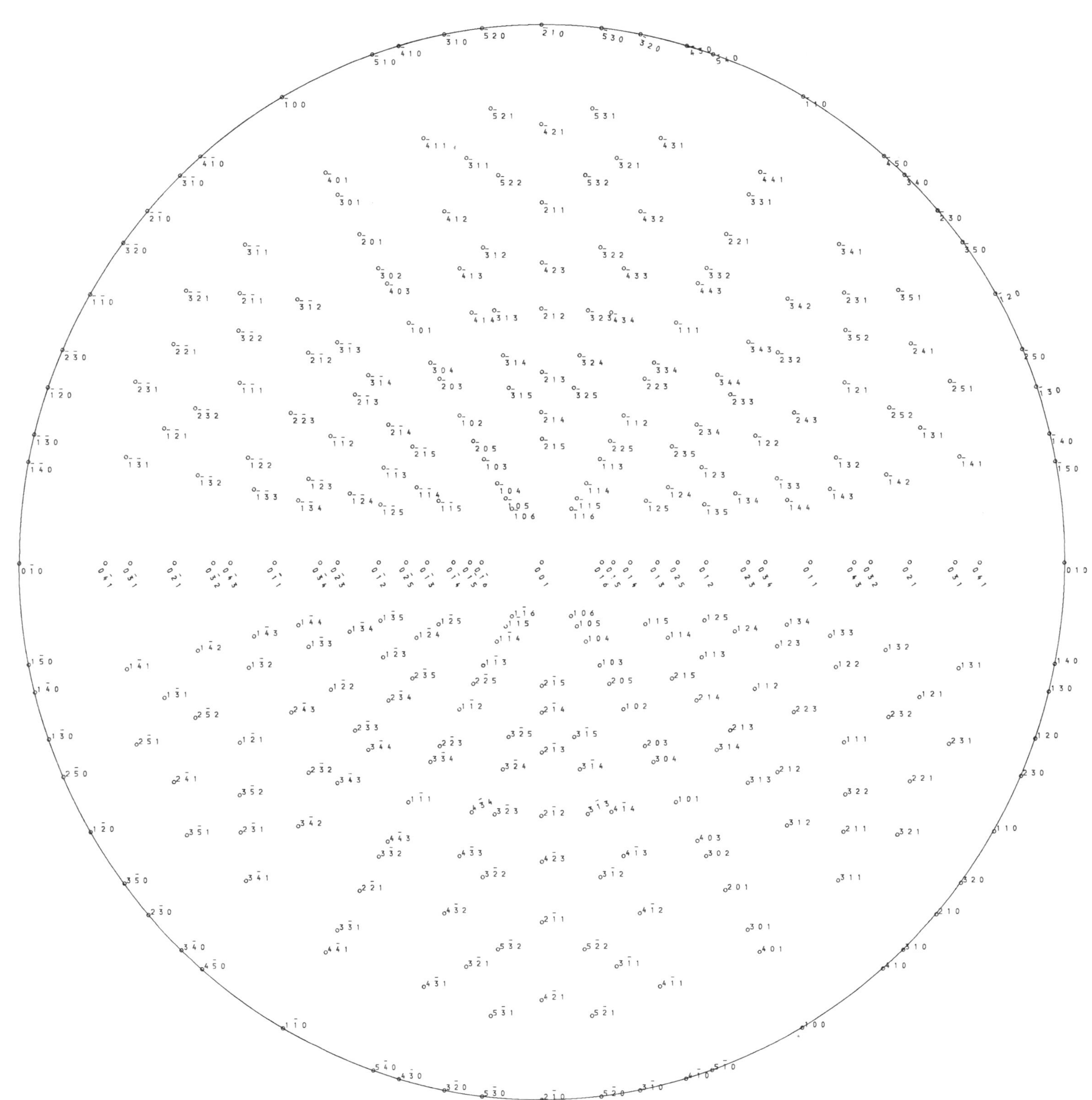

82 Aluminium nickel $AlNi_2$ Axial ratio 1.2135 010 otherwise $\bar{1}2\bar{1}0$ axis

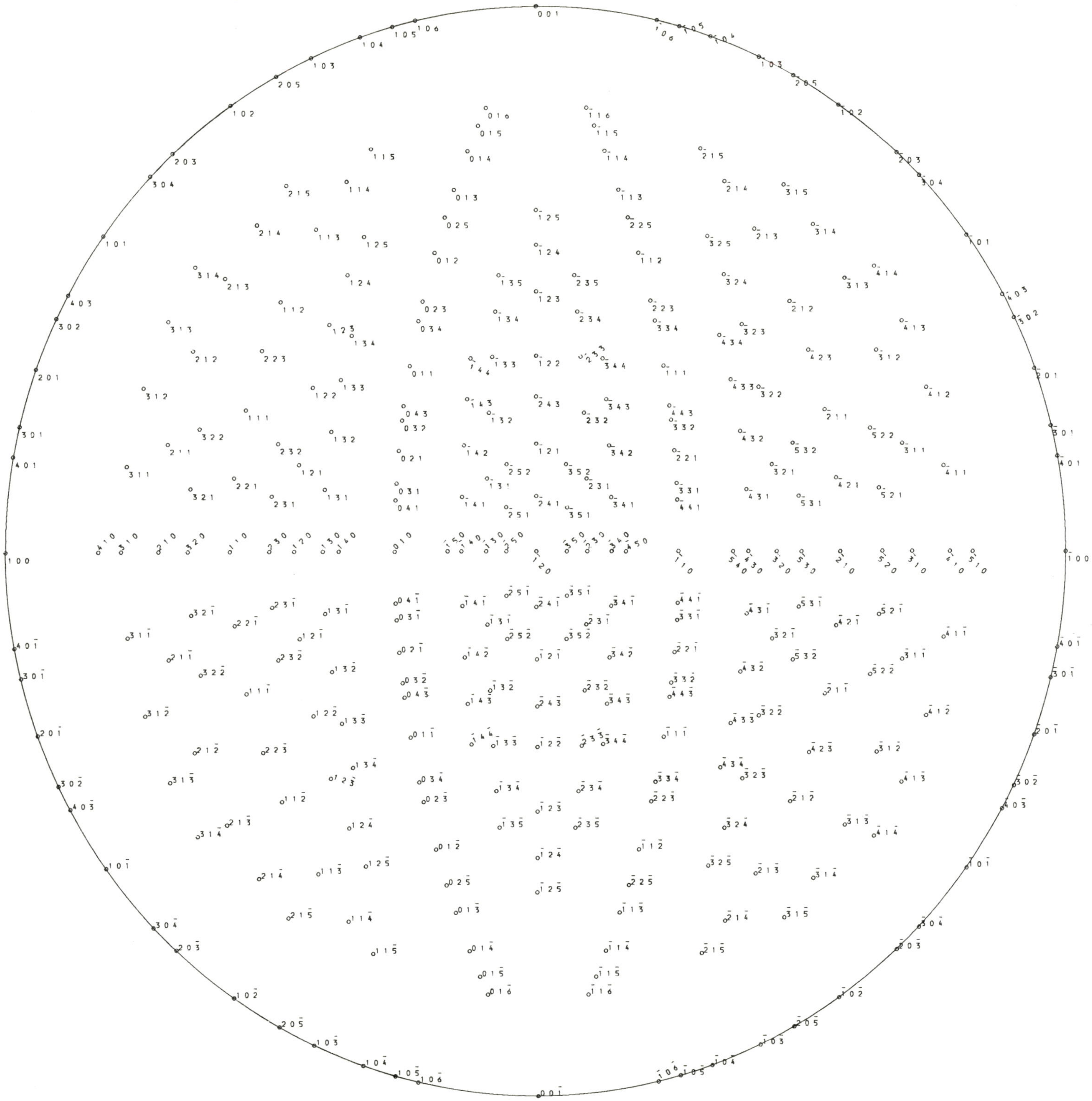

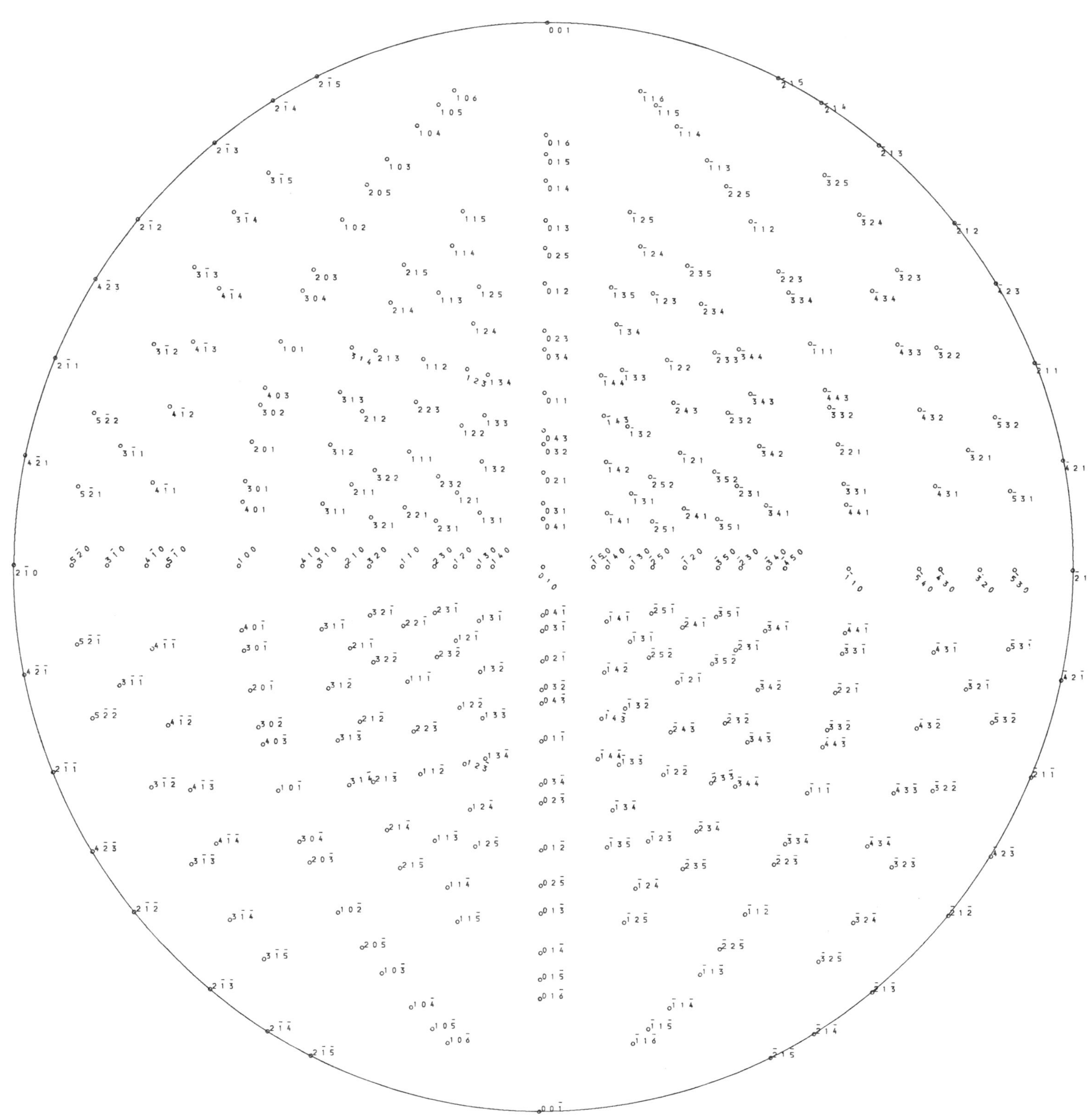

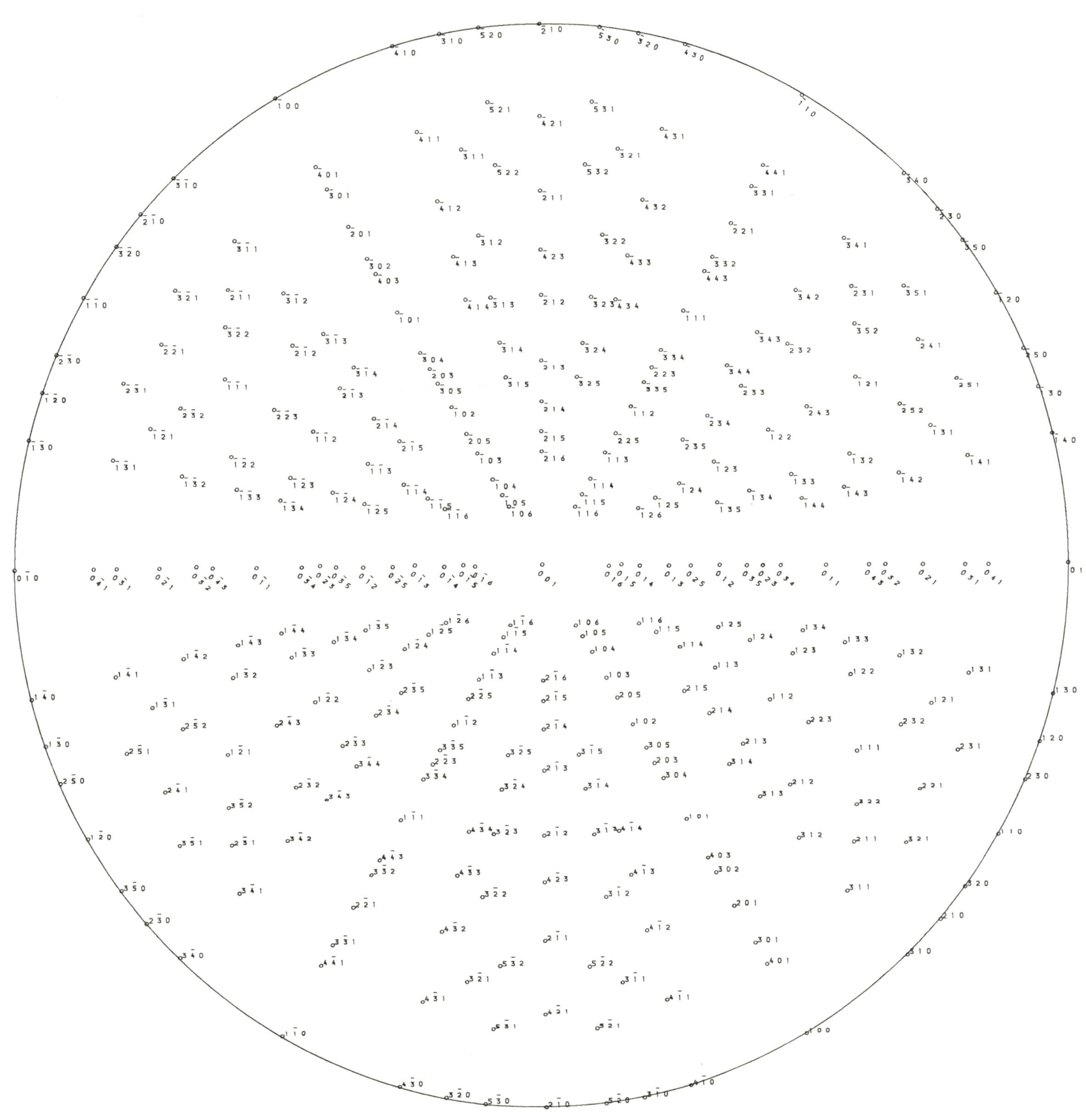

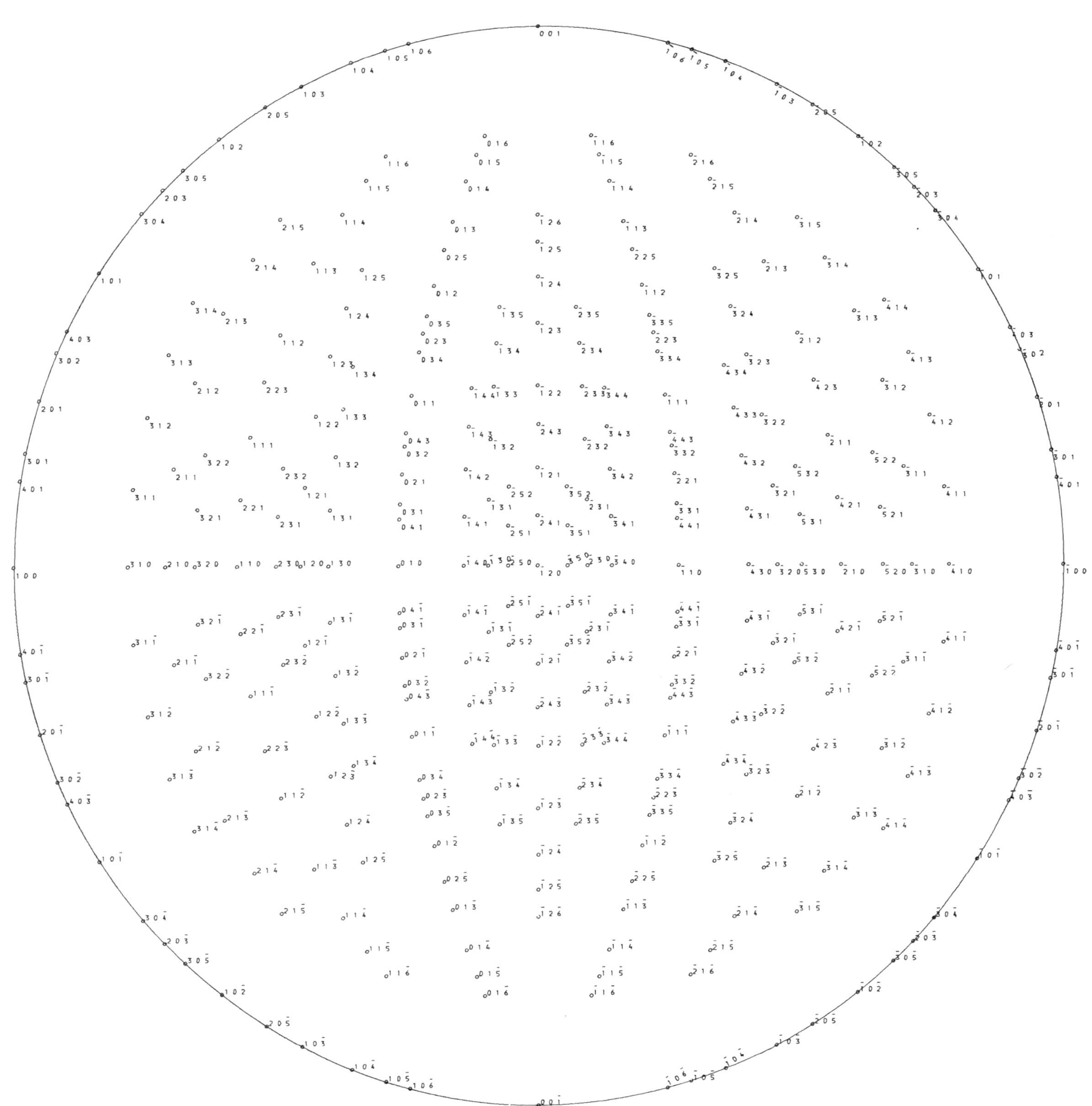

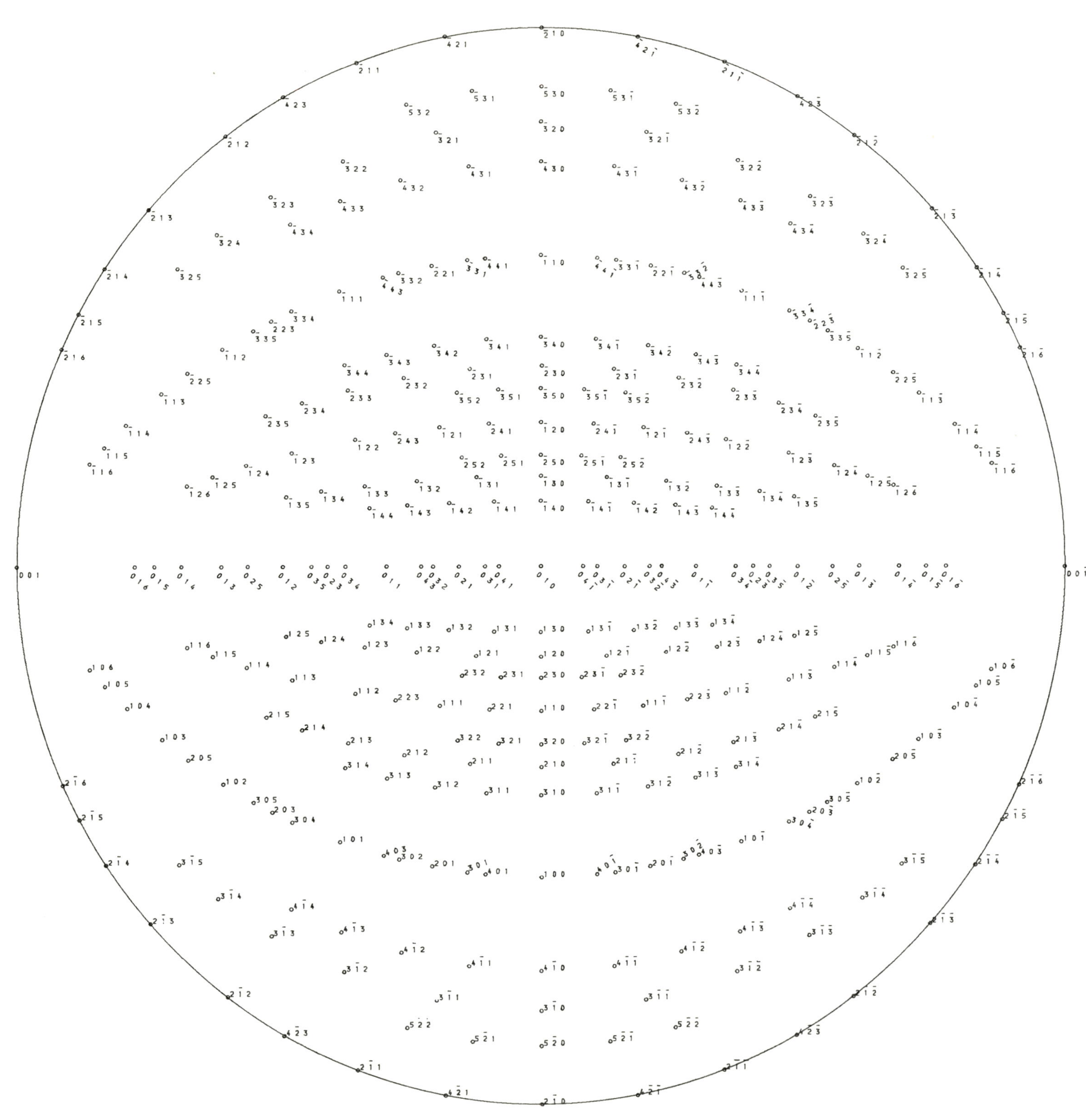

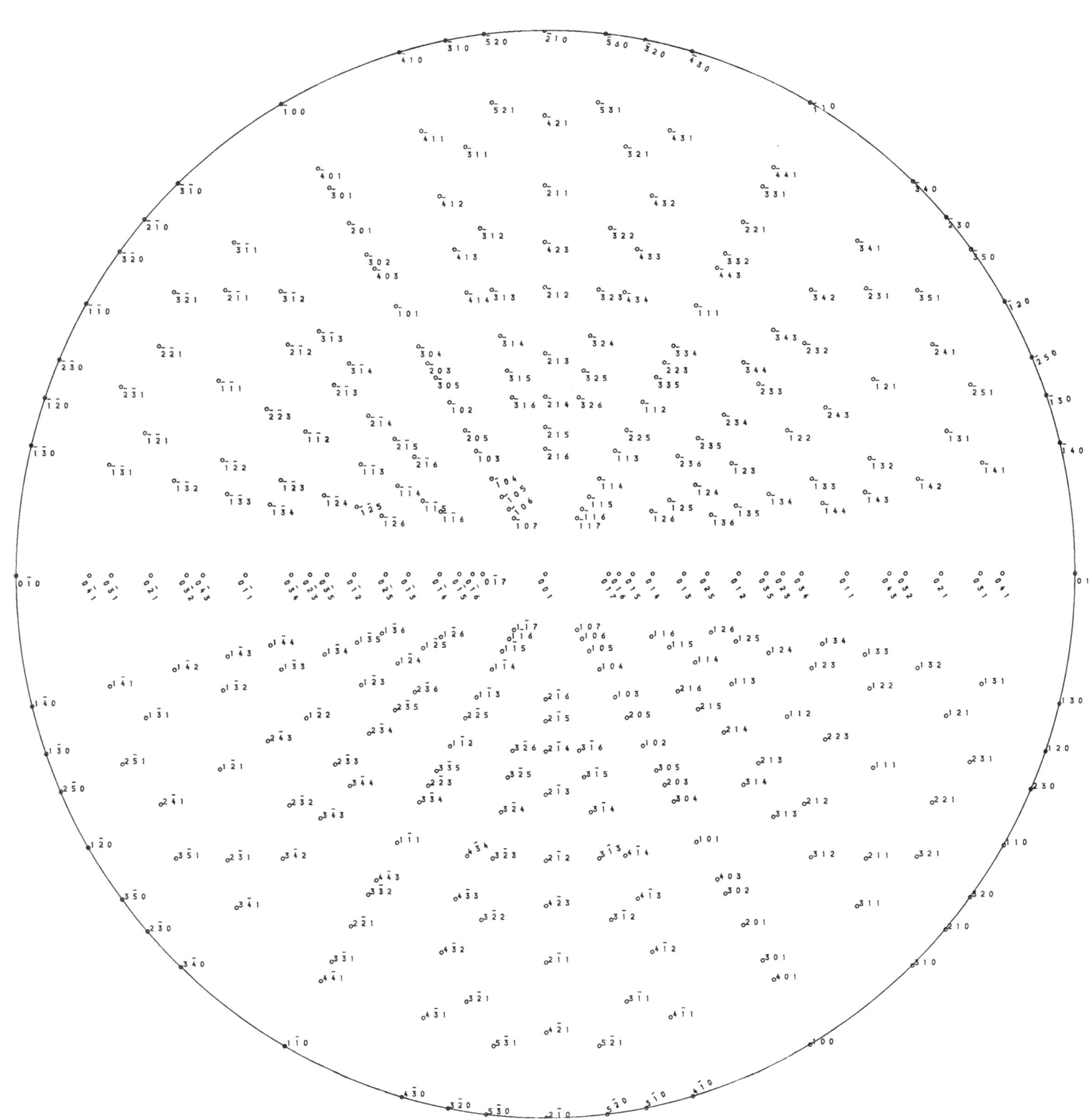

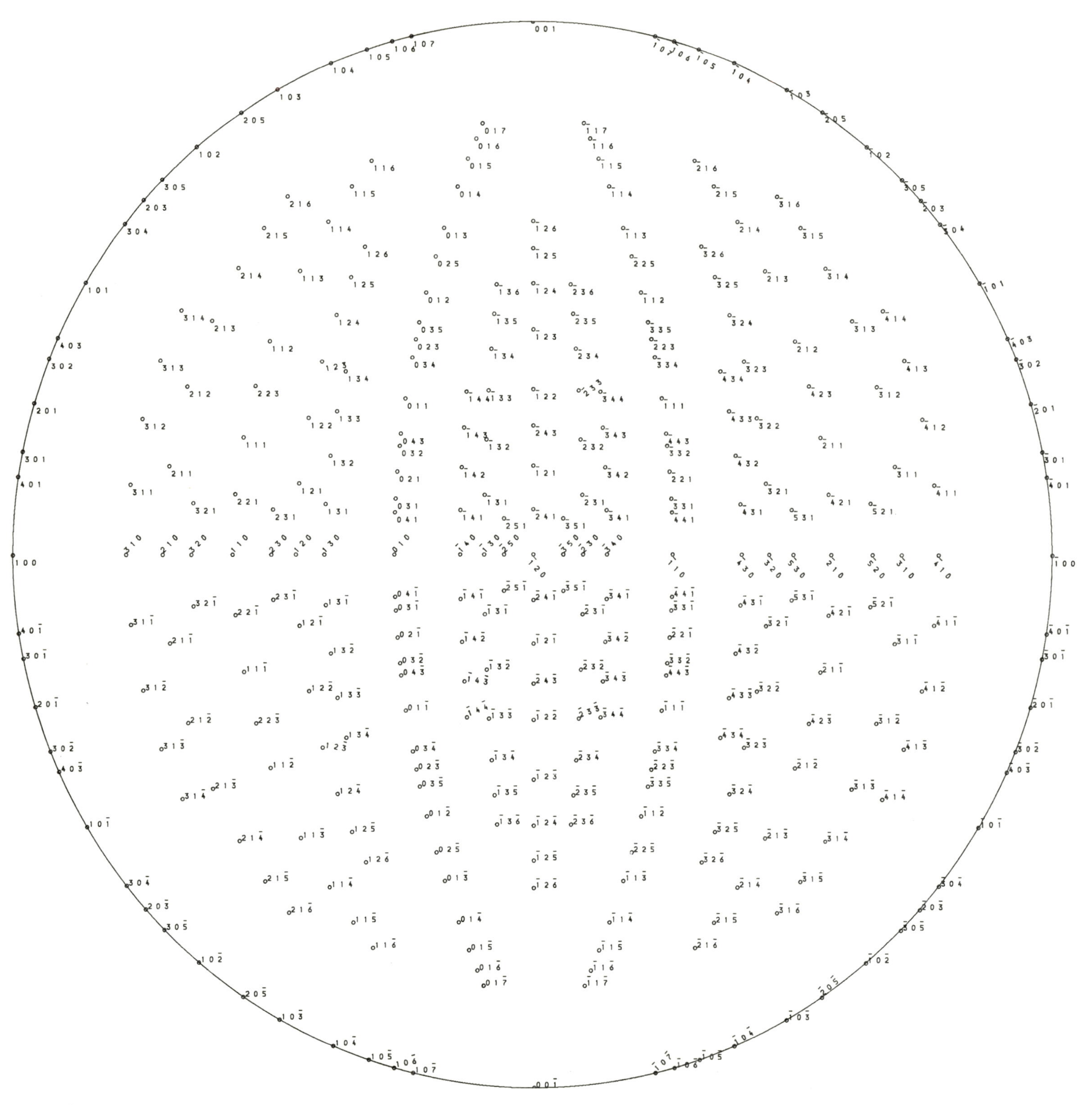

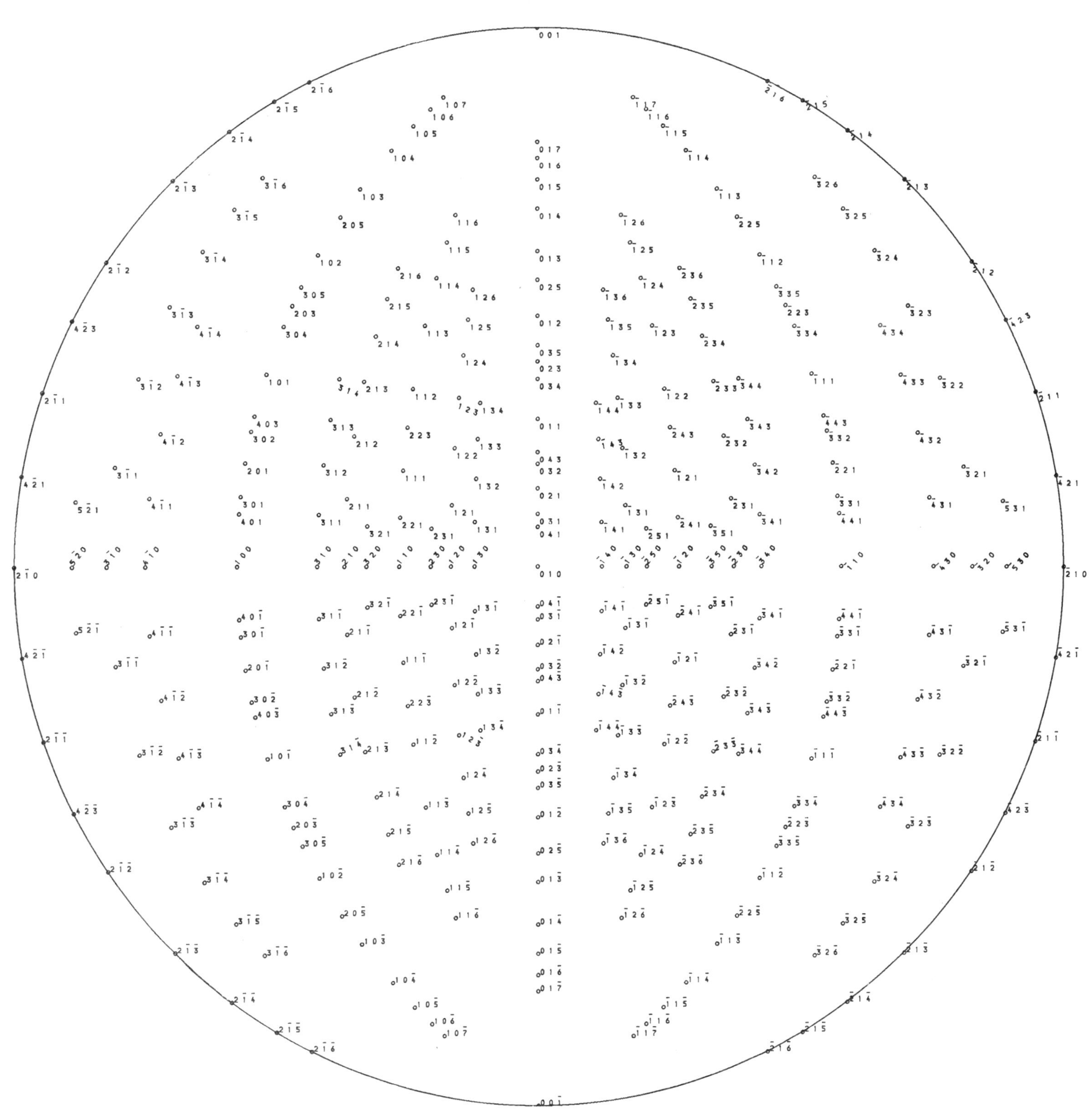

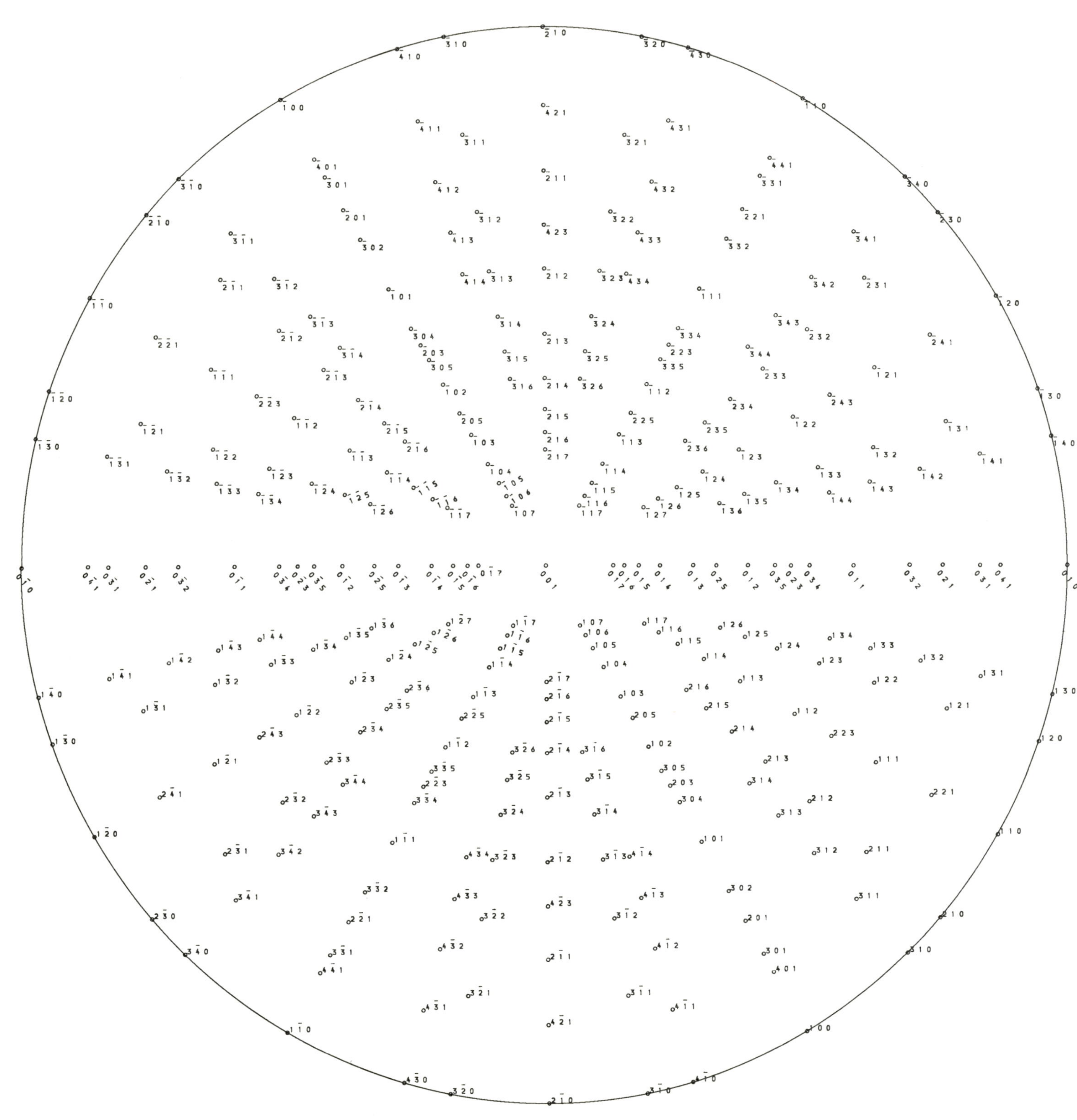

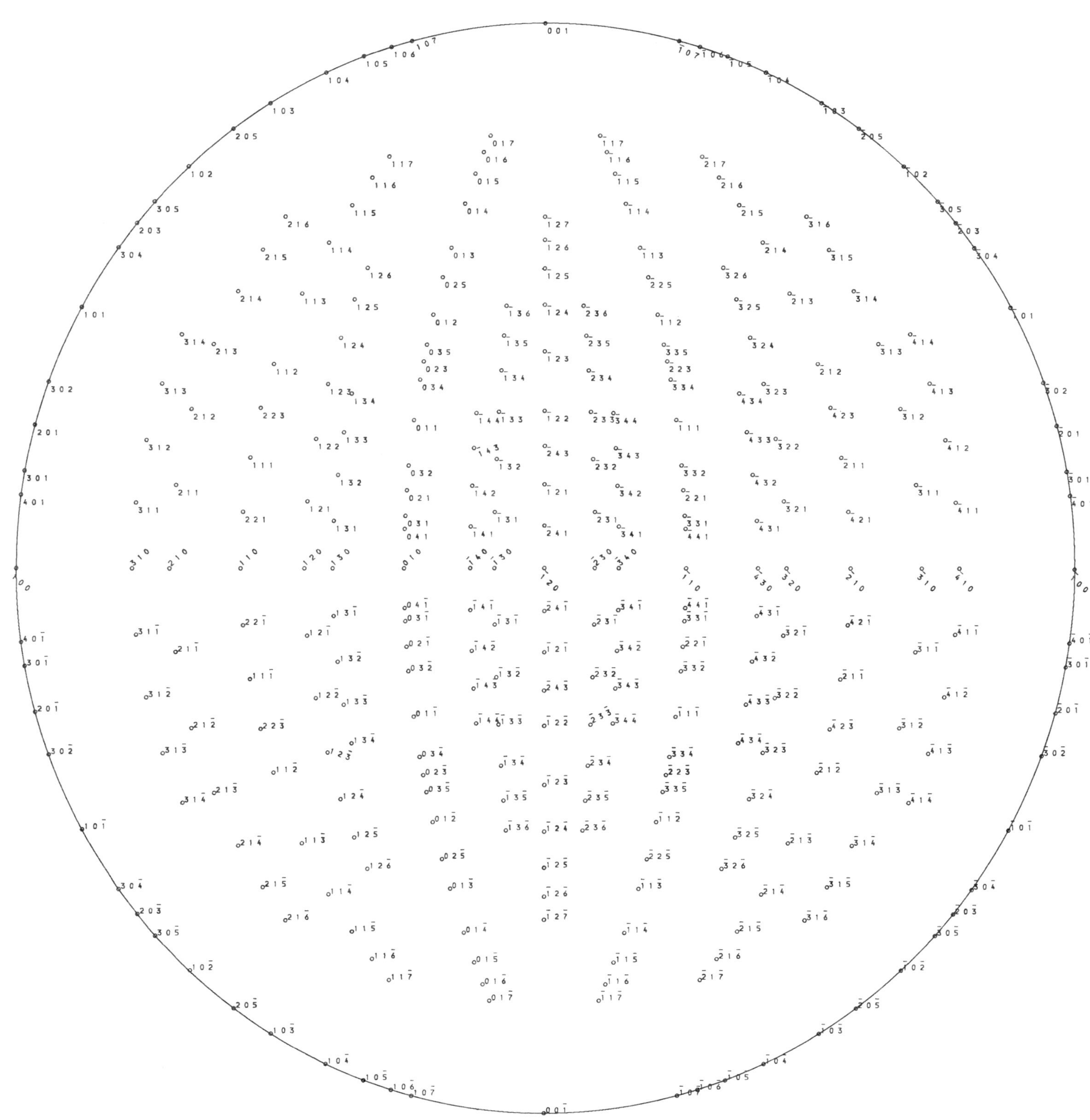

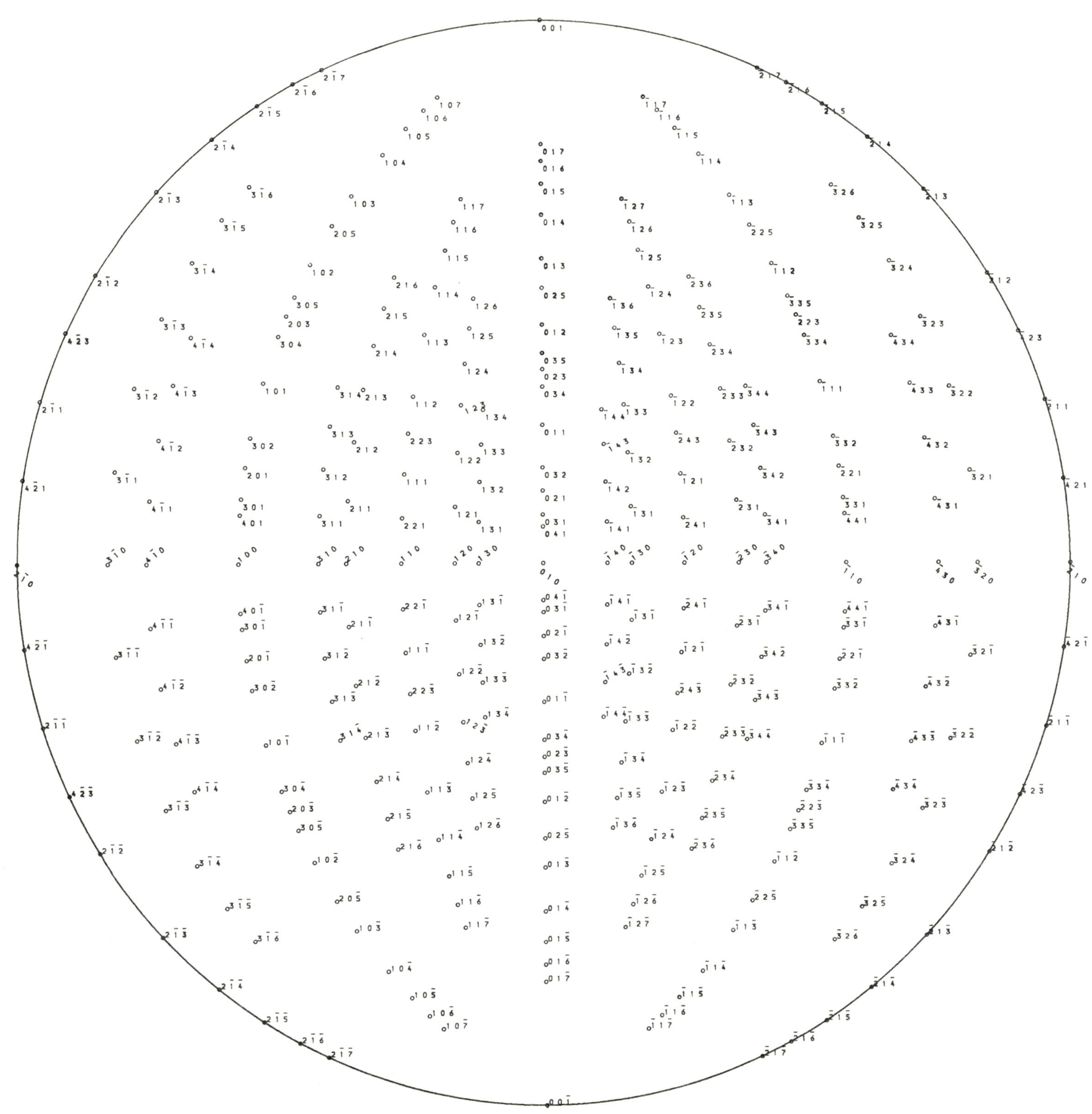

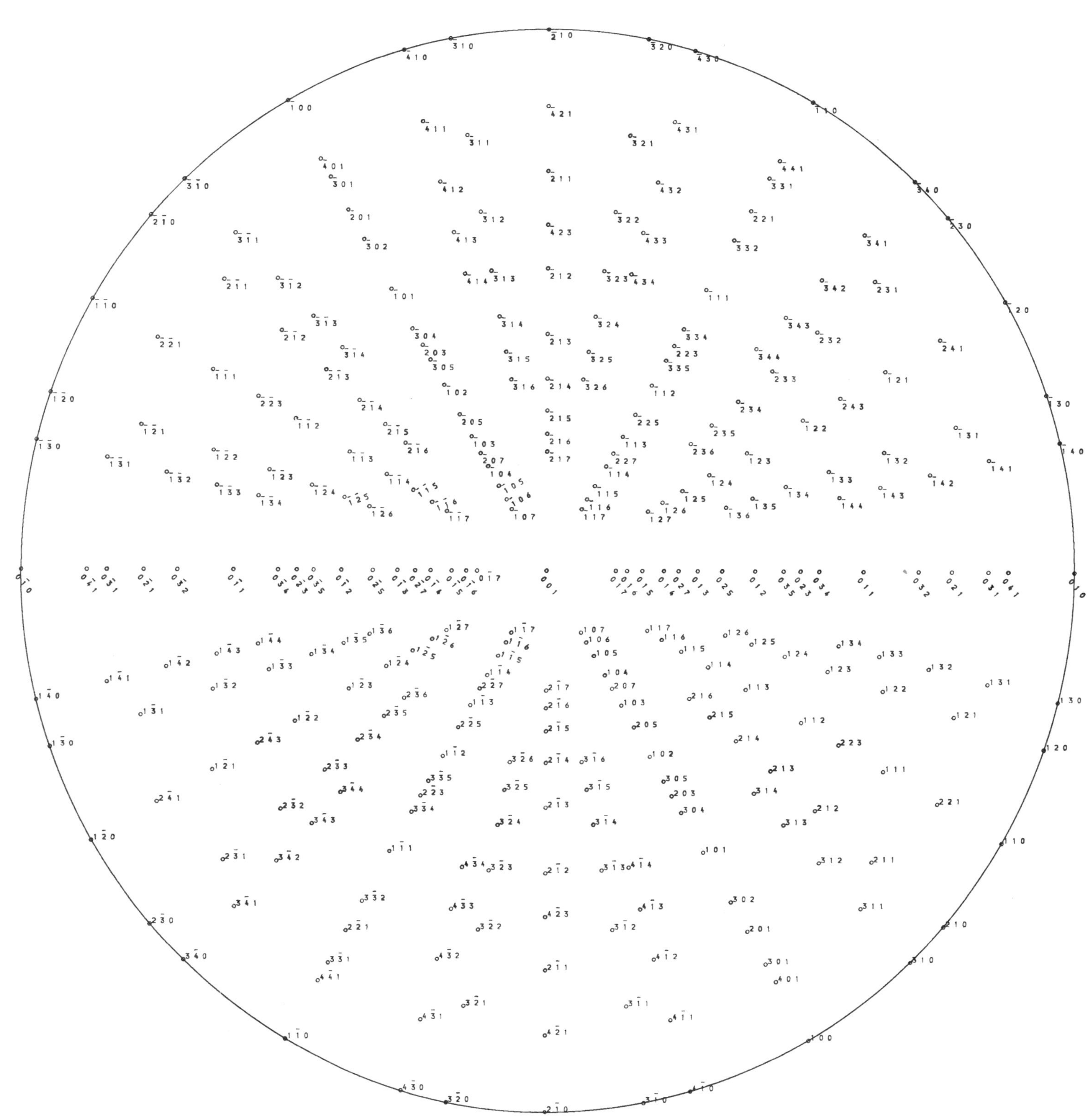

94 Cobalt Co or Magnesium Mg   Axial ratio 1.6230   010 otherwise $\bar{1}2\bar{1}0$ axis

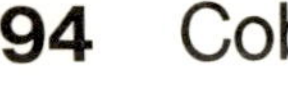

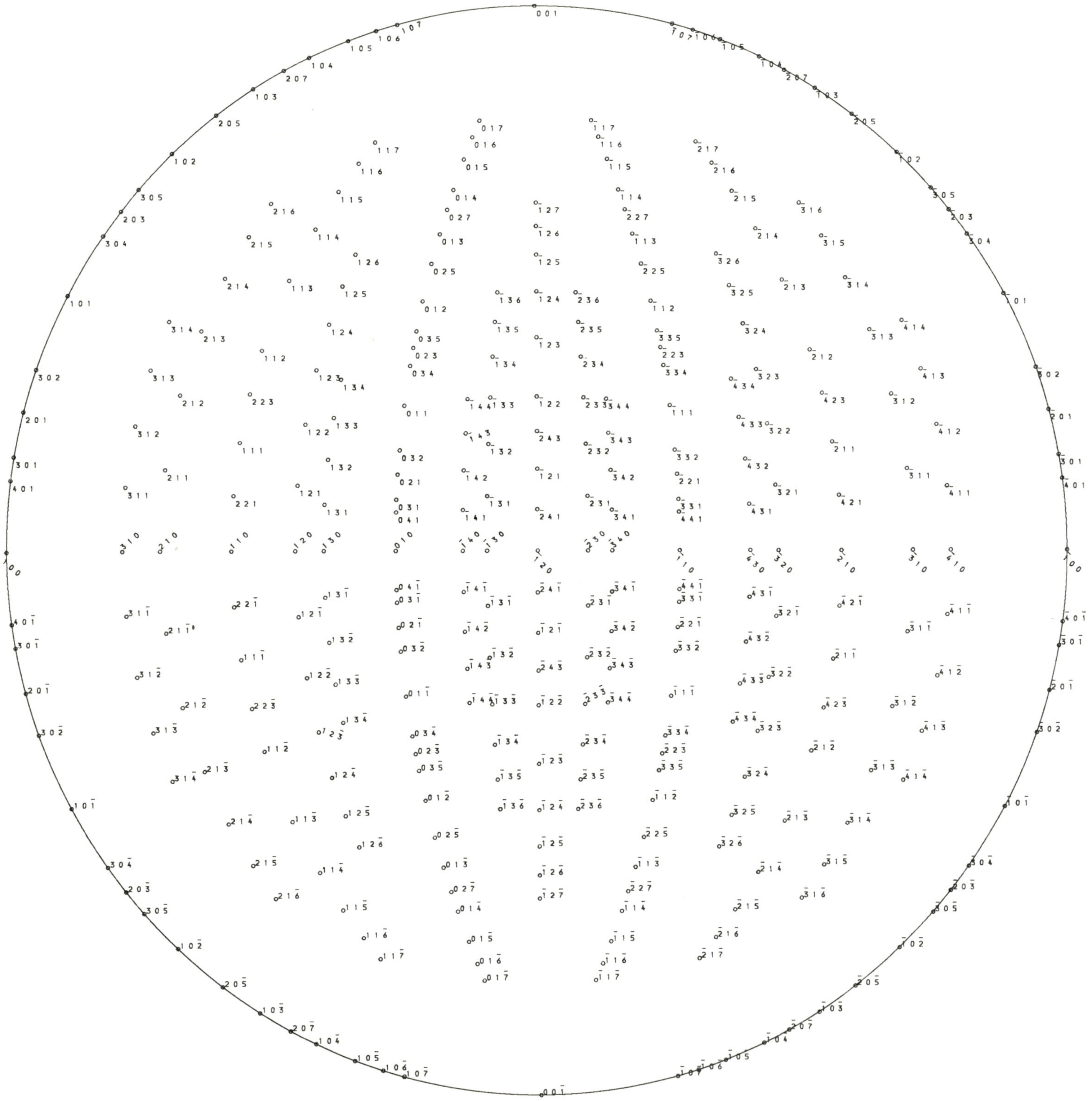

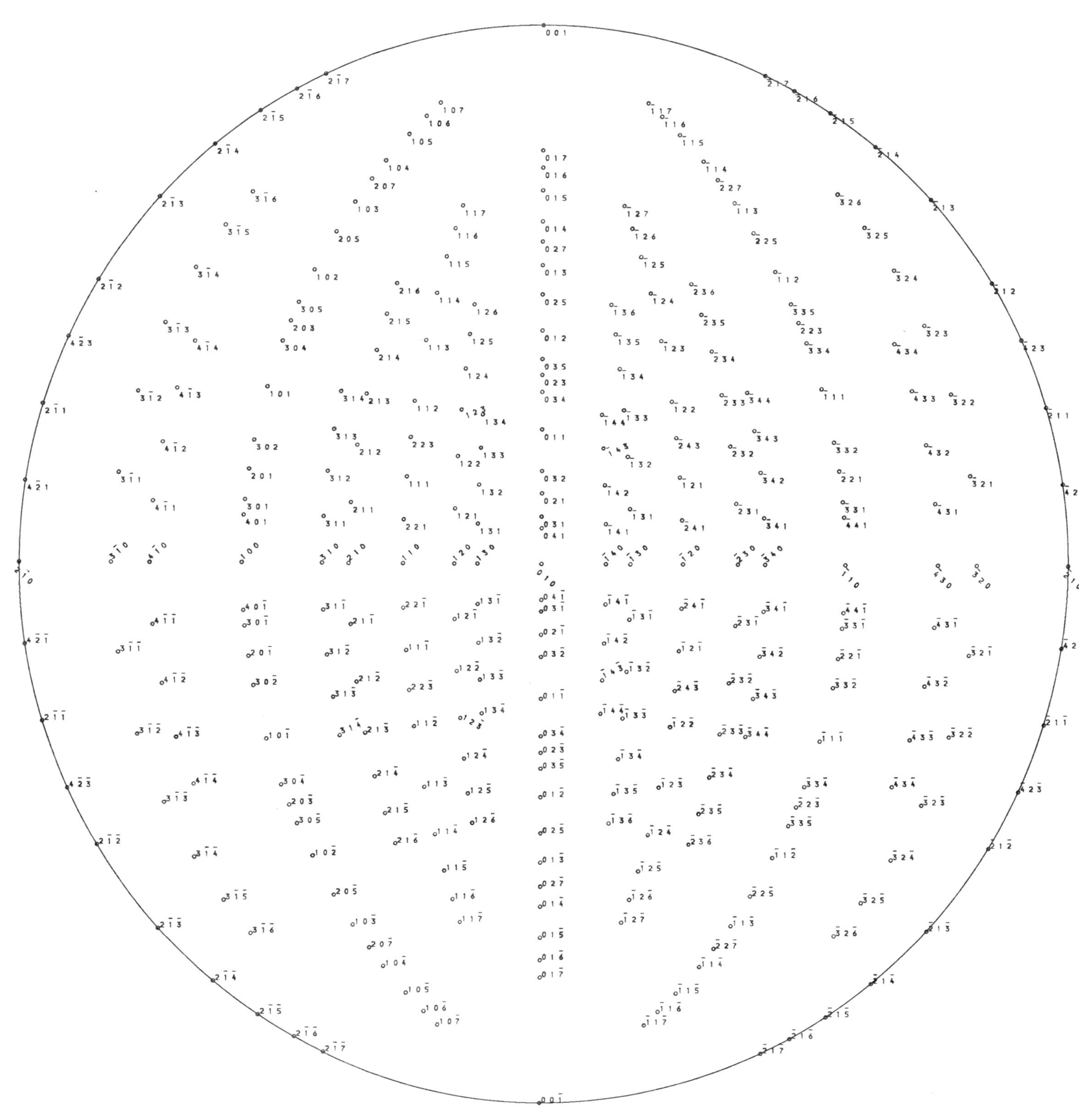

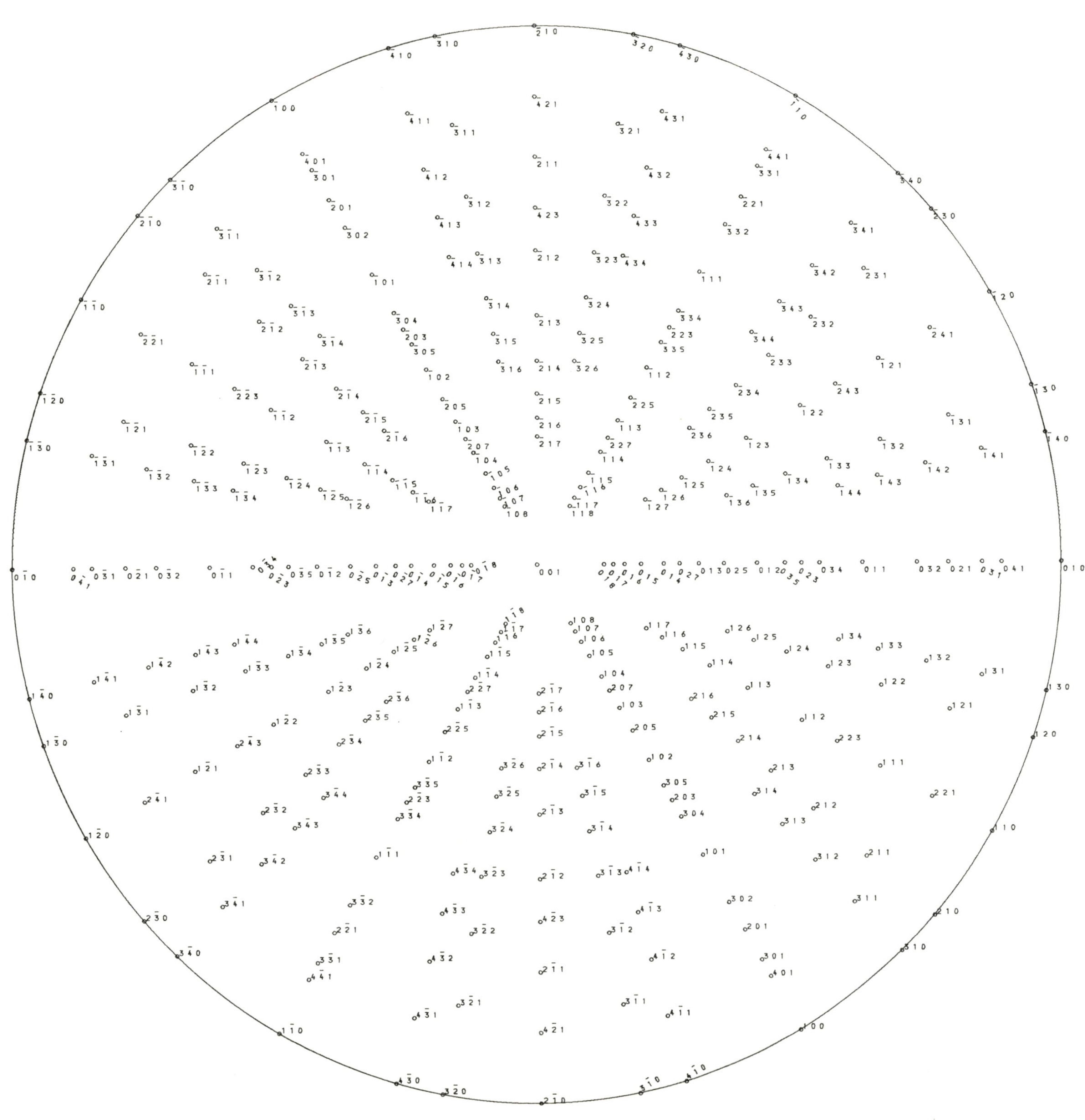

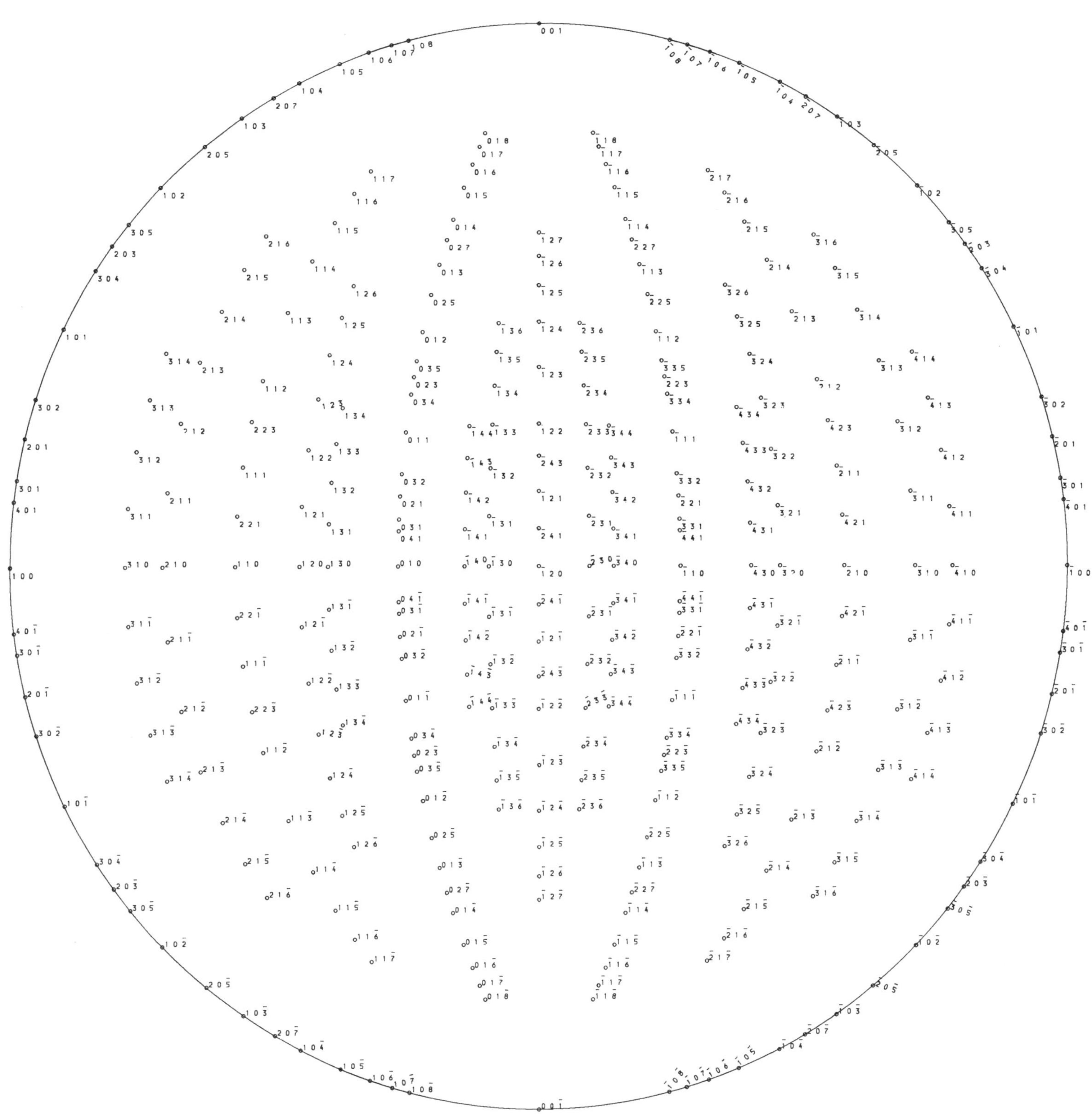

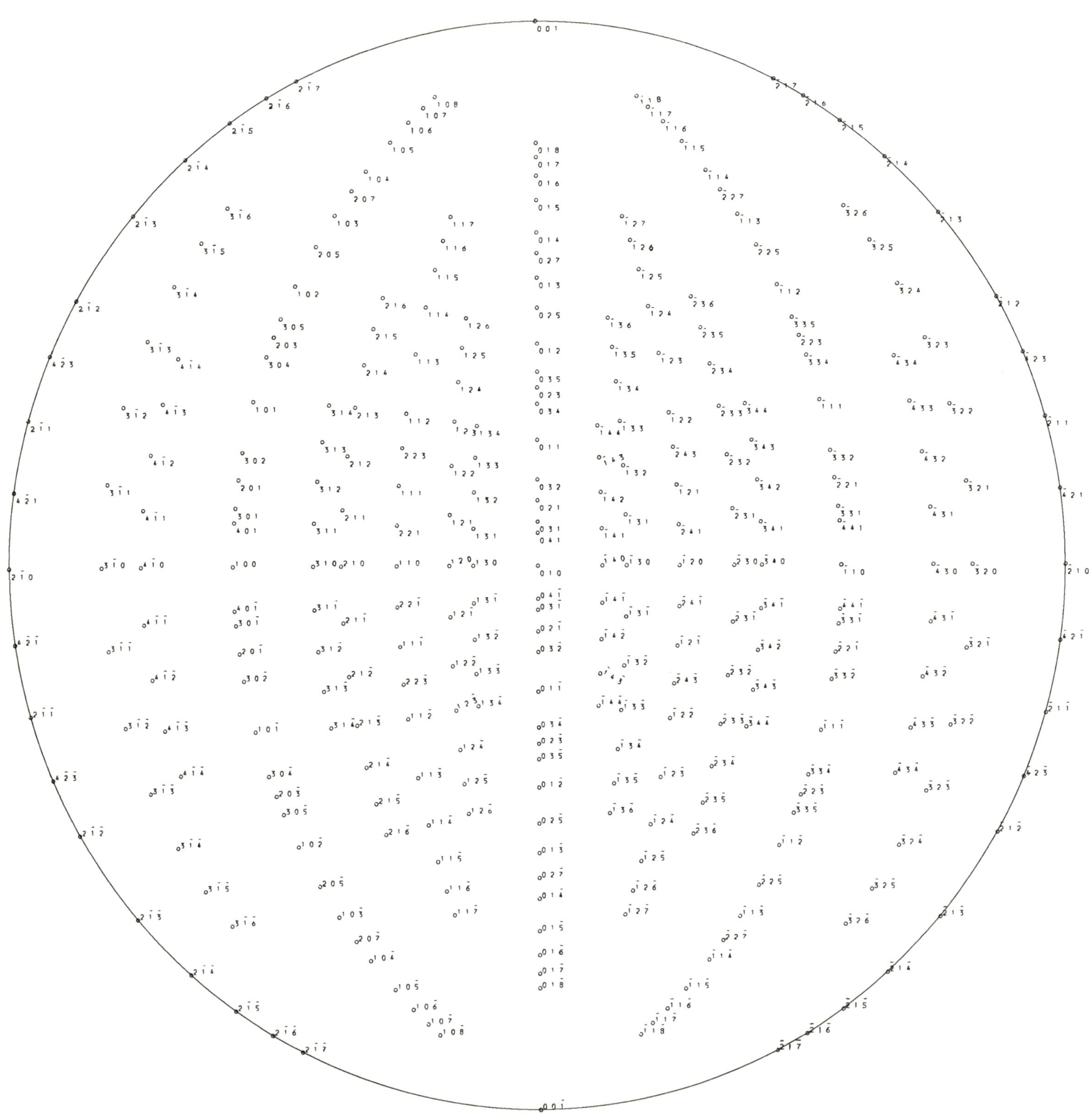

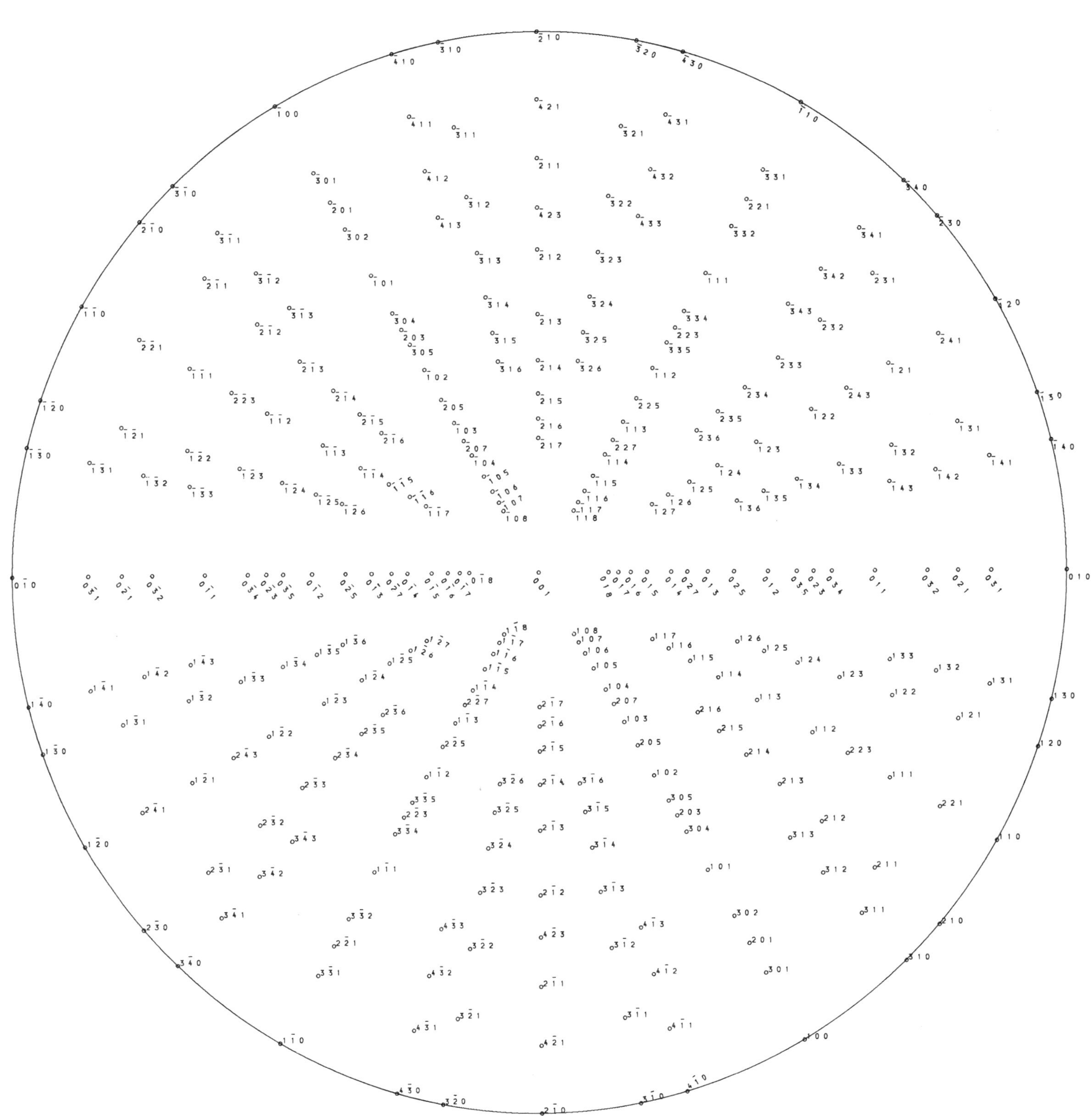

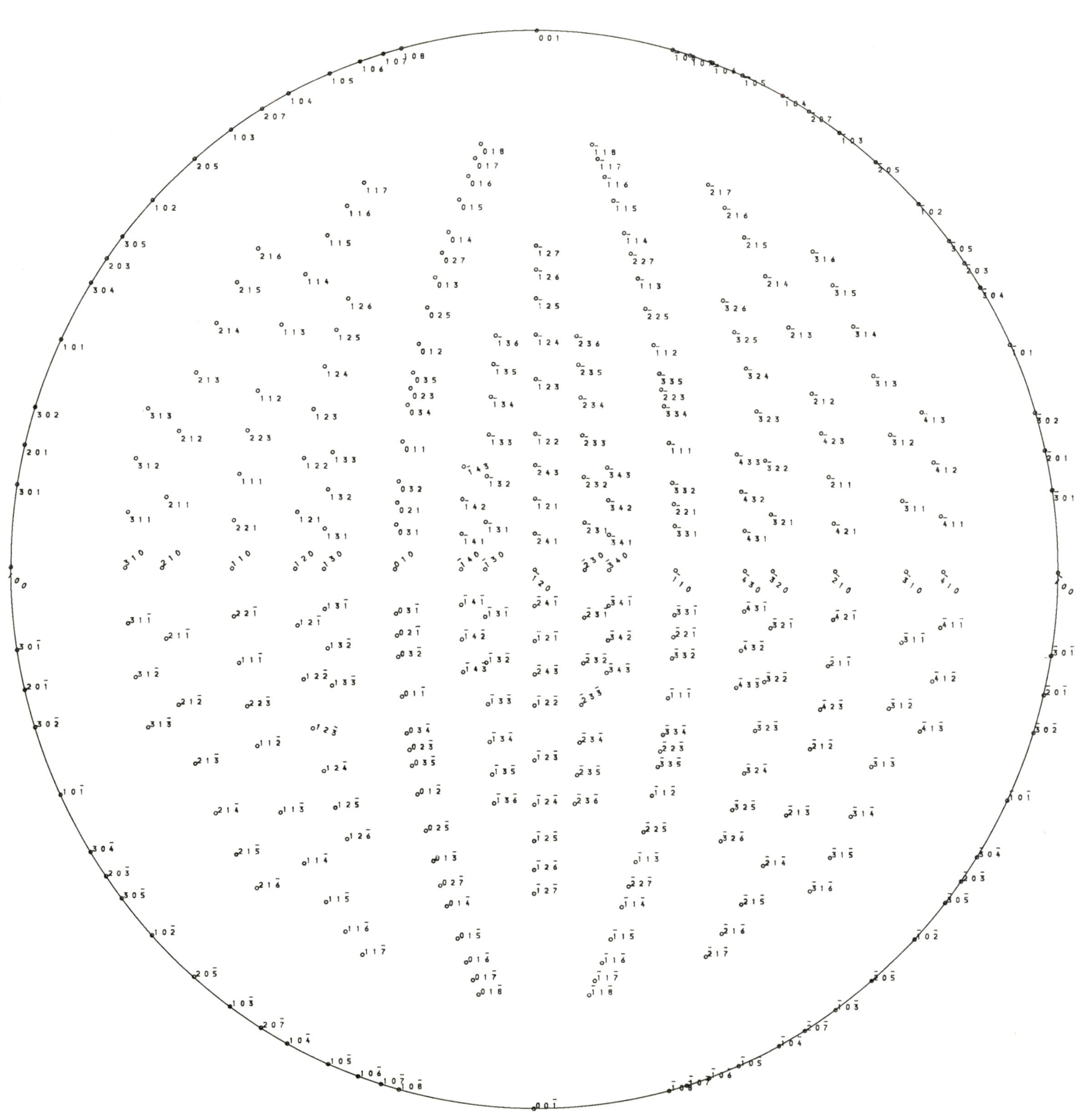

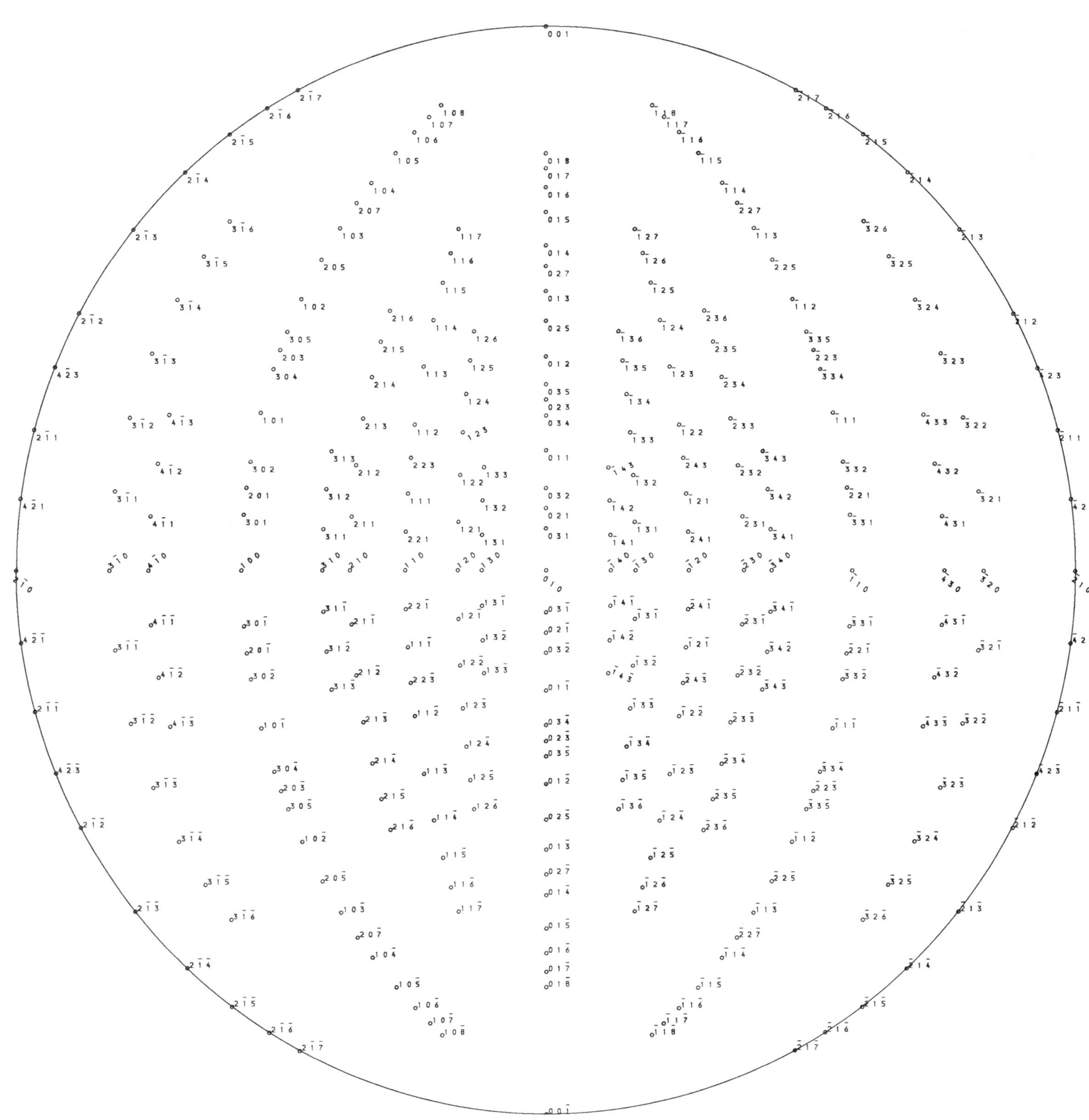

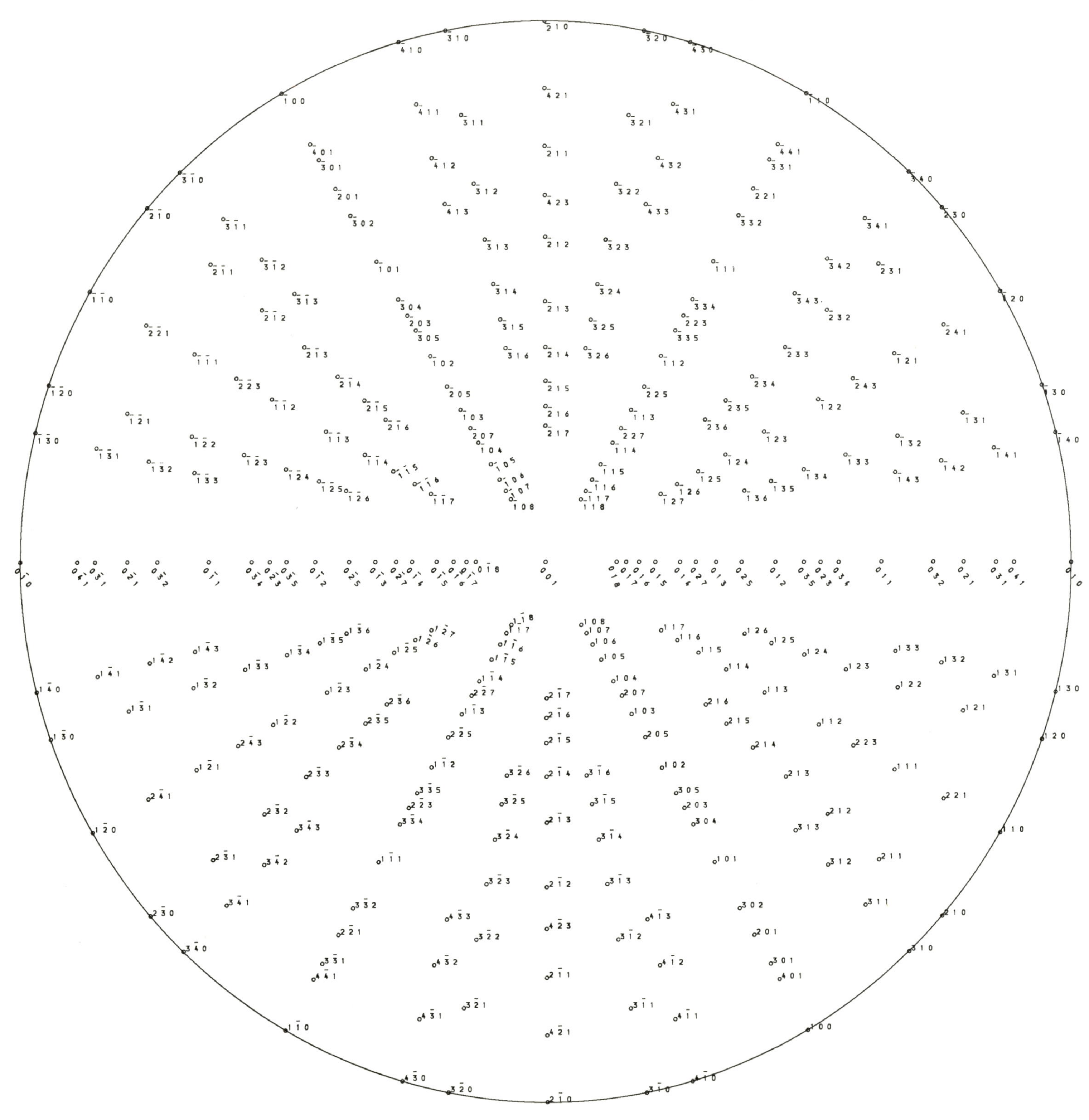

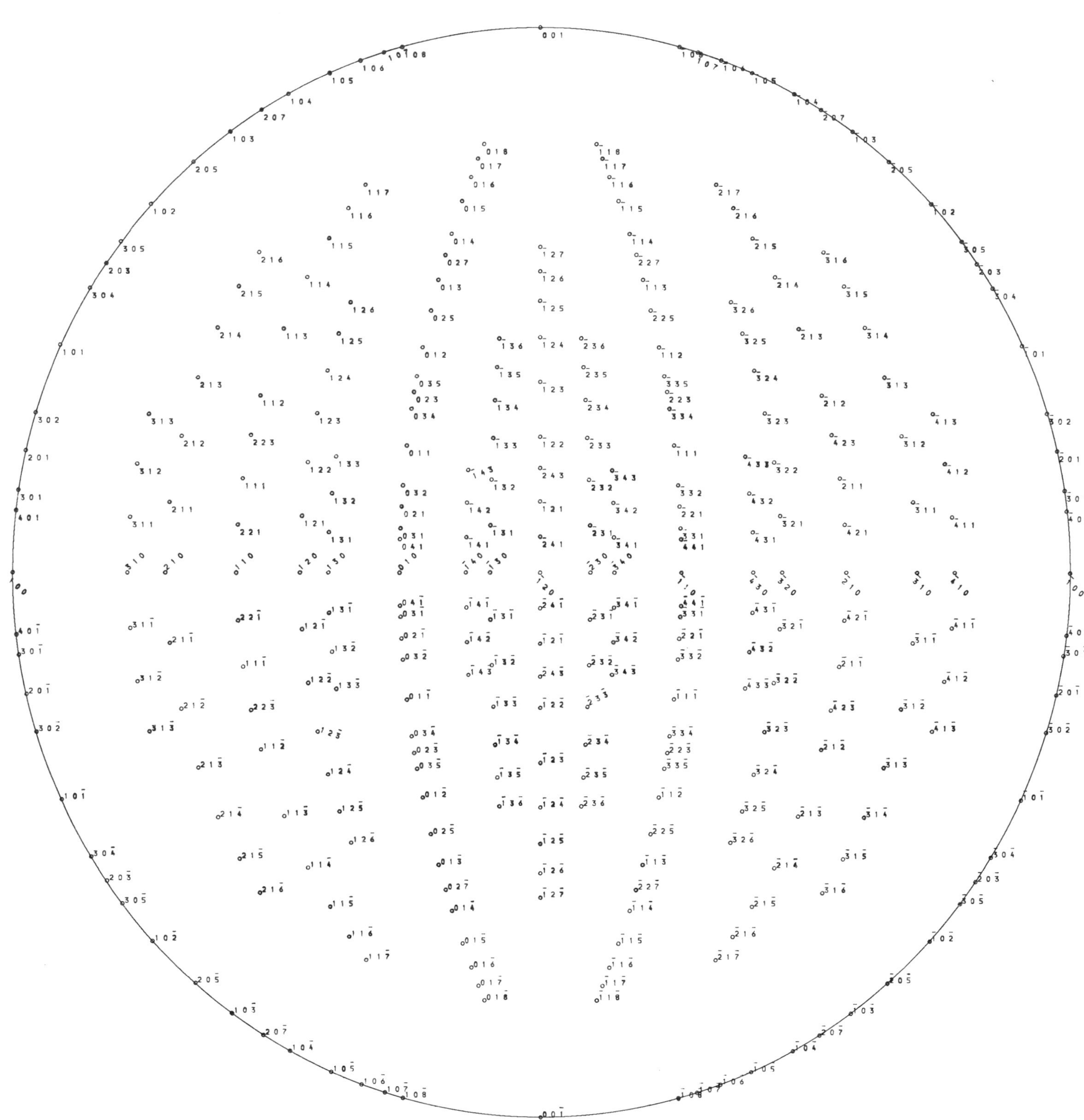

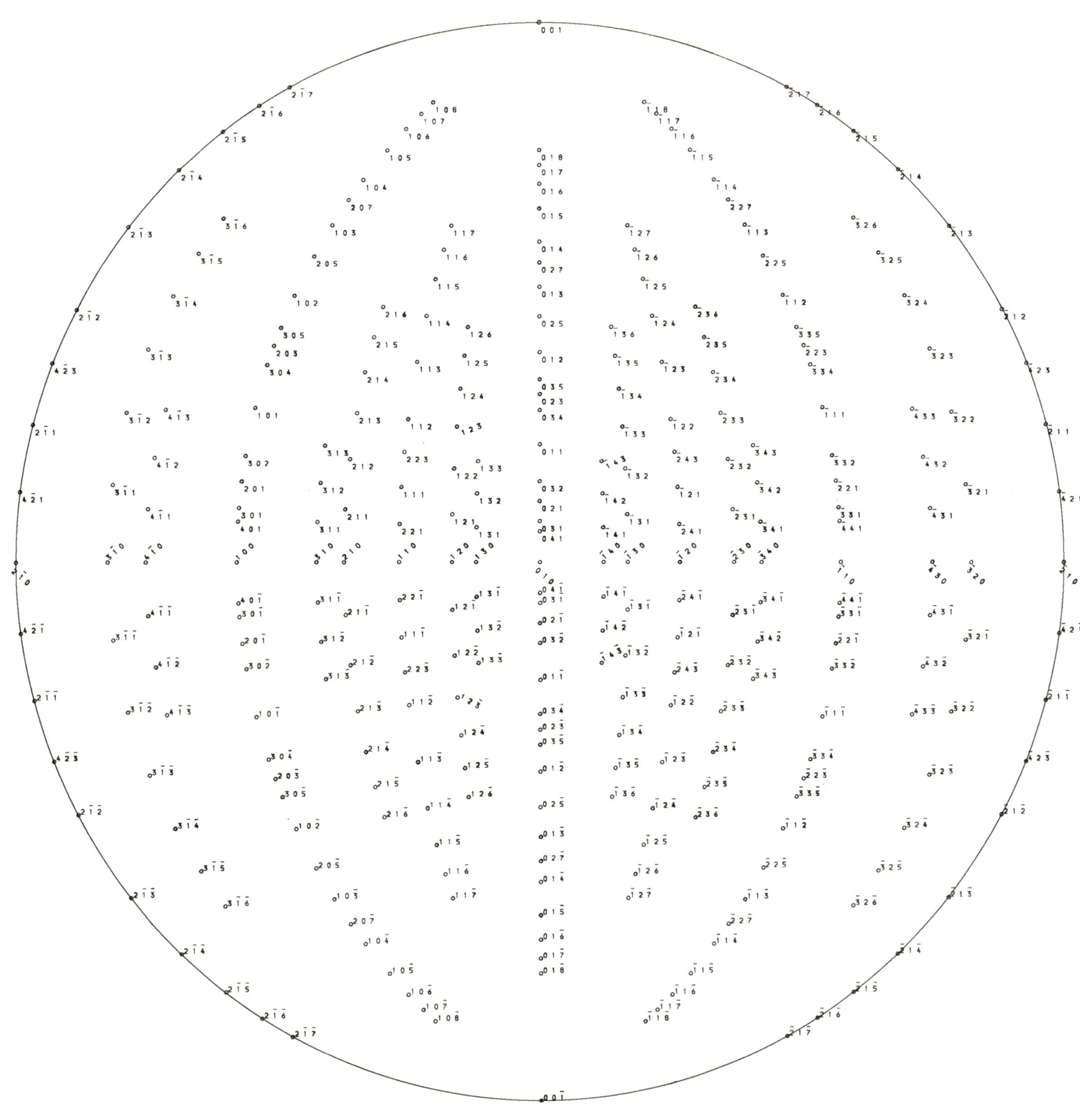

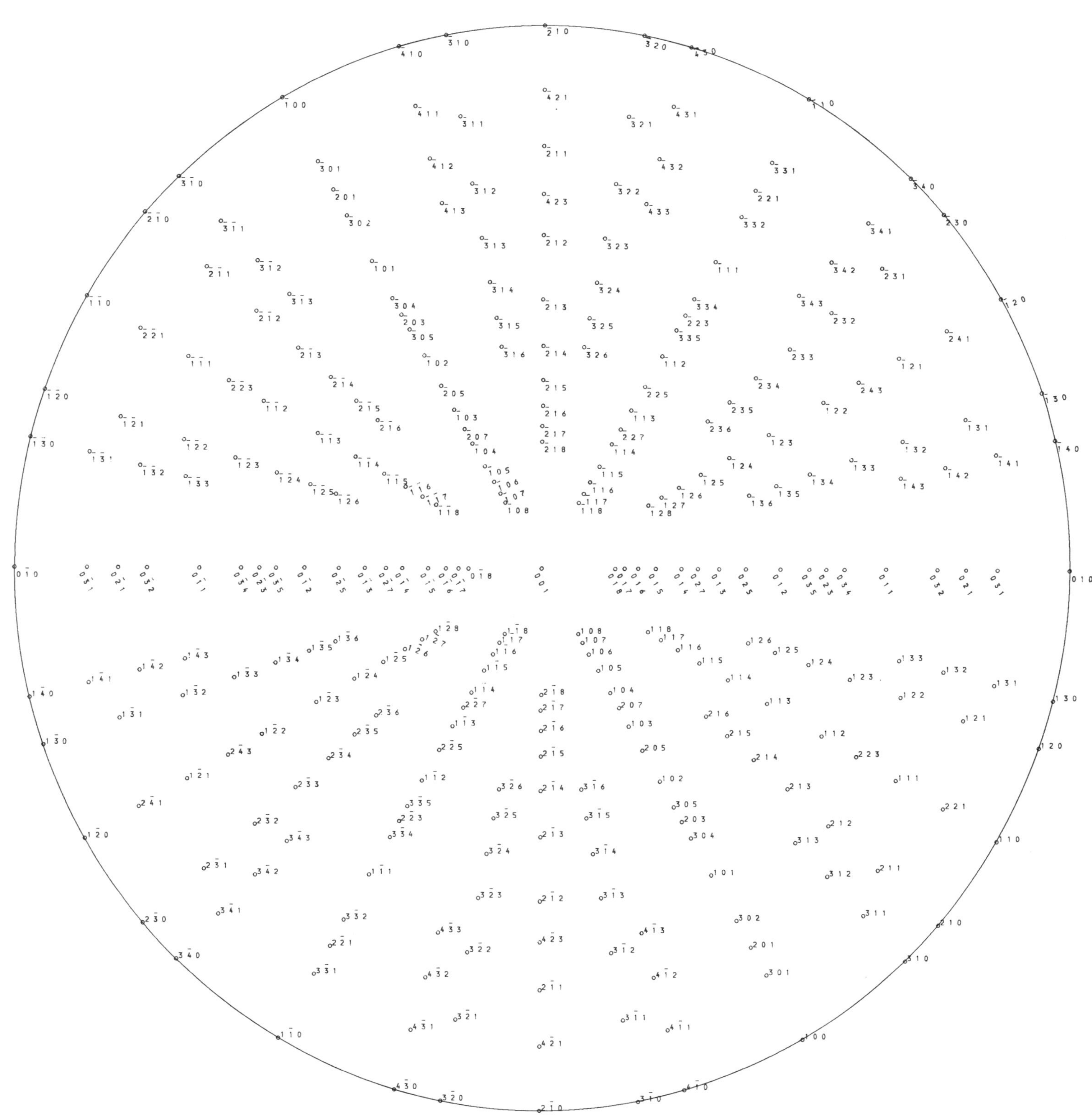

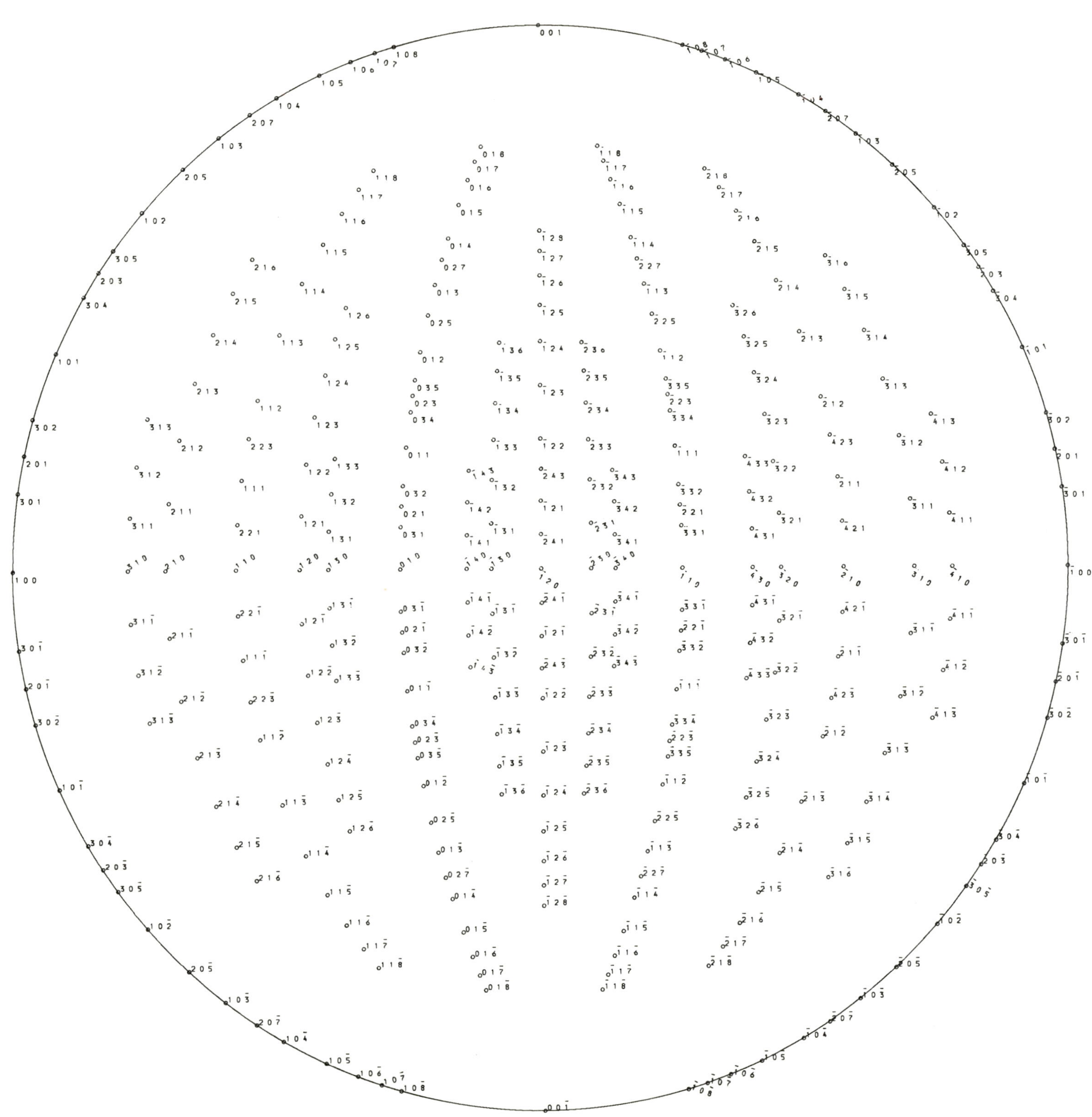

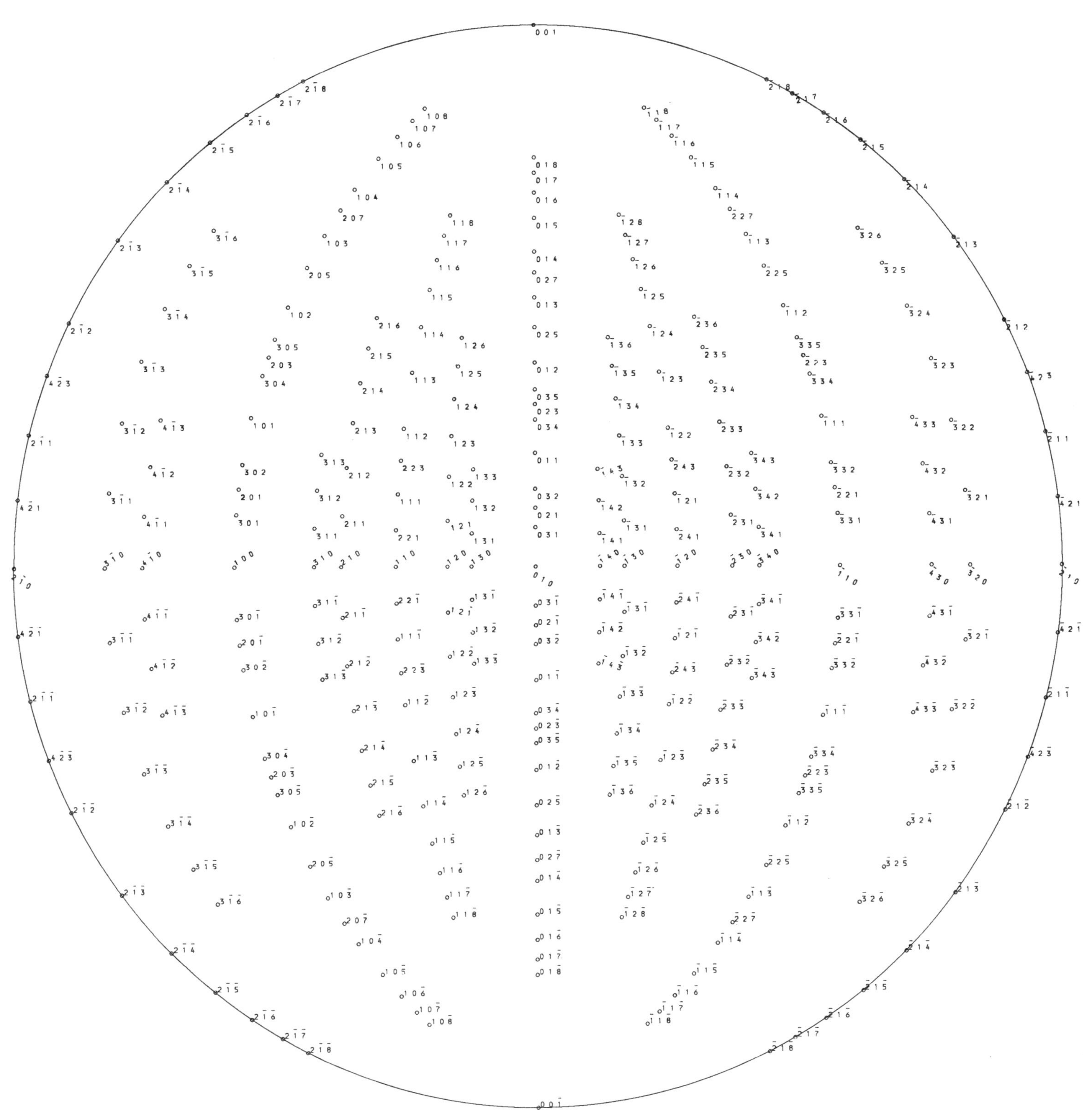

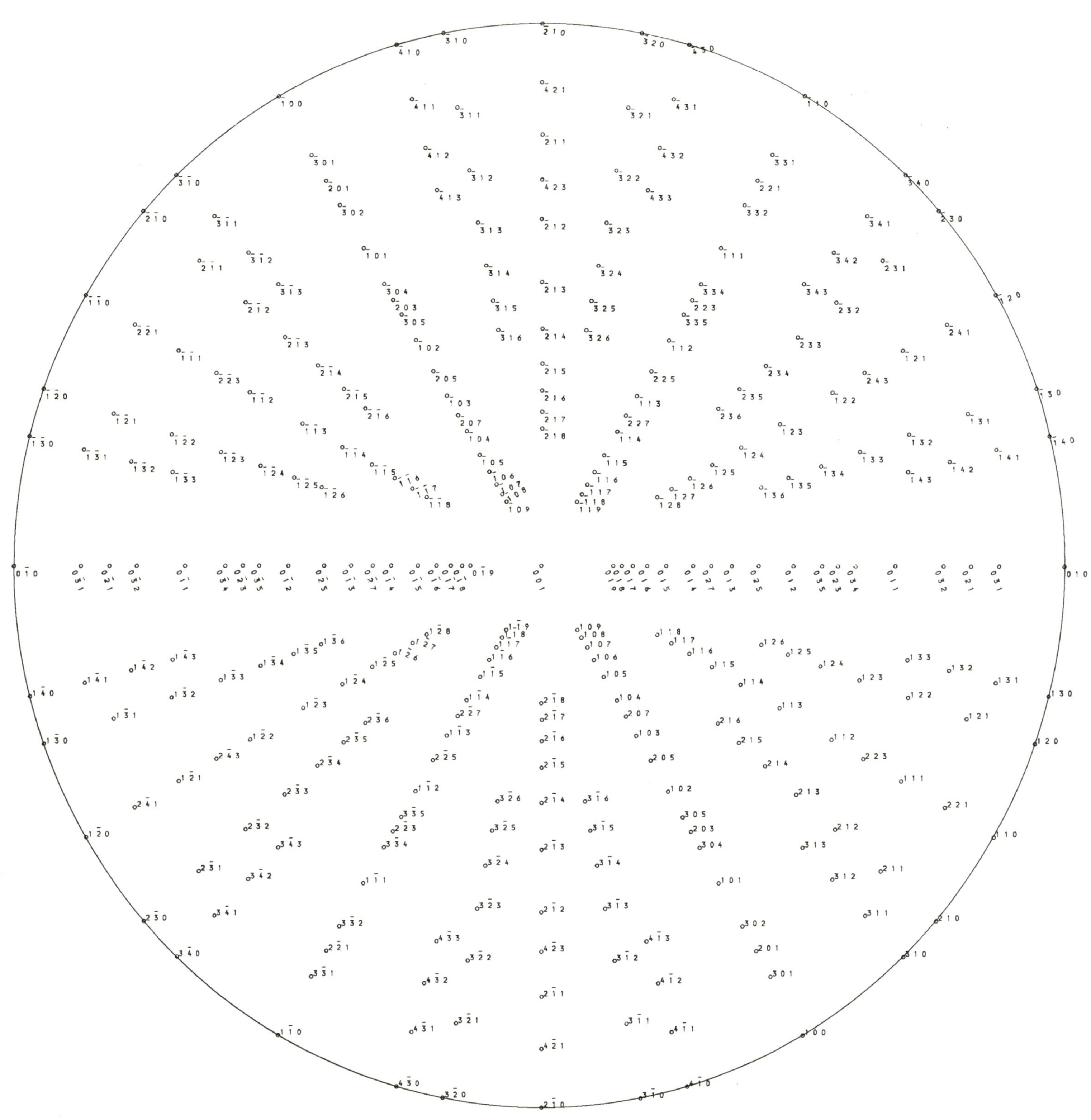

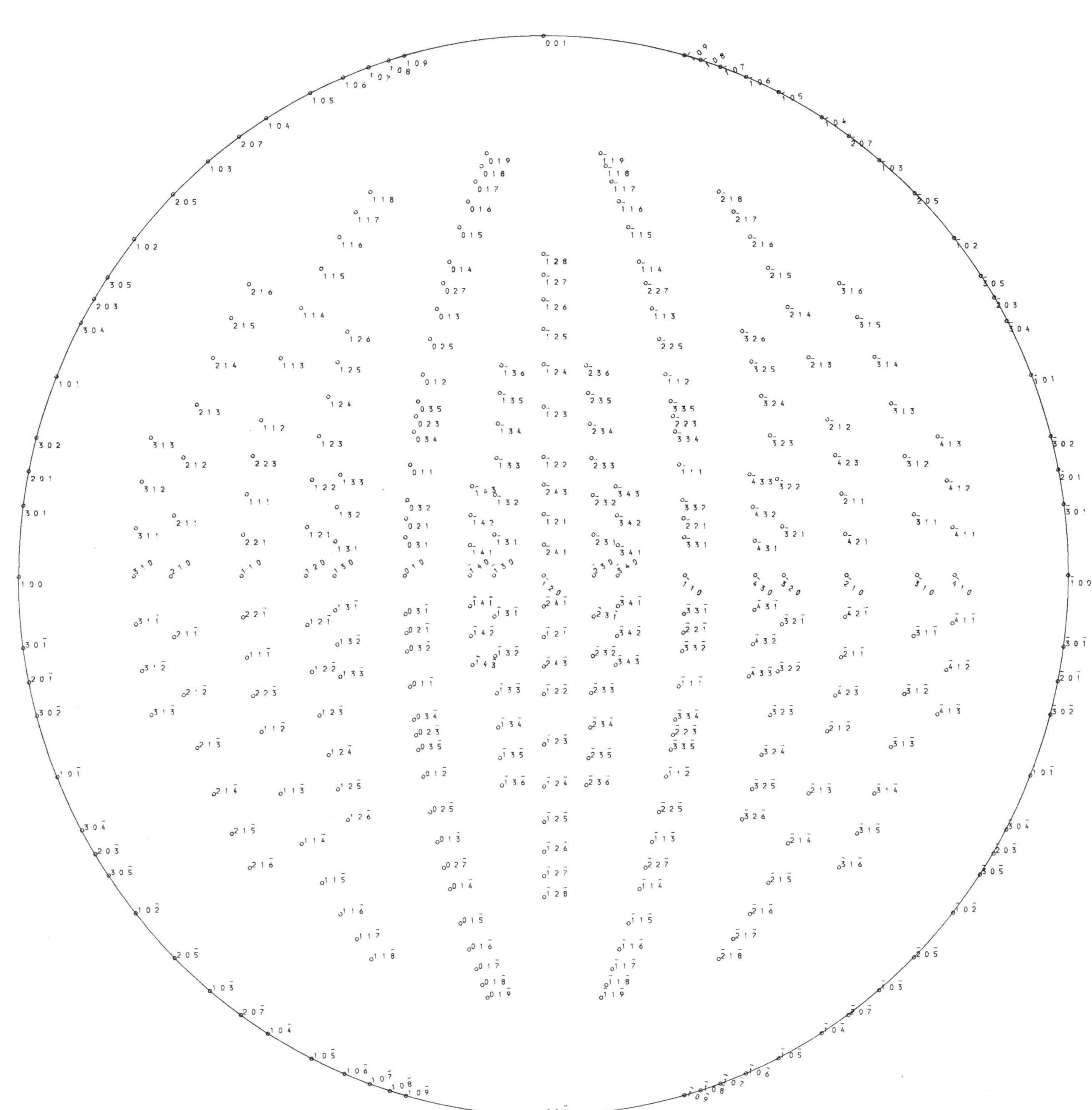

110 Boron carbide B$_4$C   Axial ratio 2.1643   120 otherwise 01$\bar{1}$0 axis

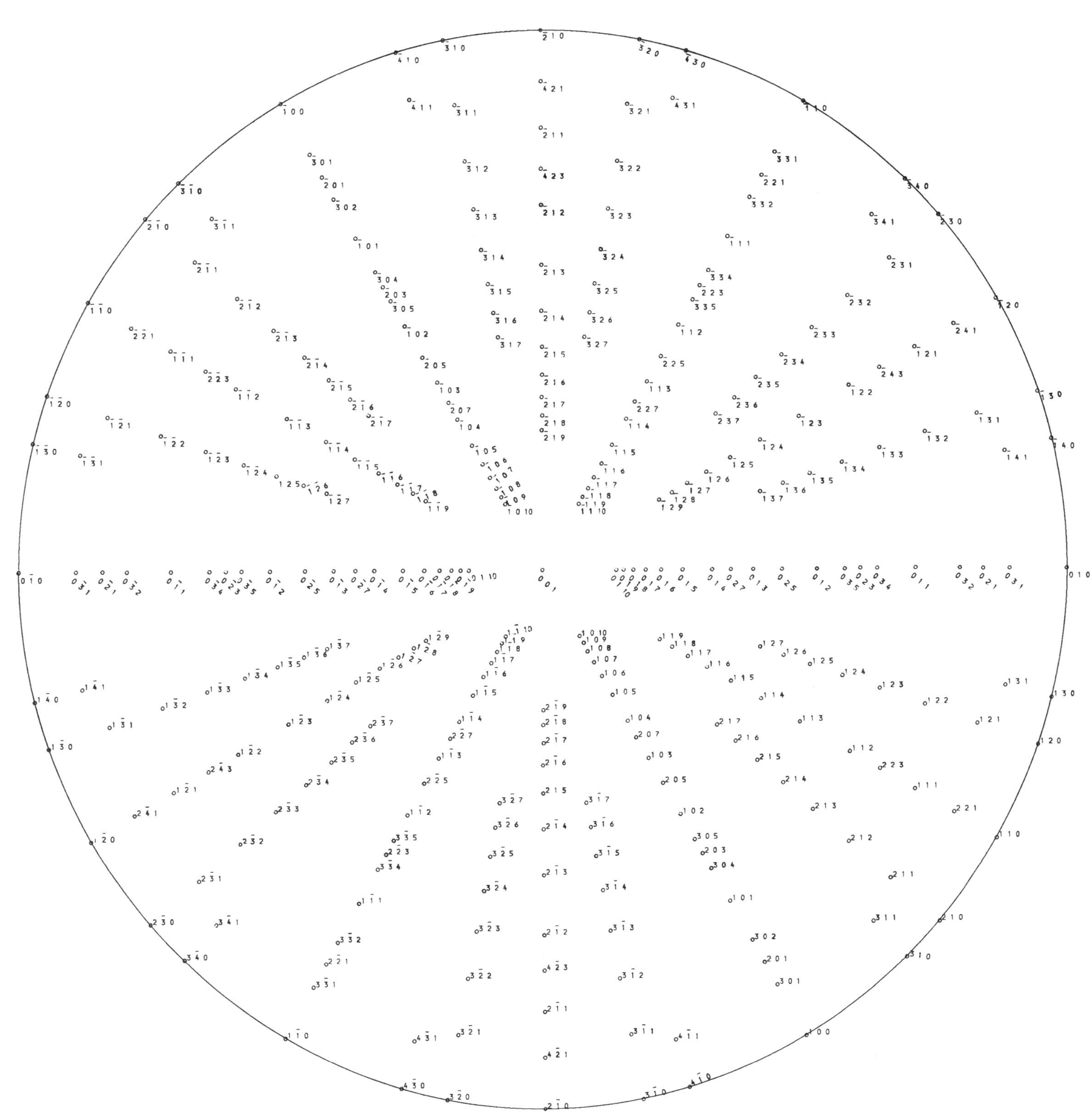

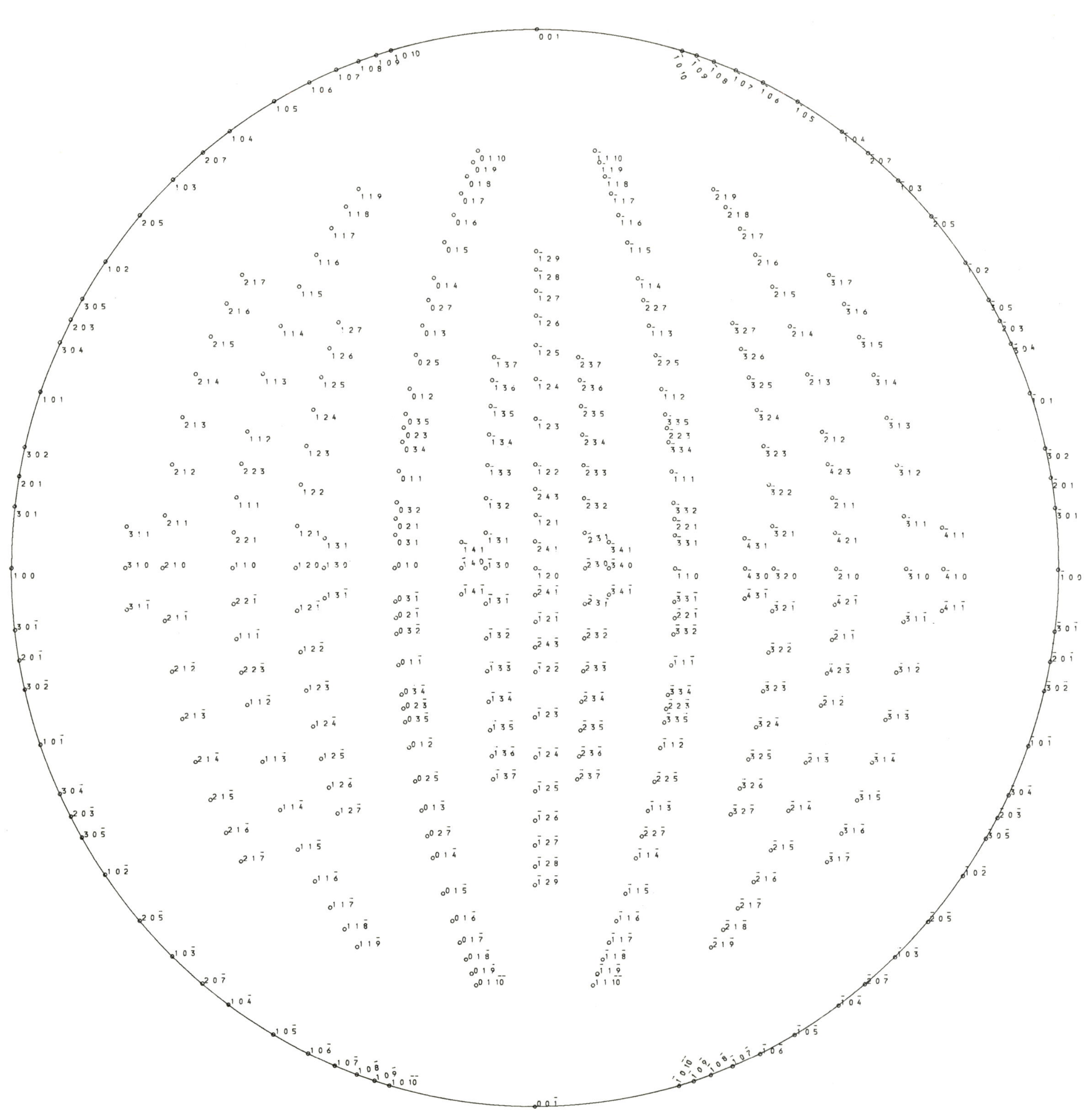

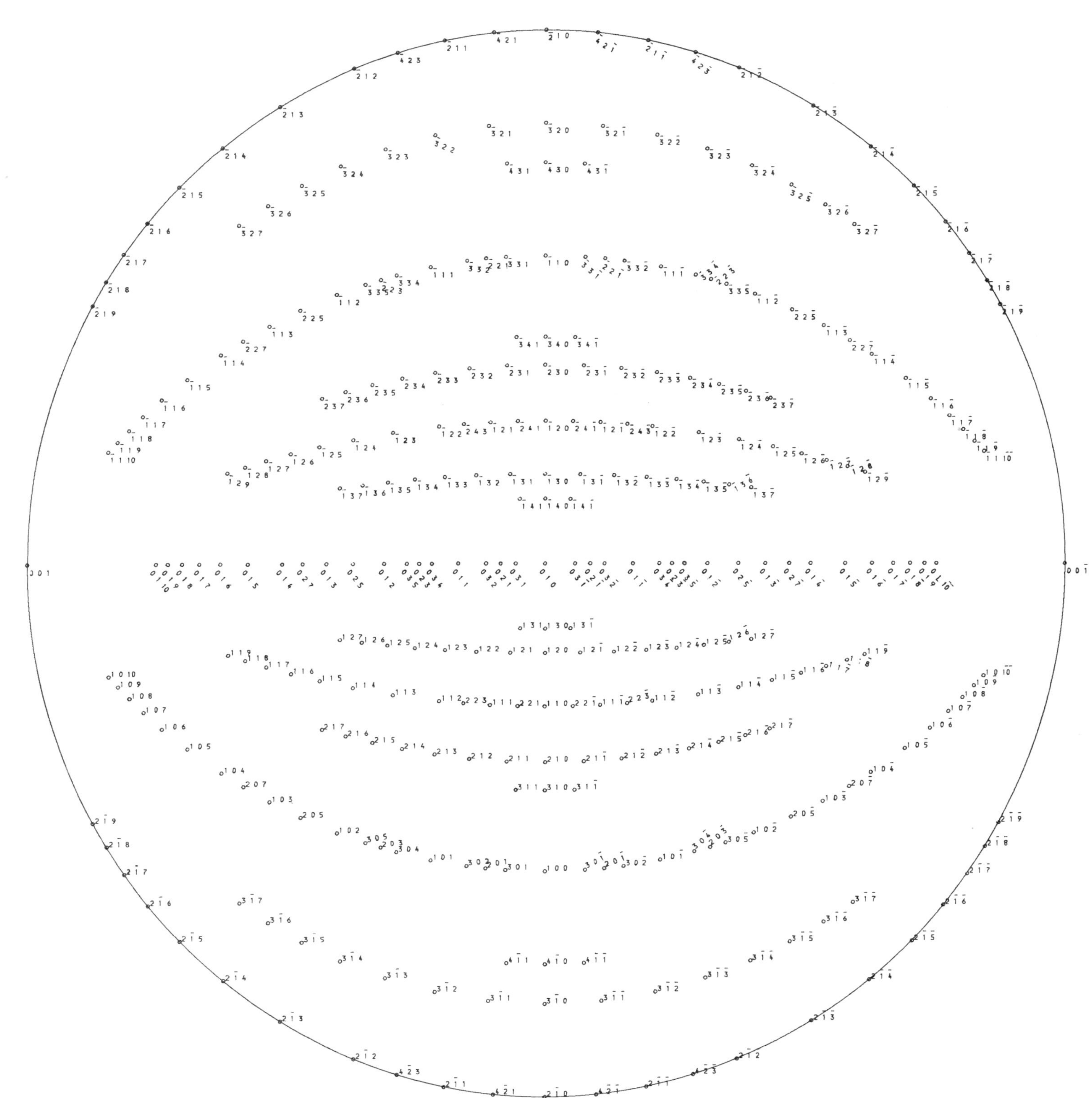

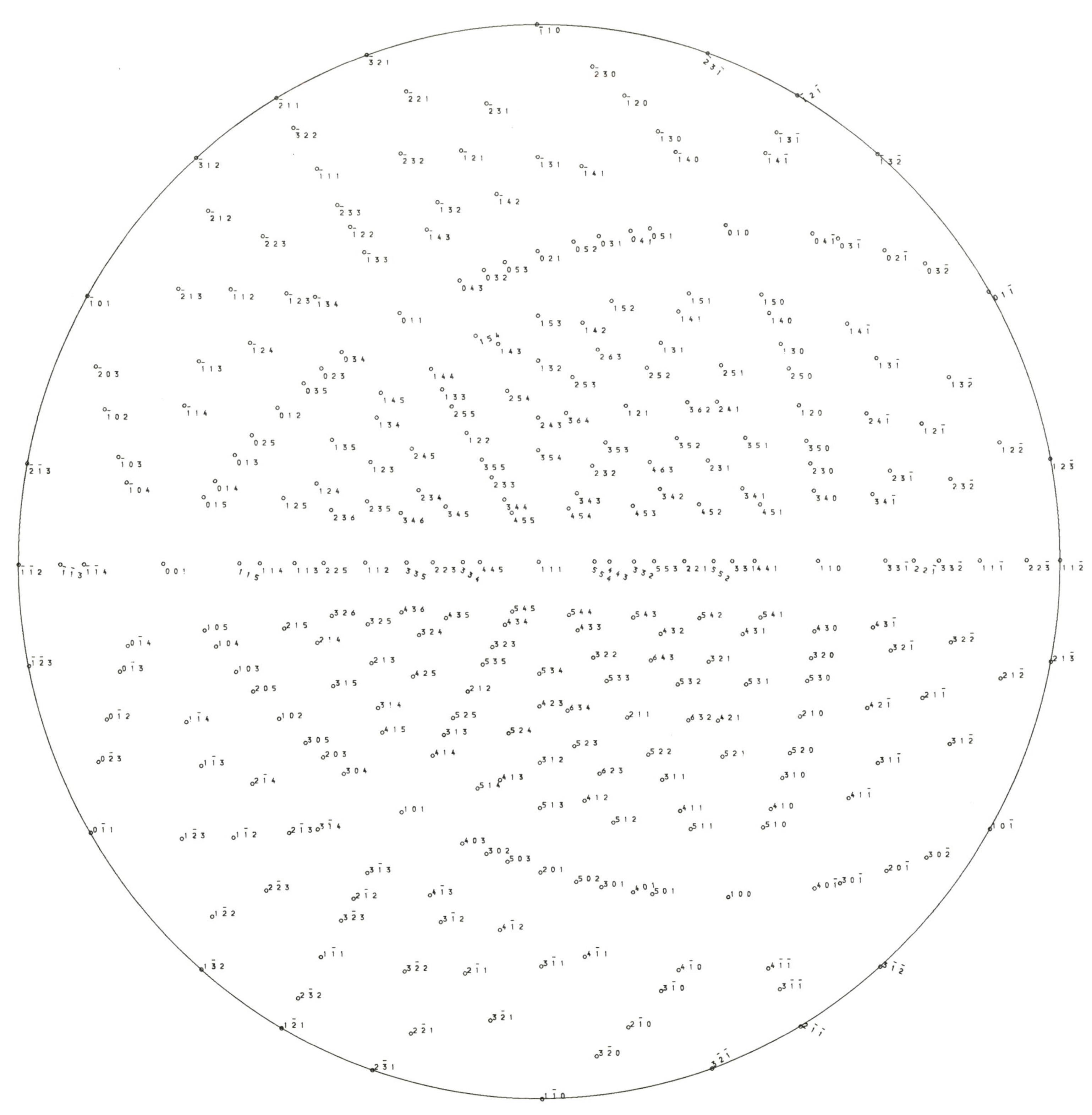

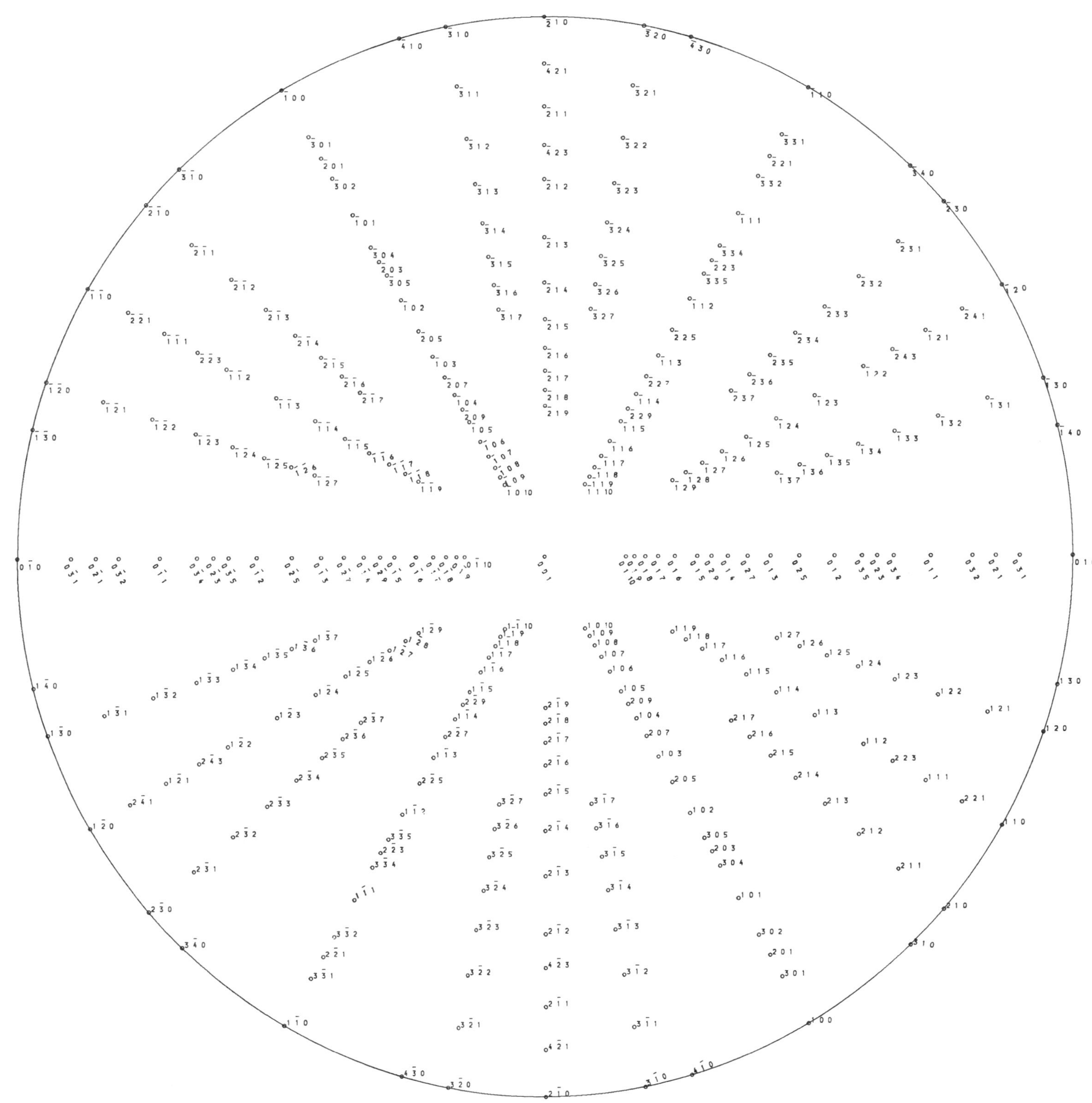

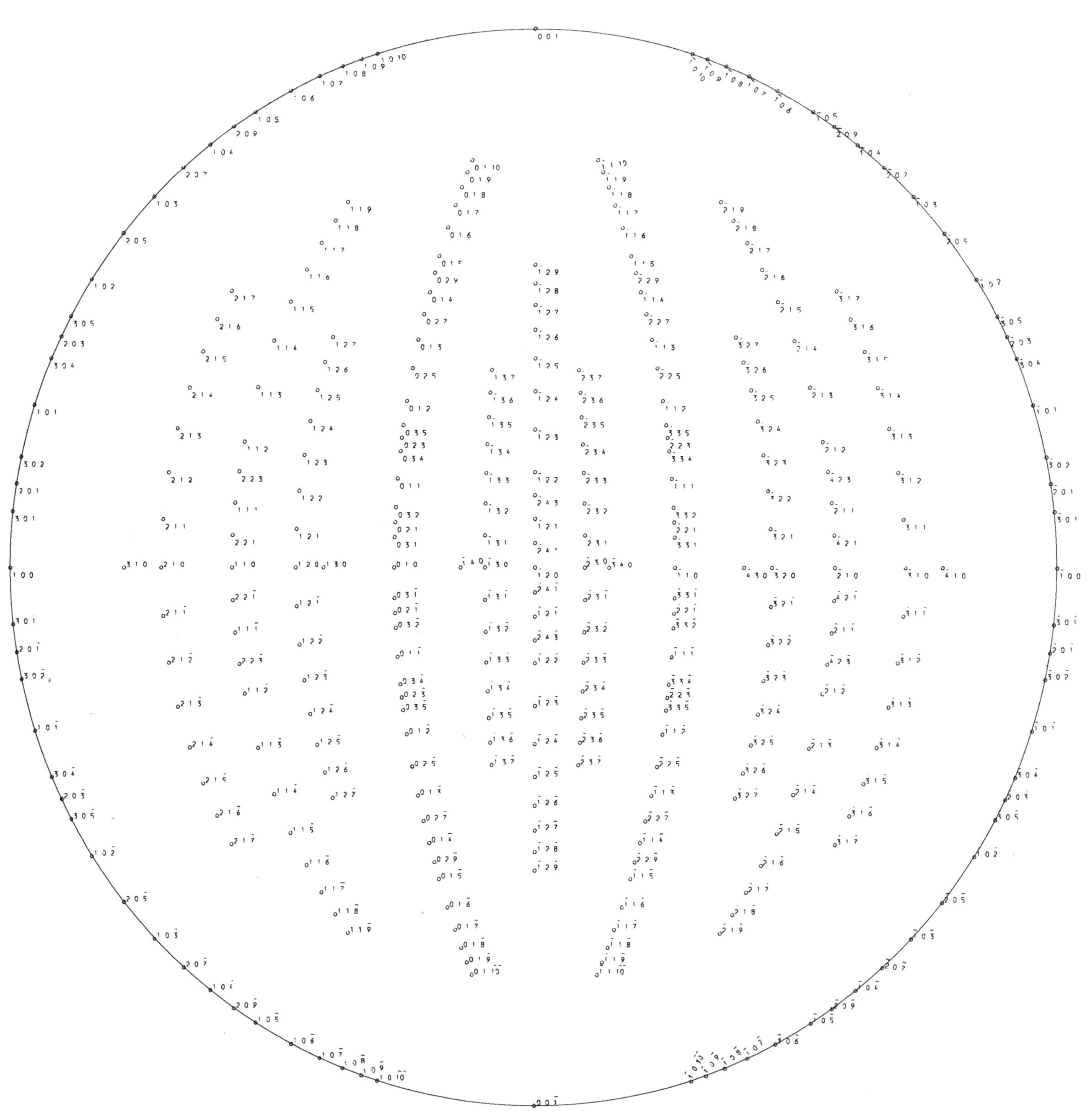

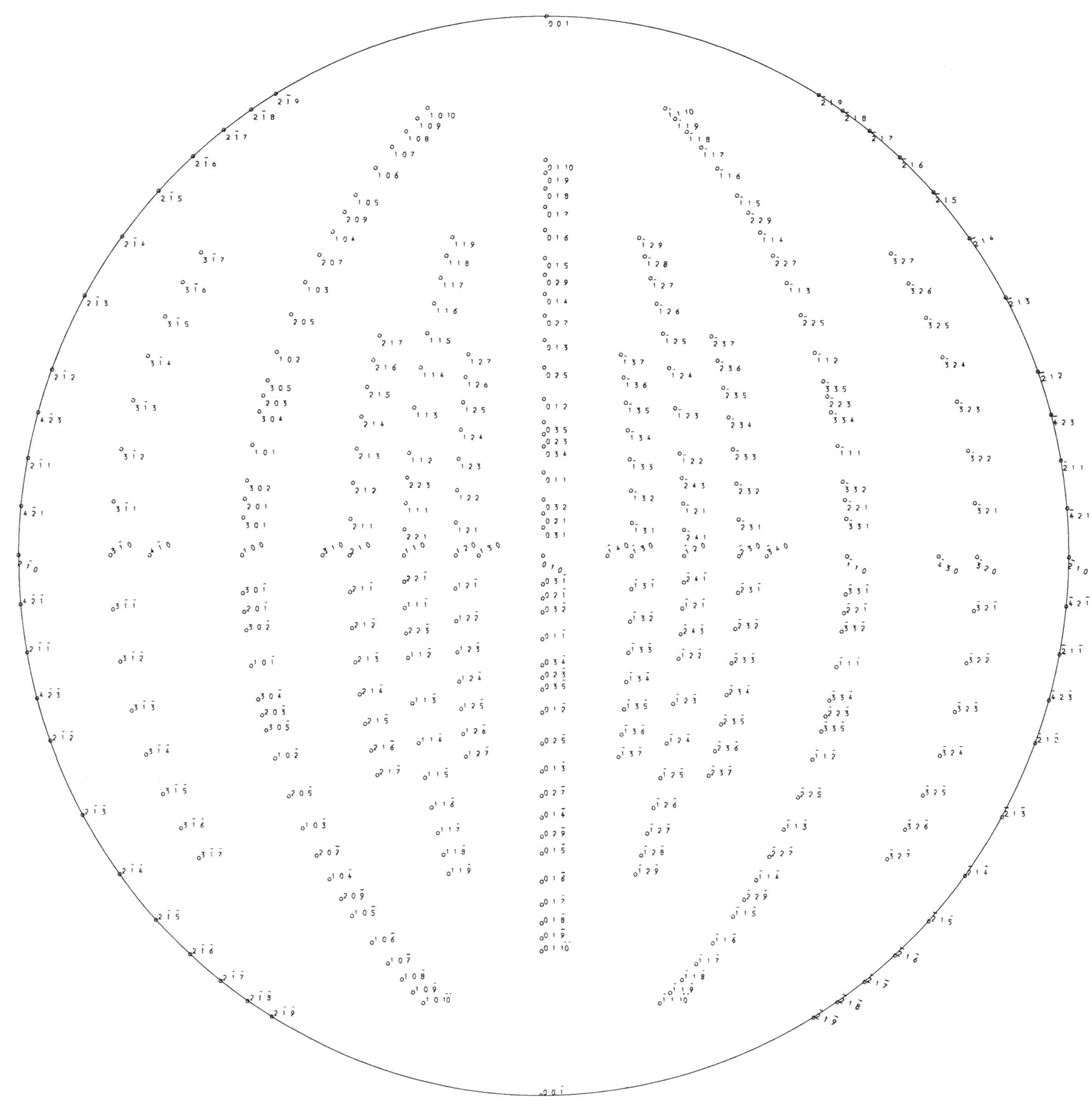

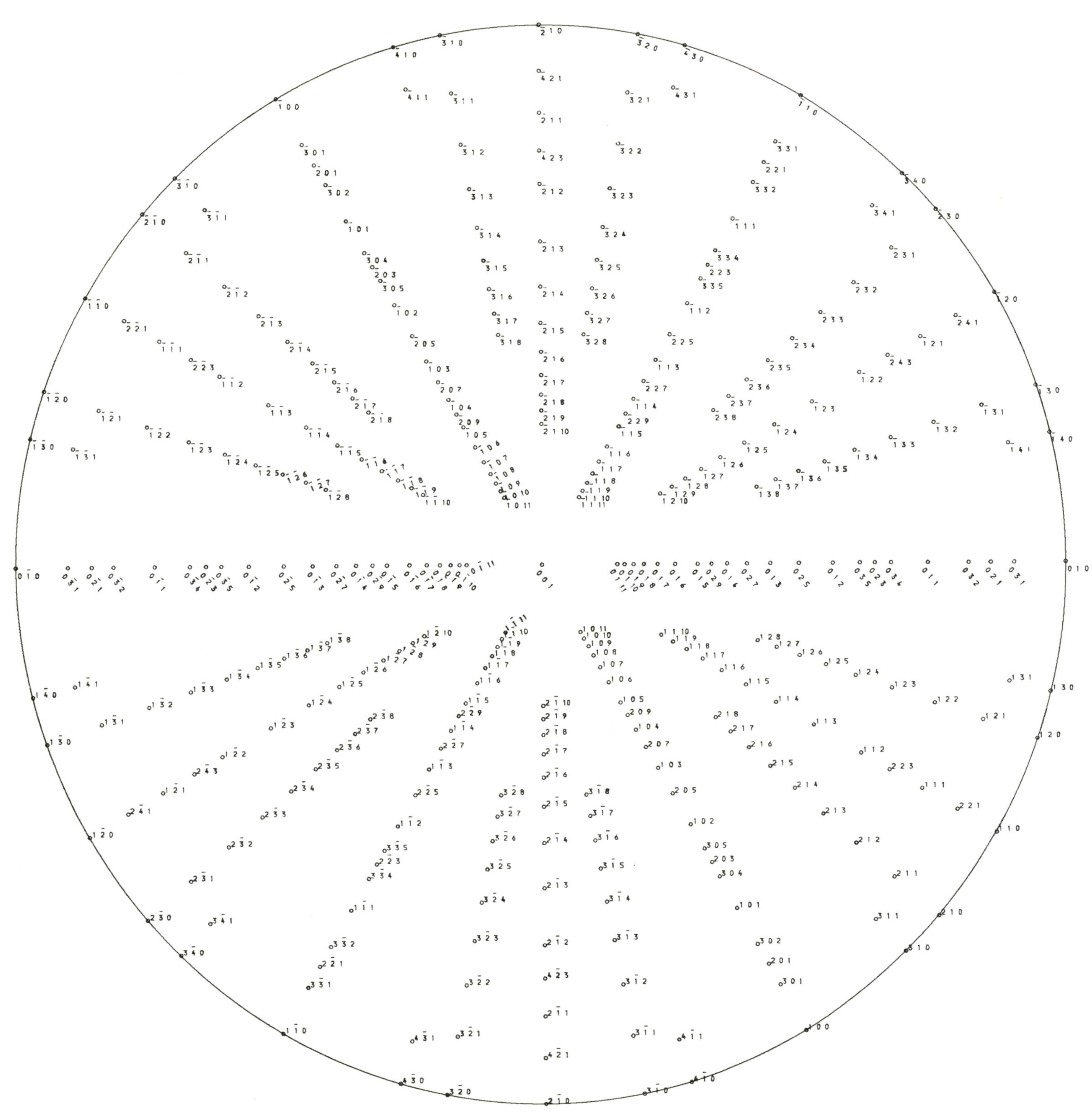

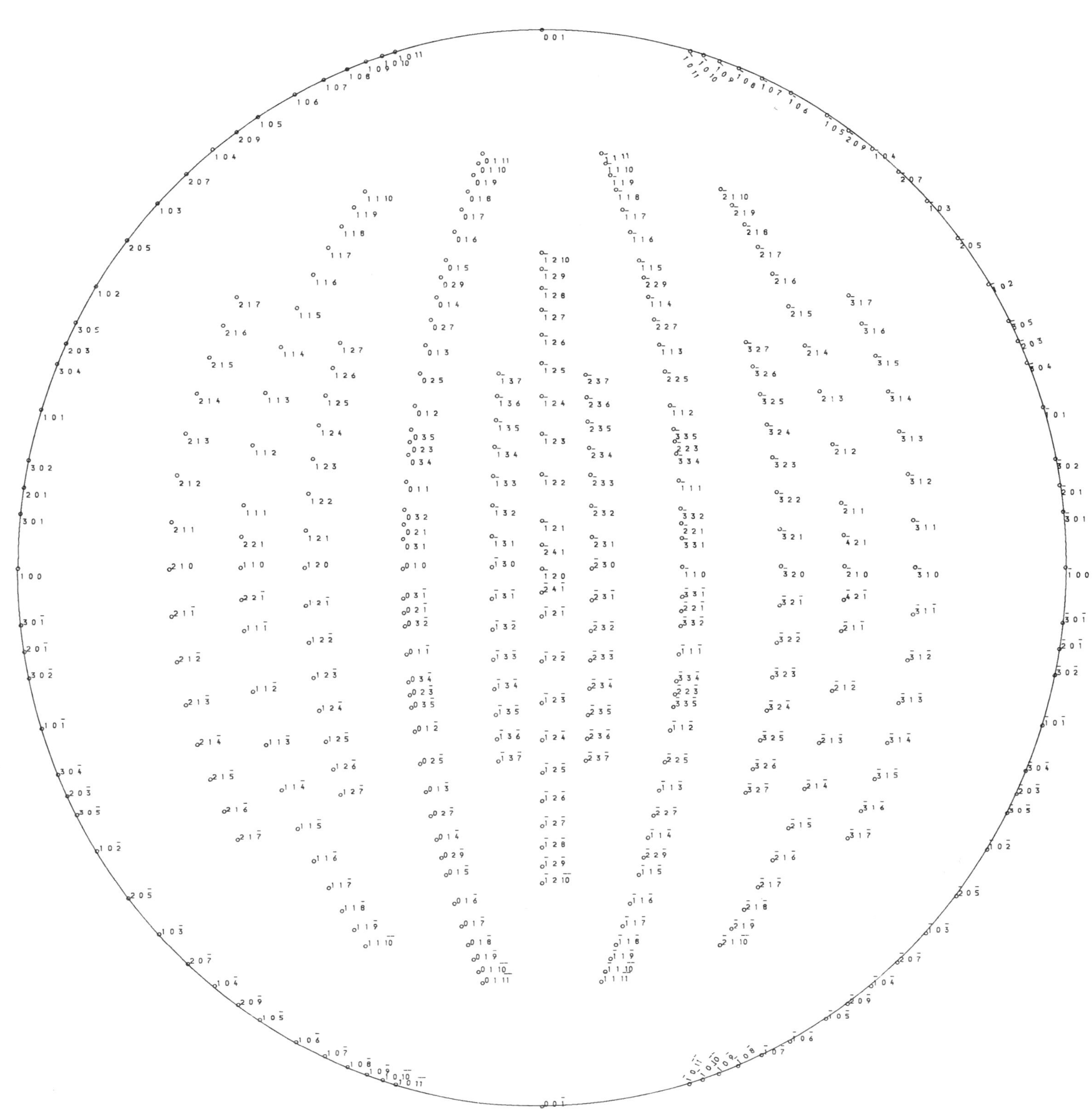

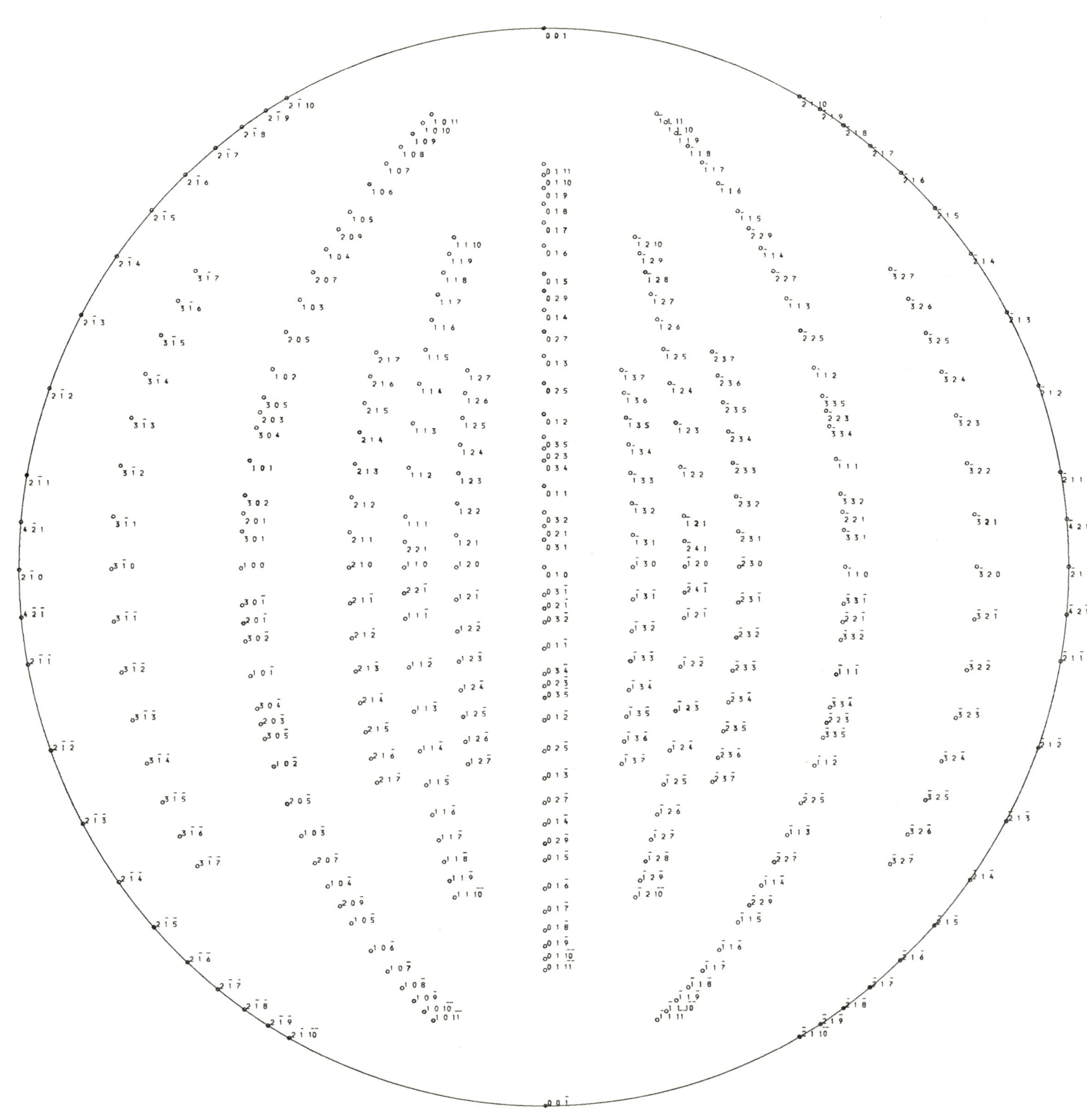

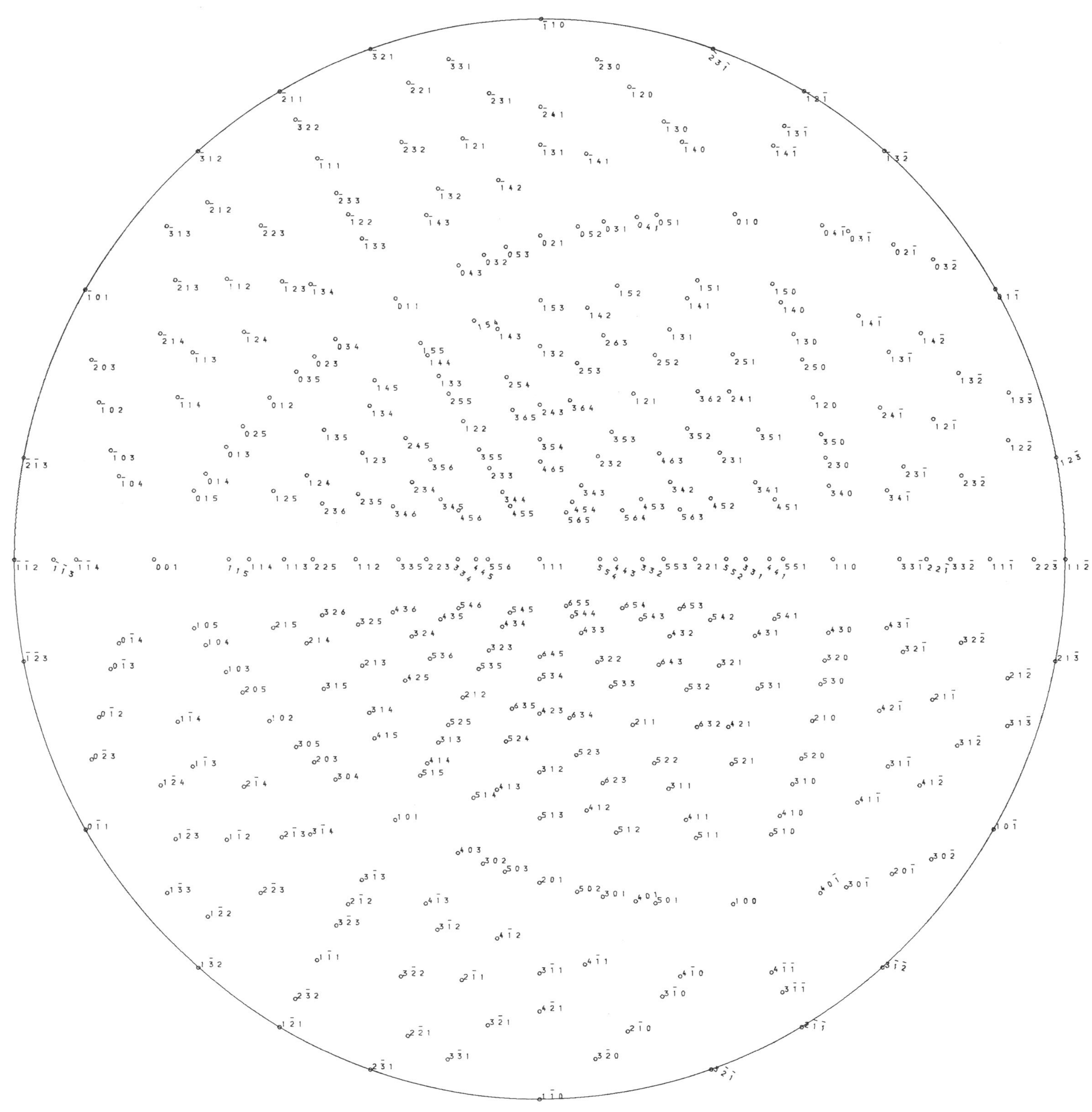

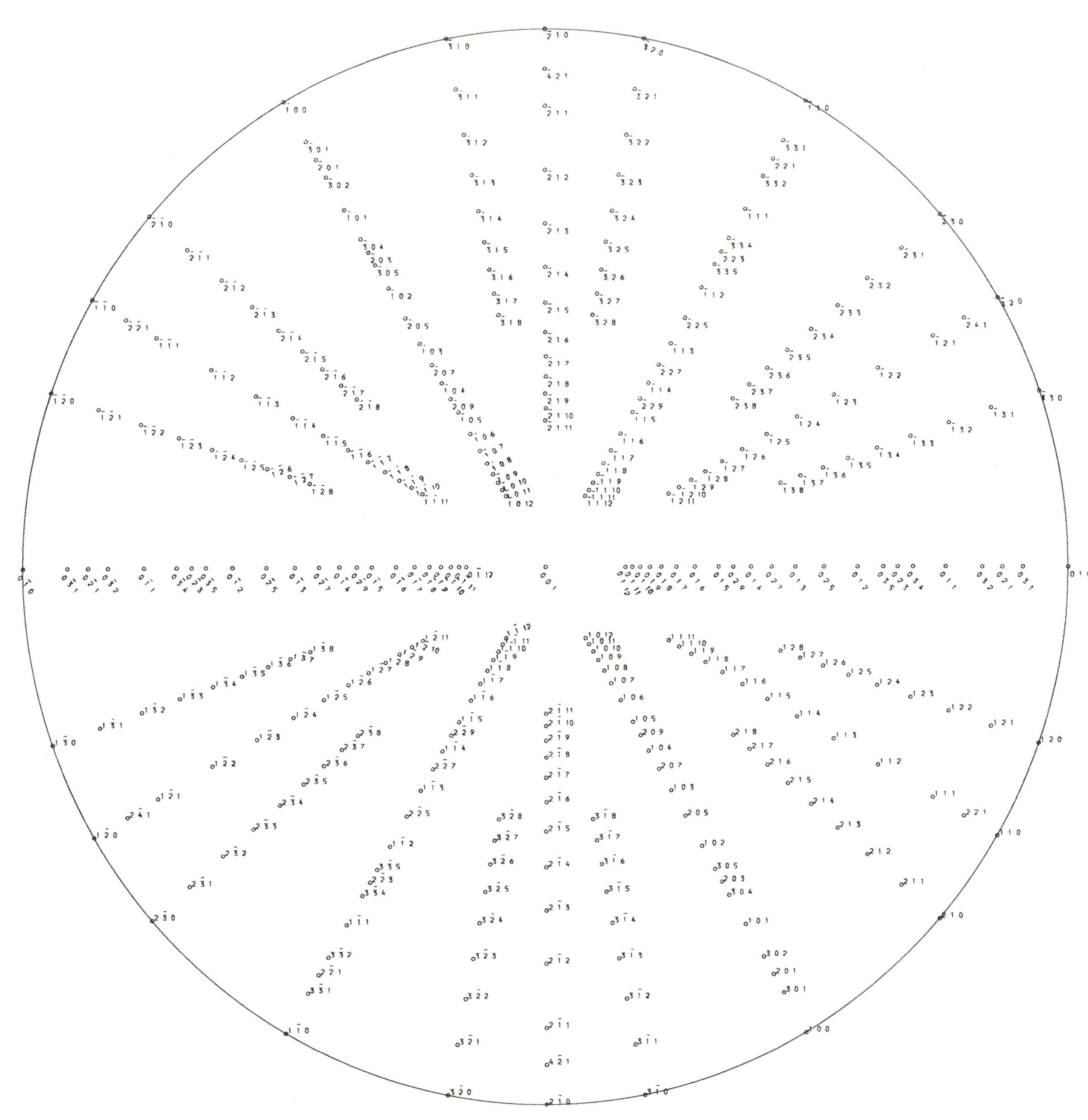

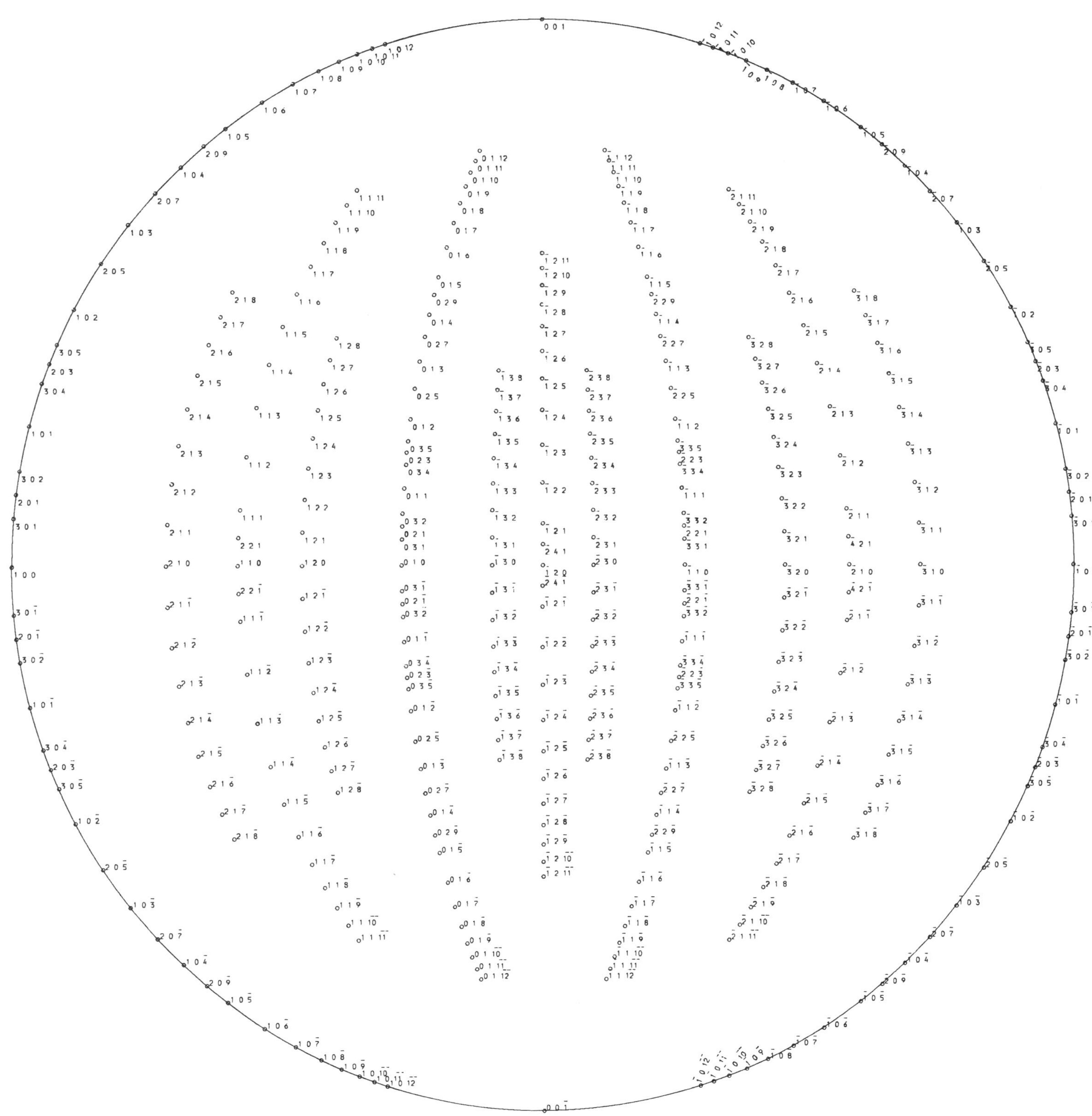

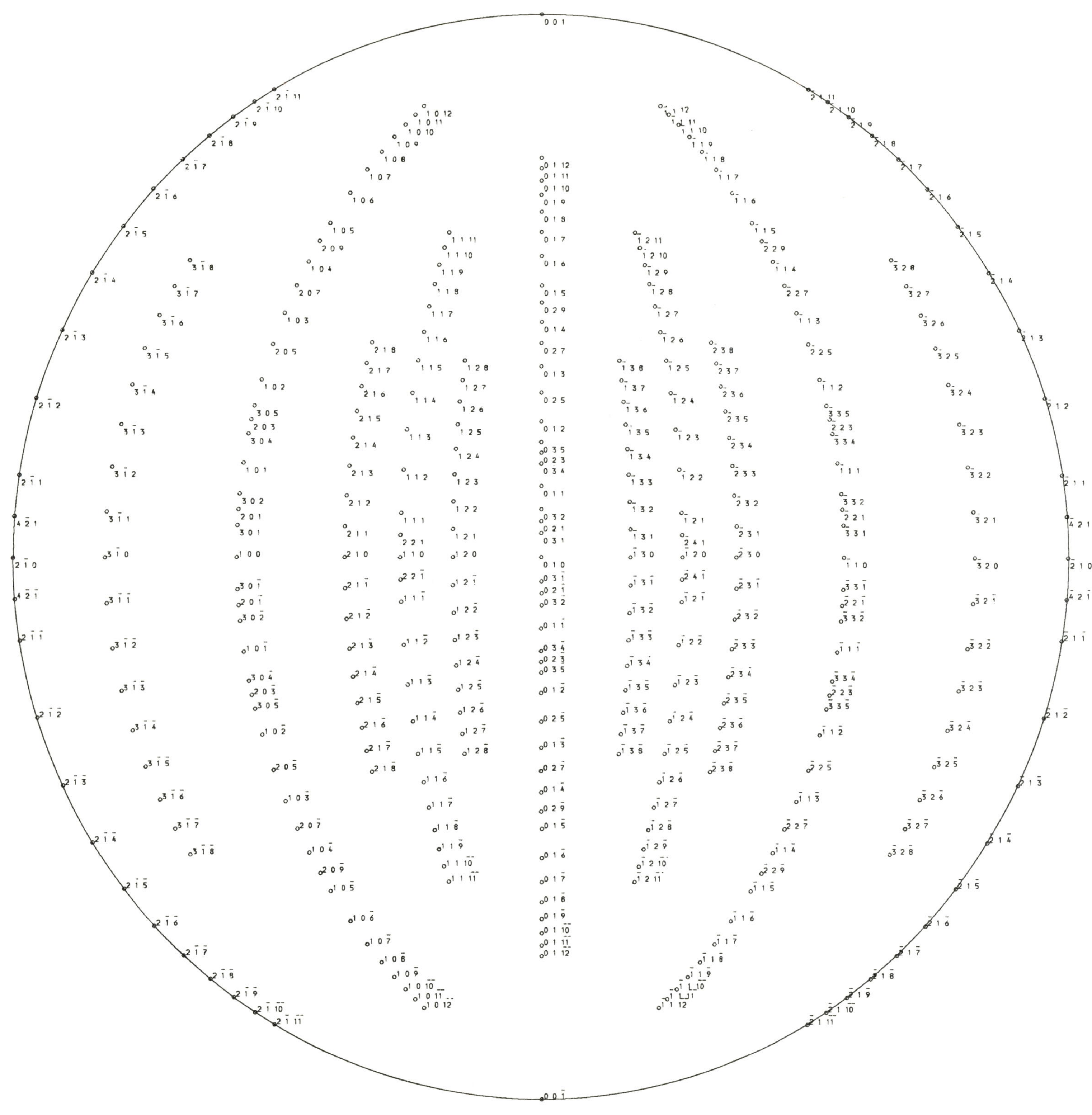

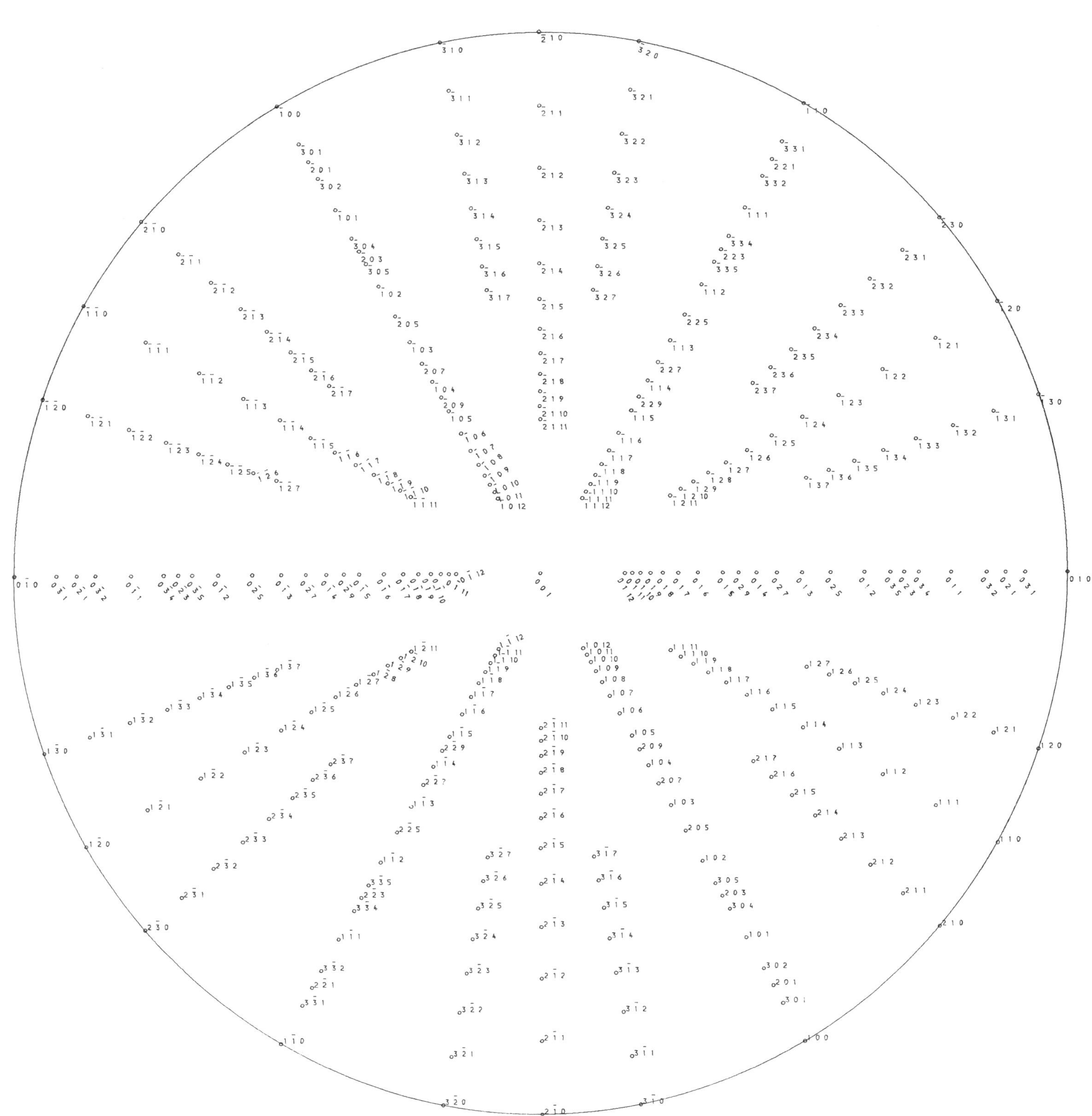

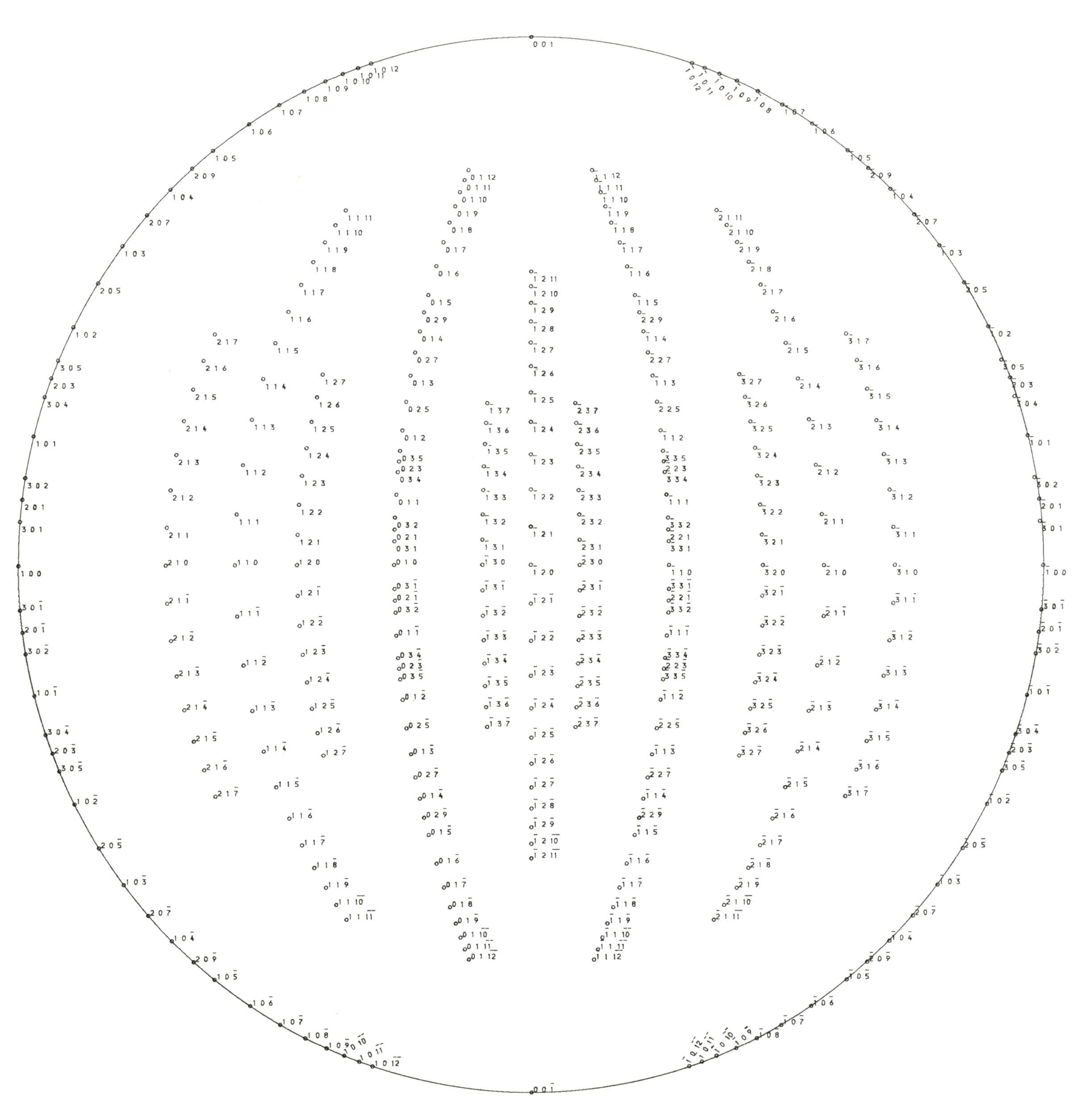

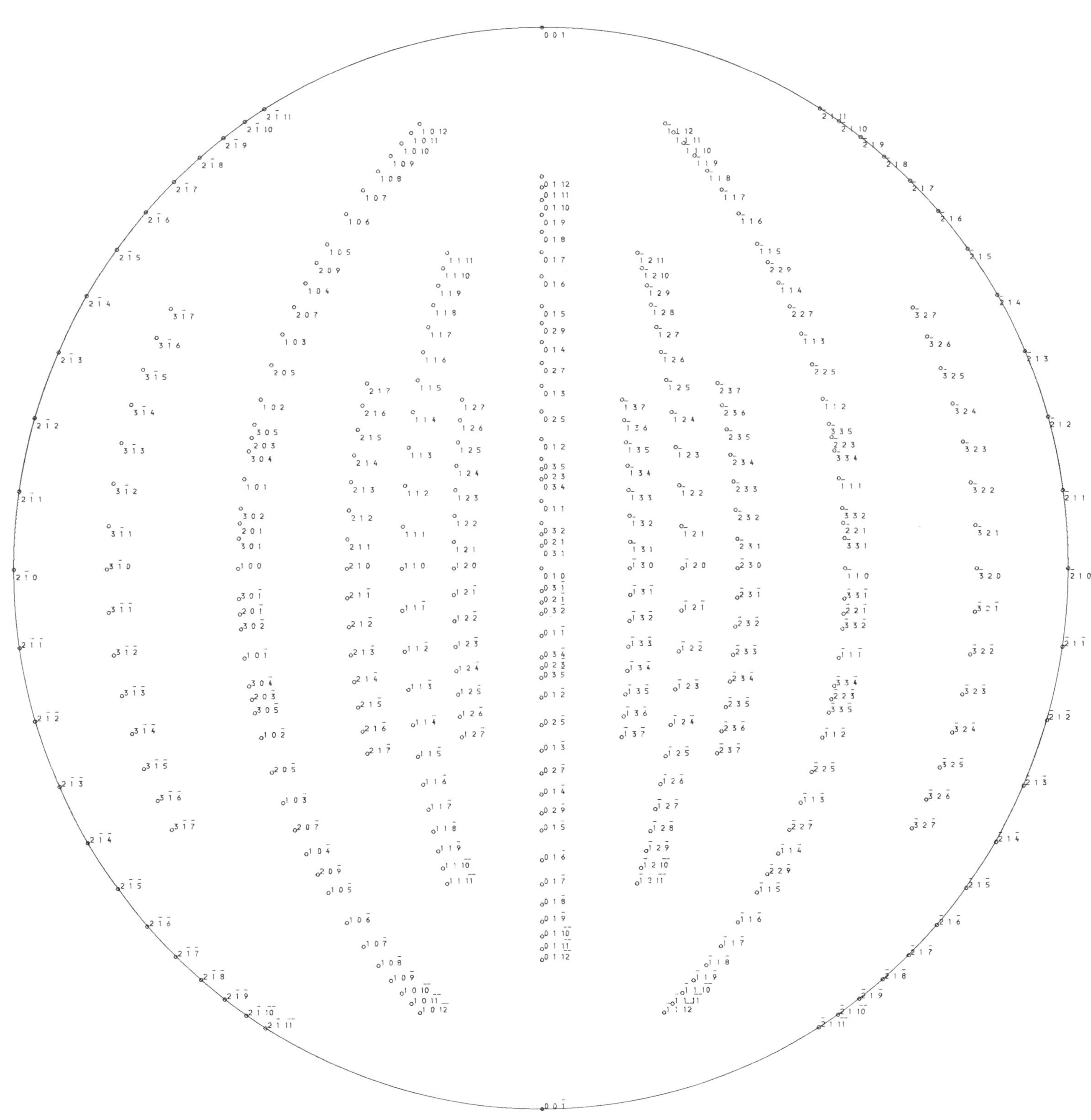

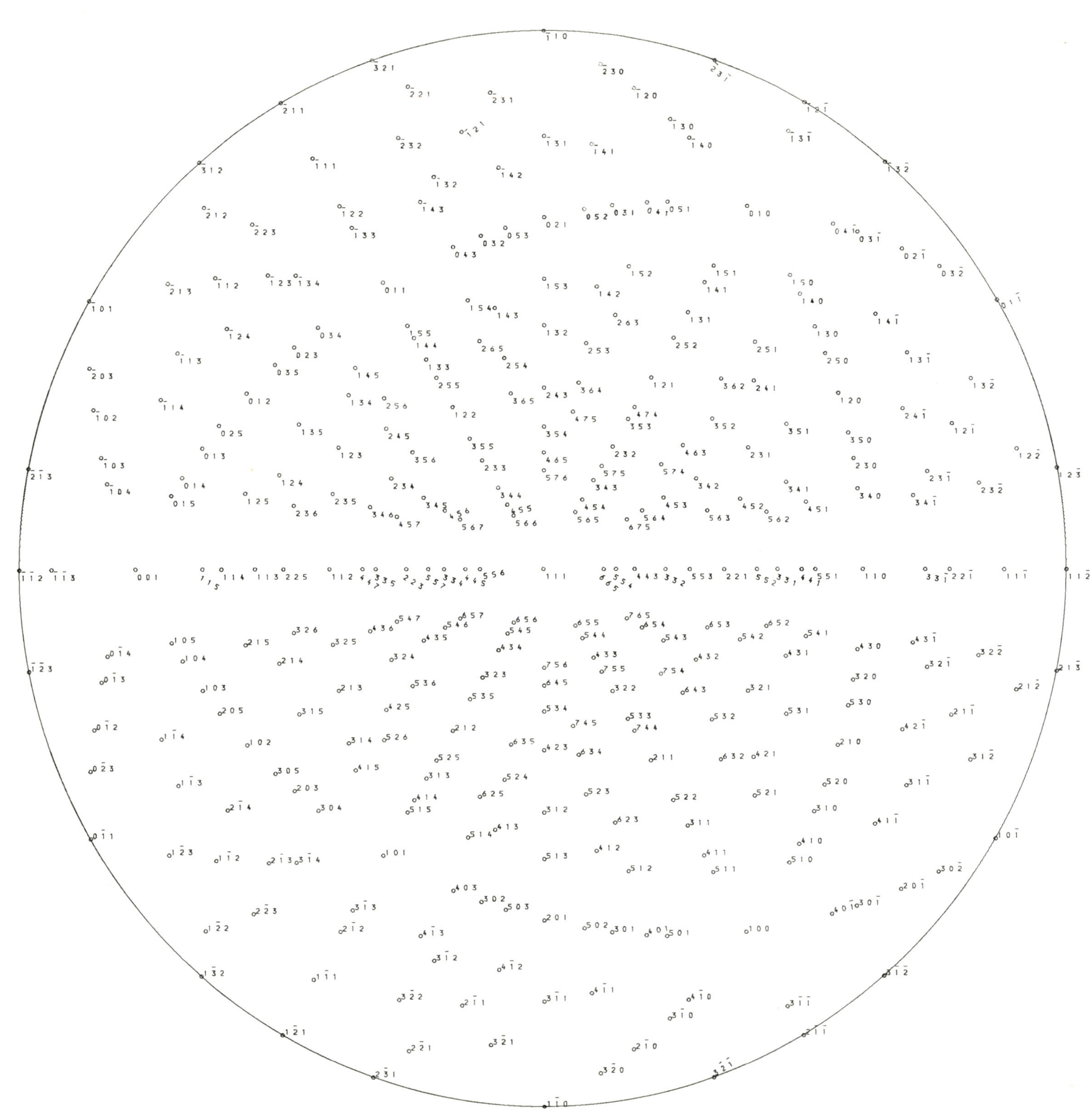

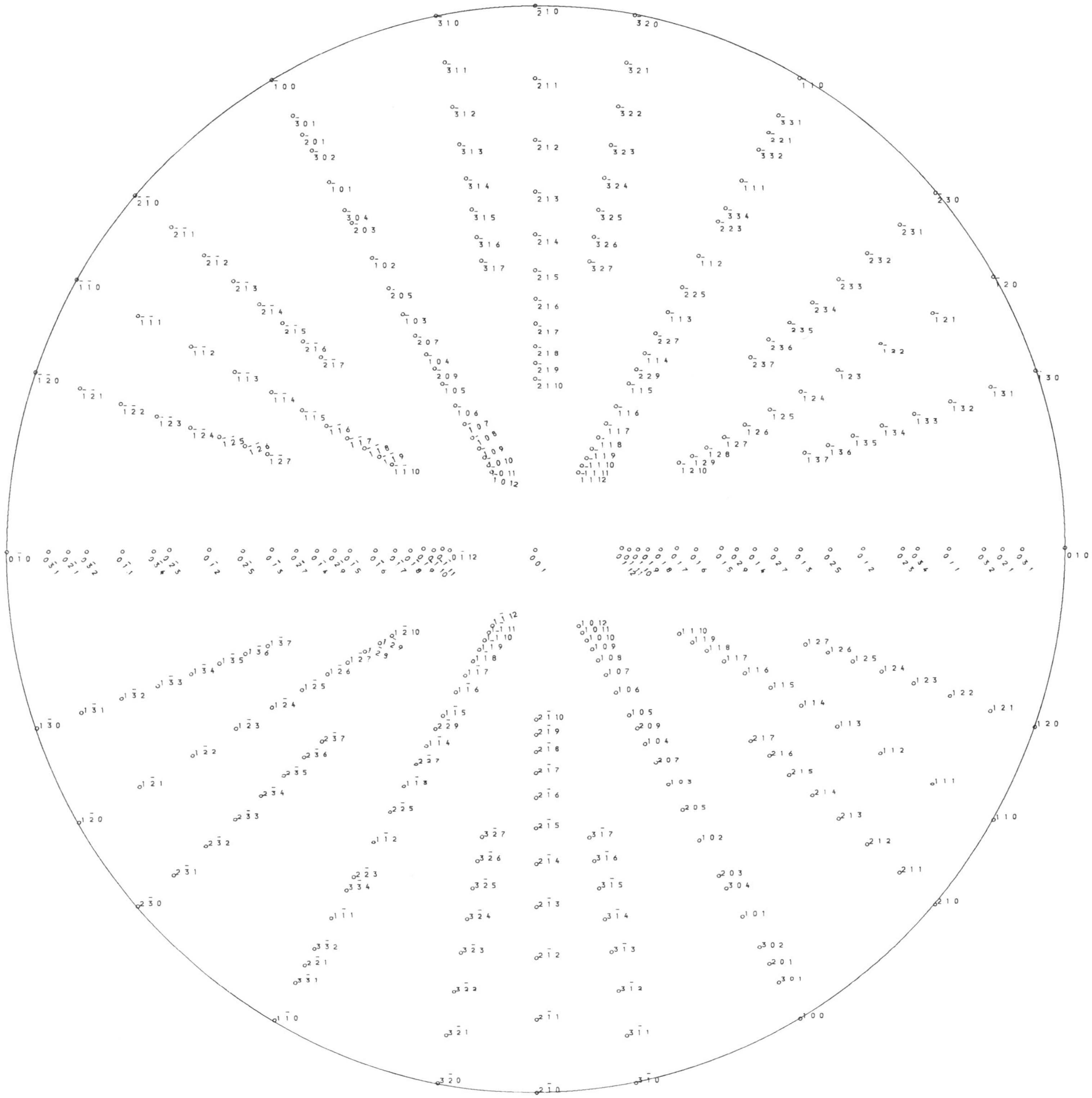

130 Y phase Axial ratio 3.4900 010 otherwise $\bar{1}2\bar{1}0$ axis

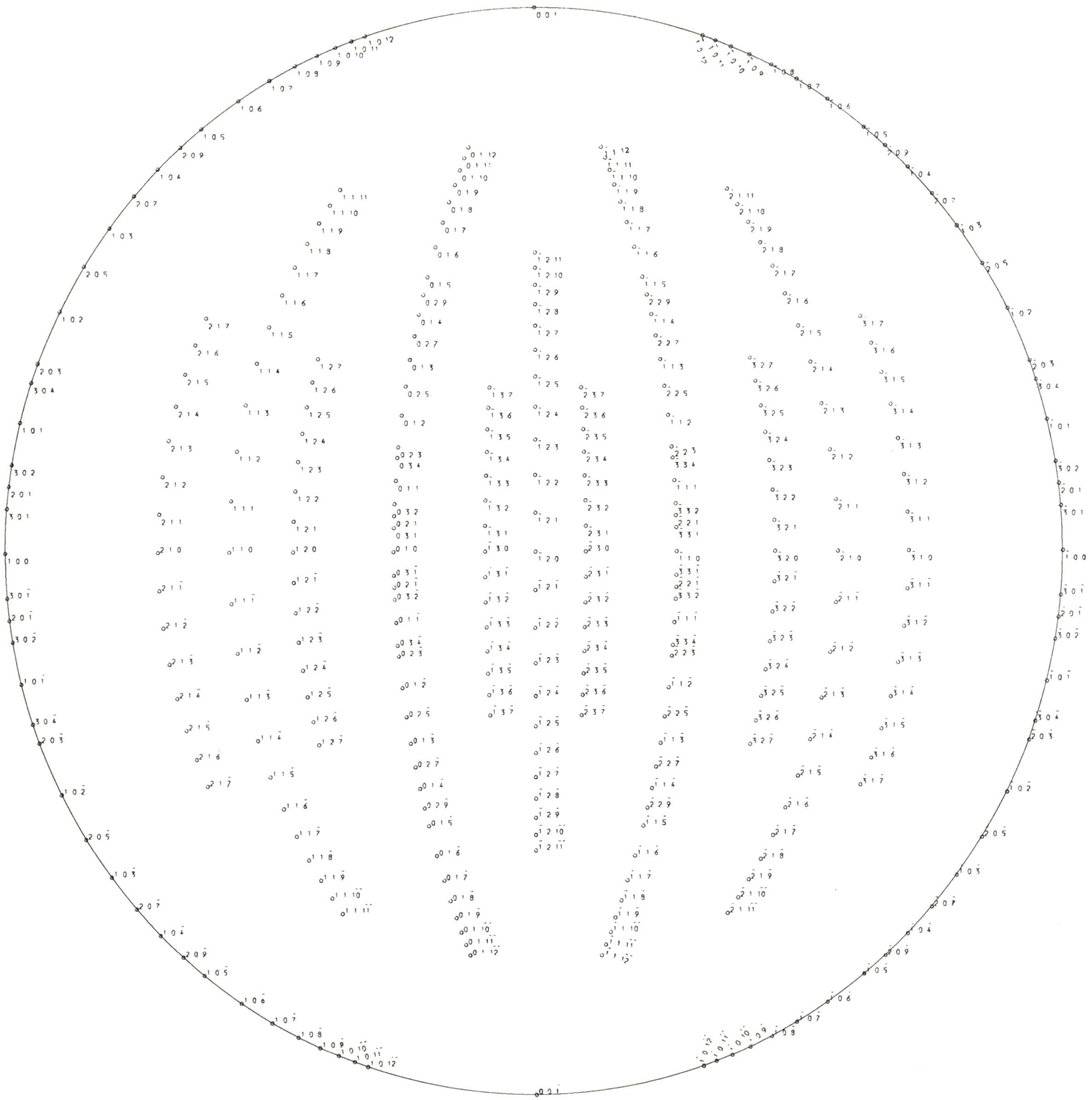

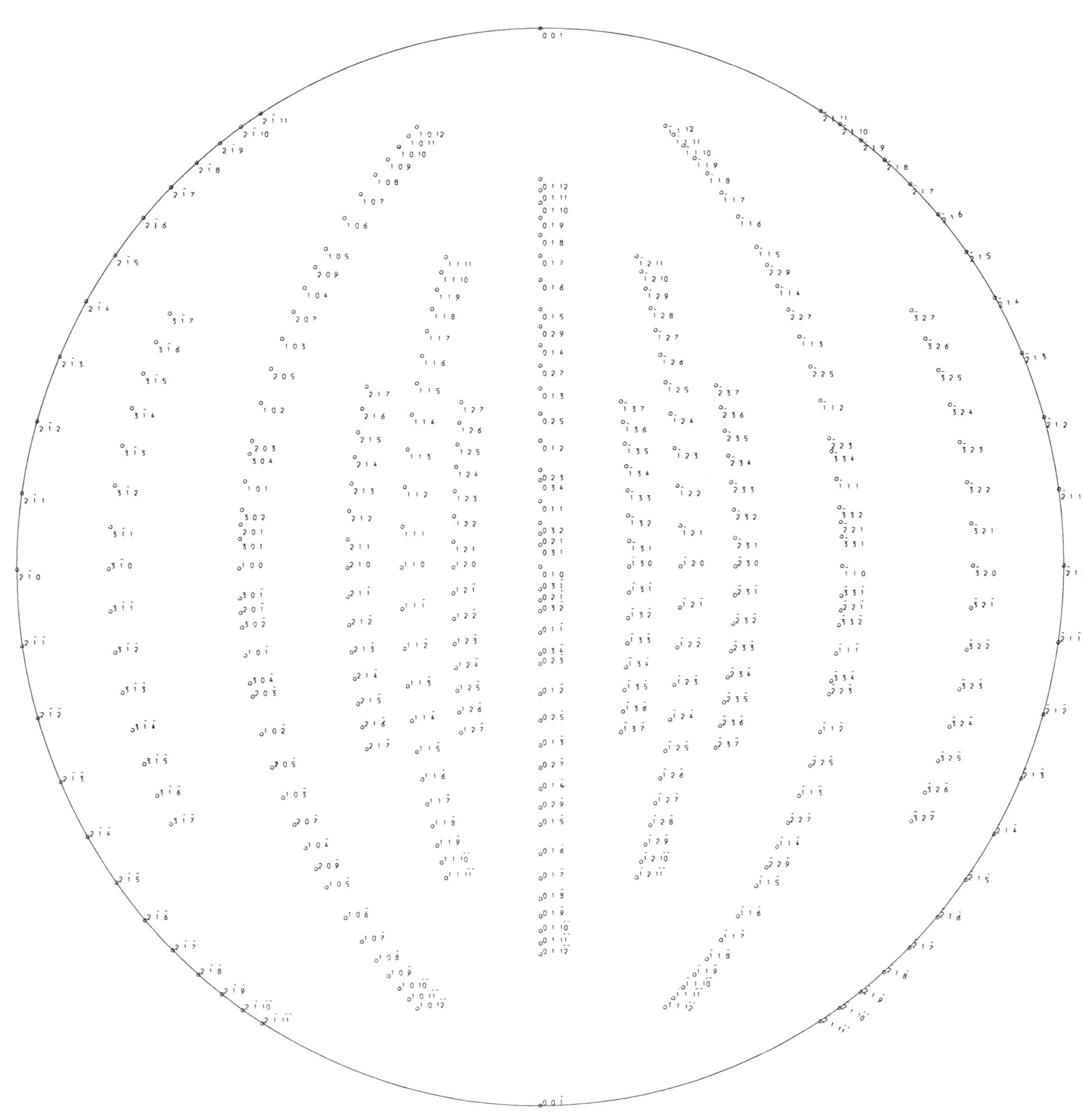

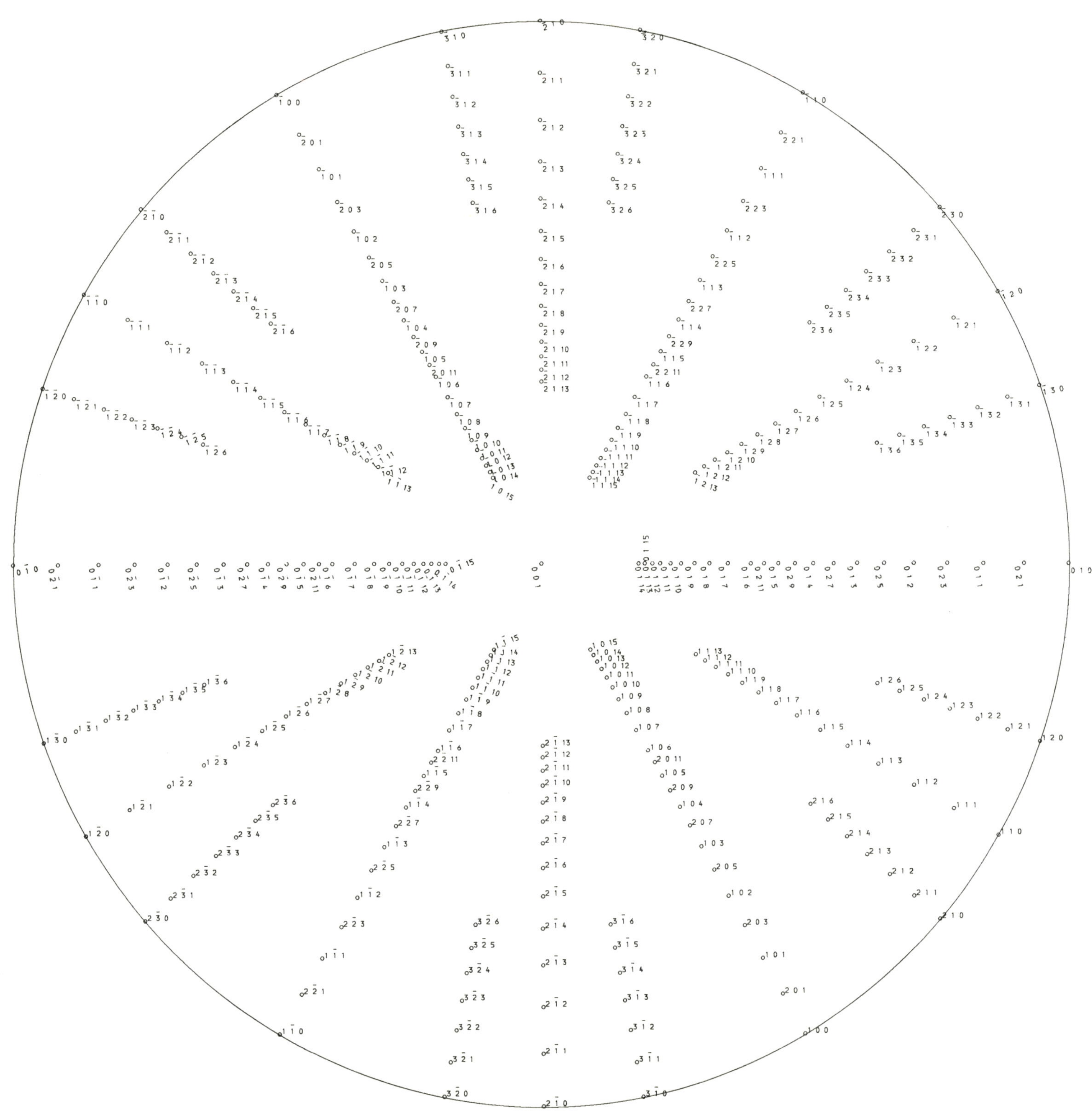

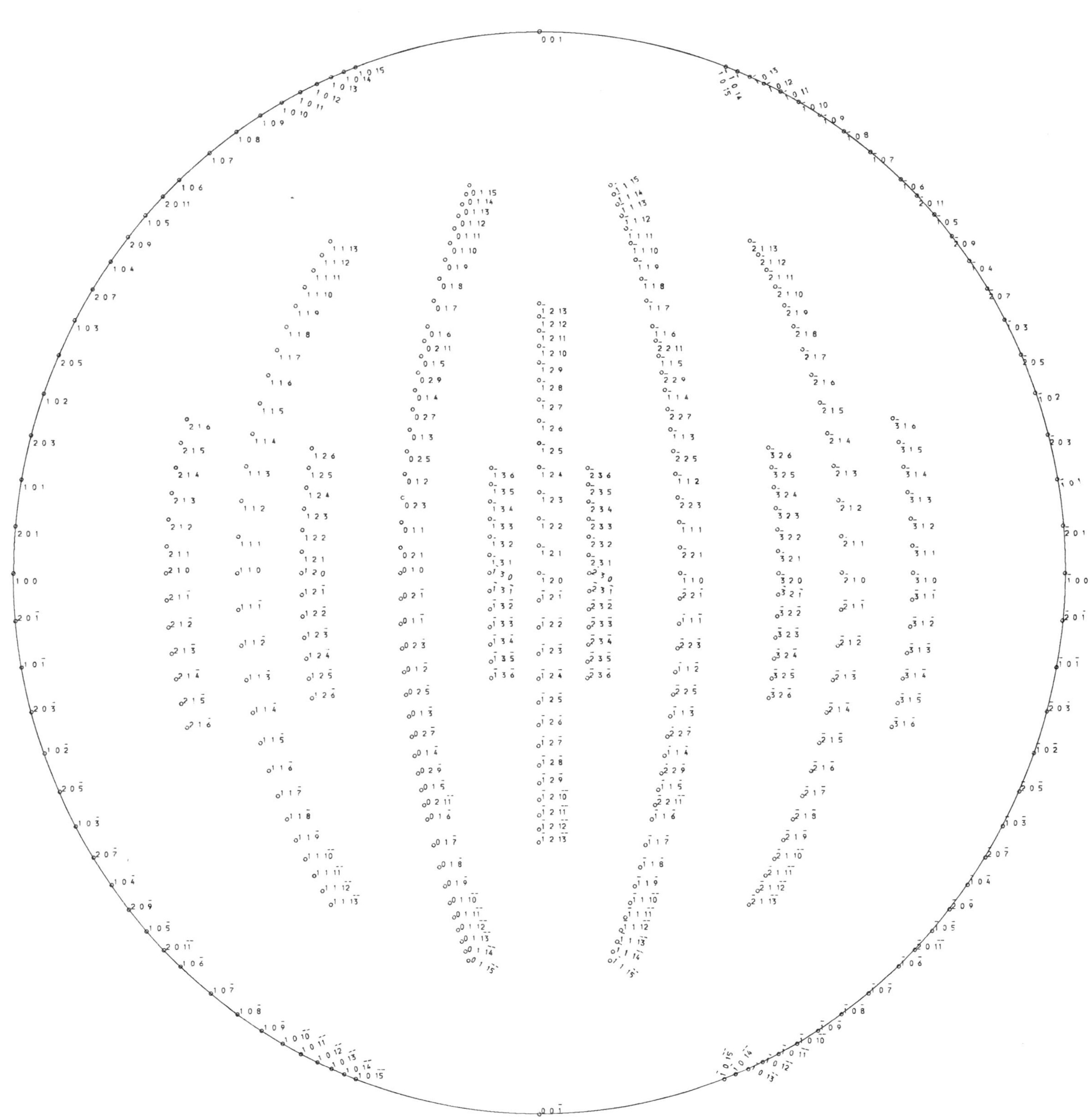

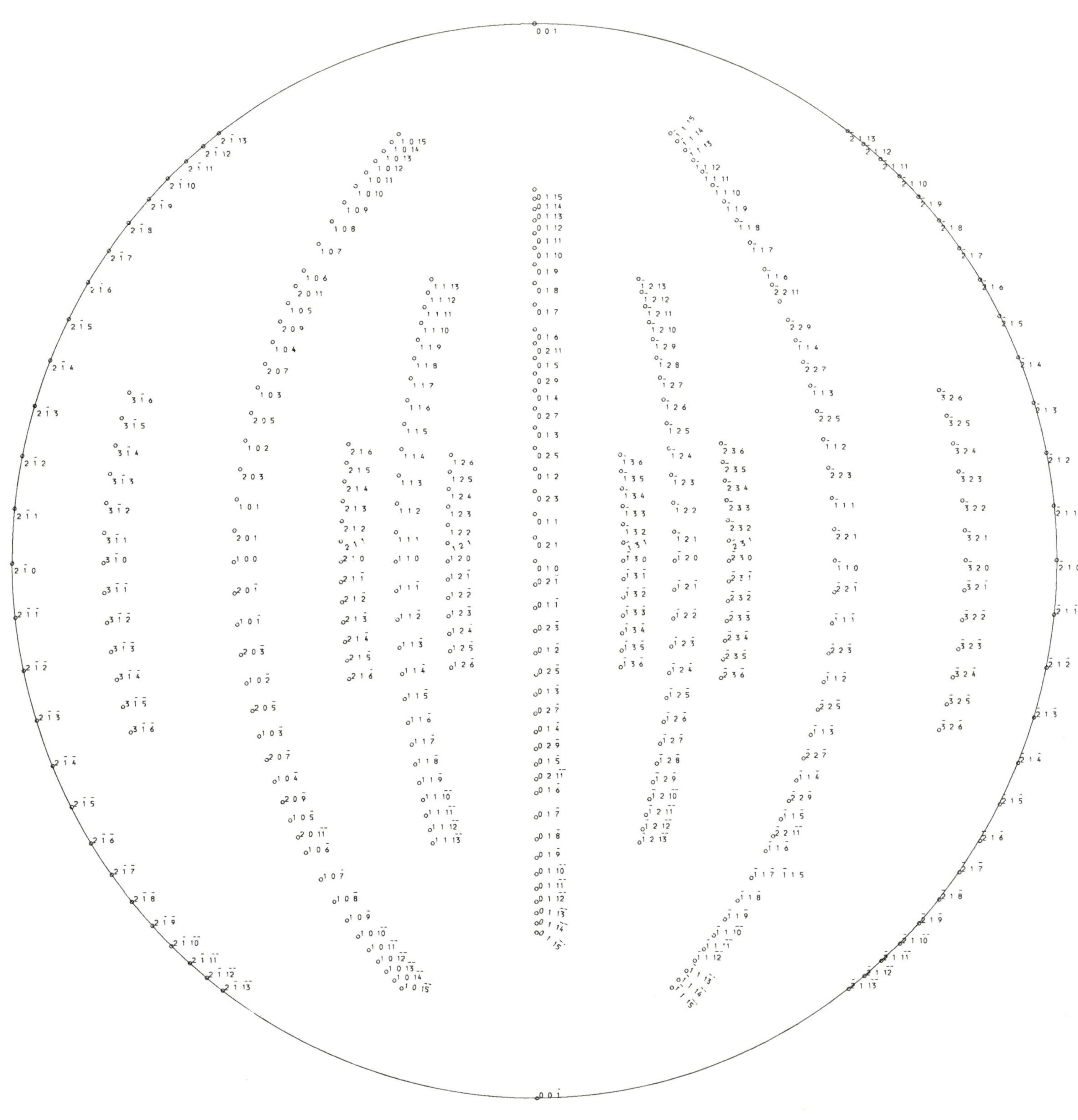

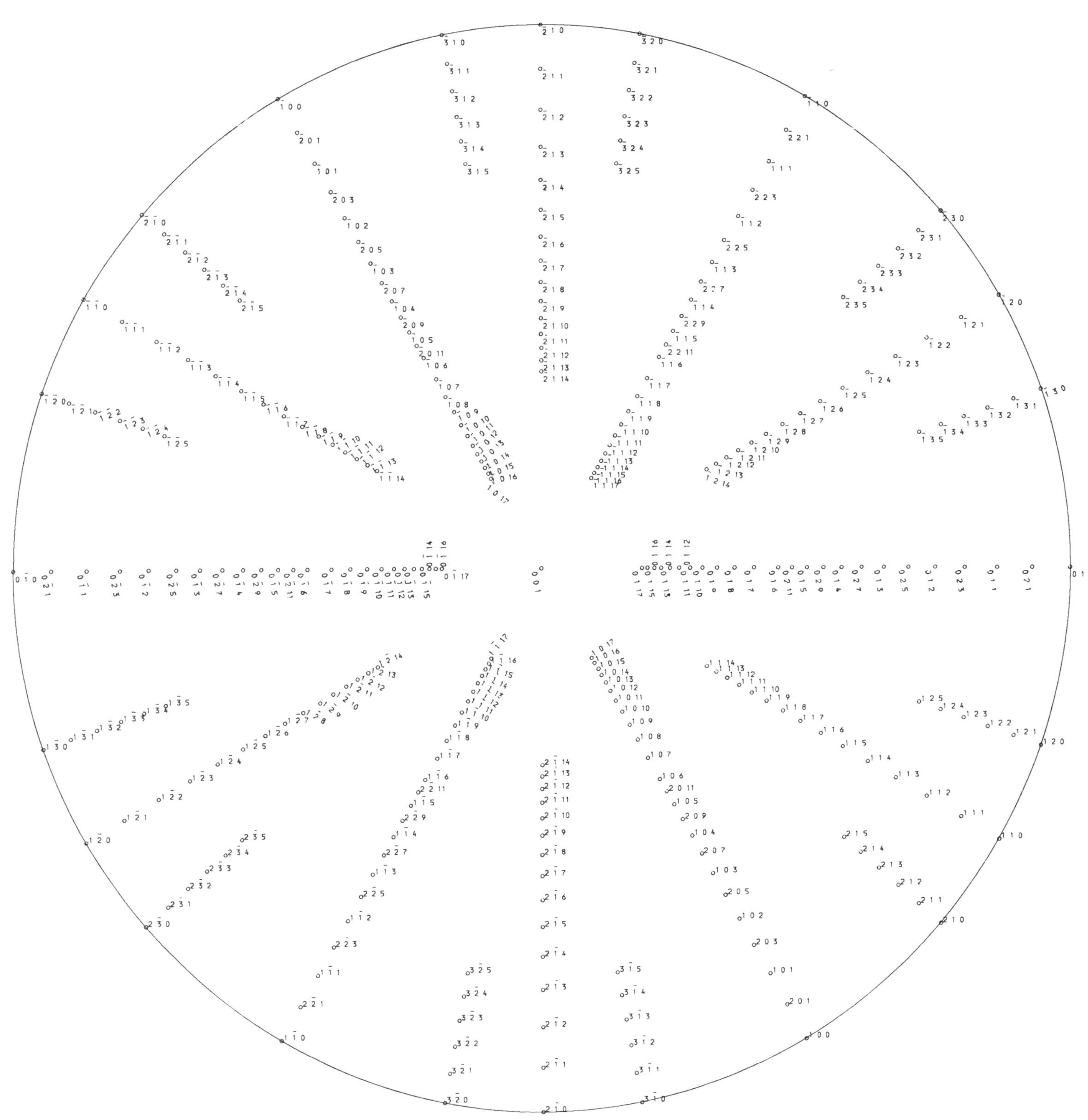

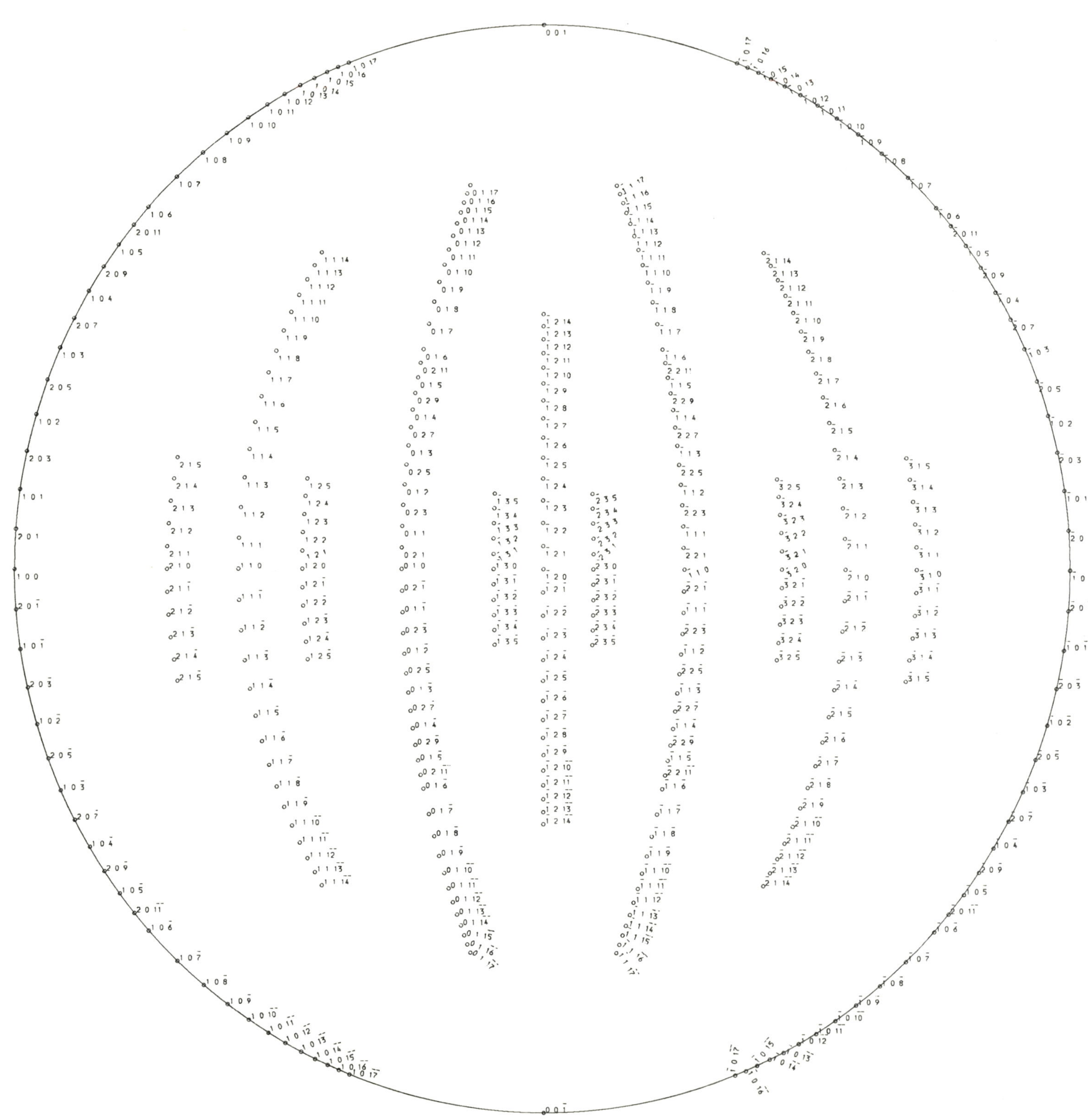

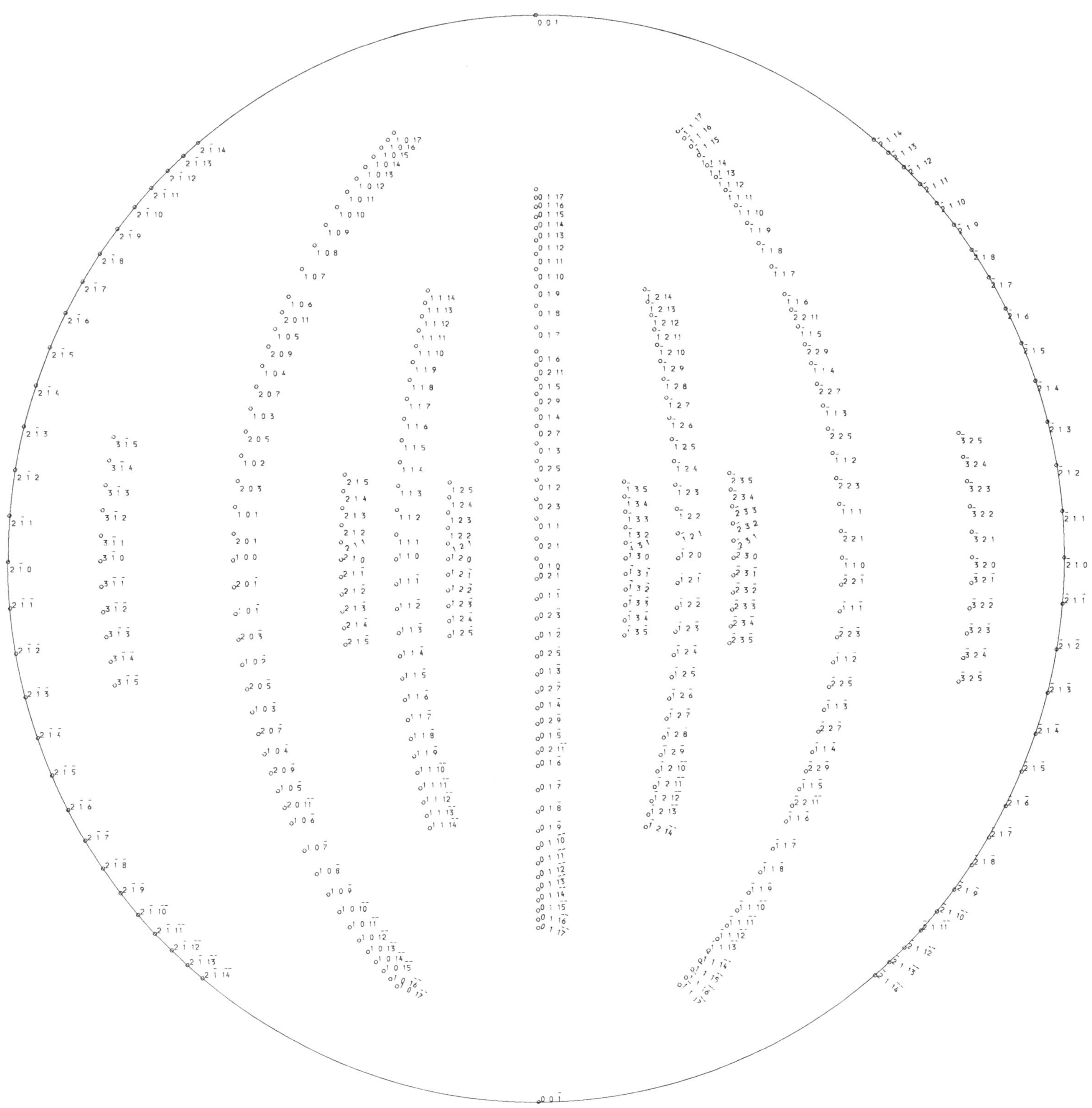

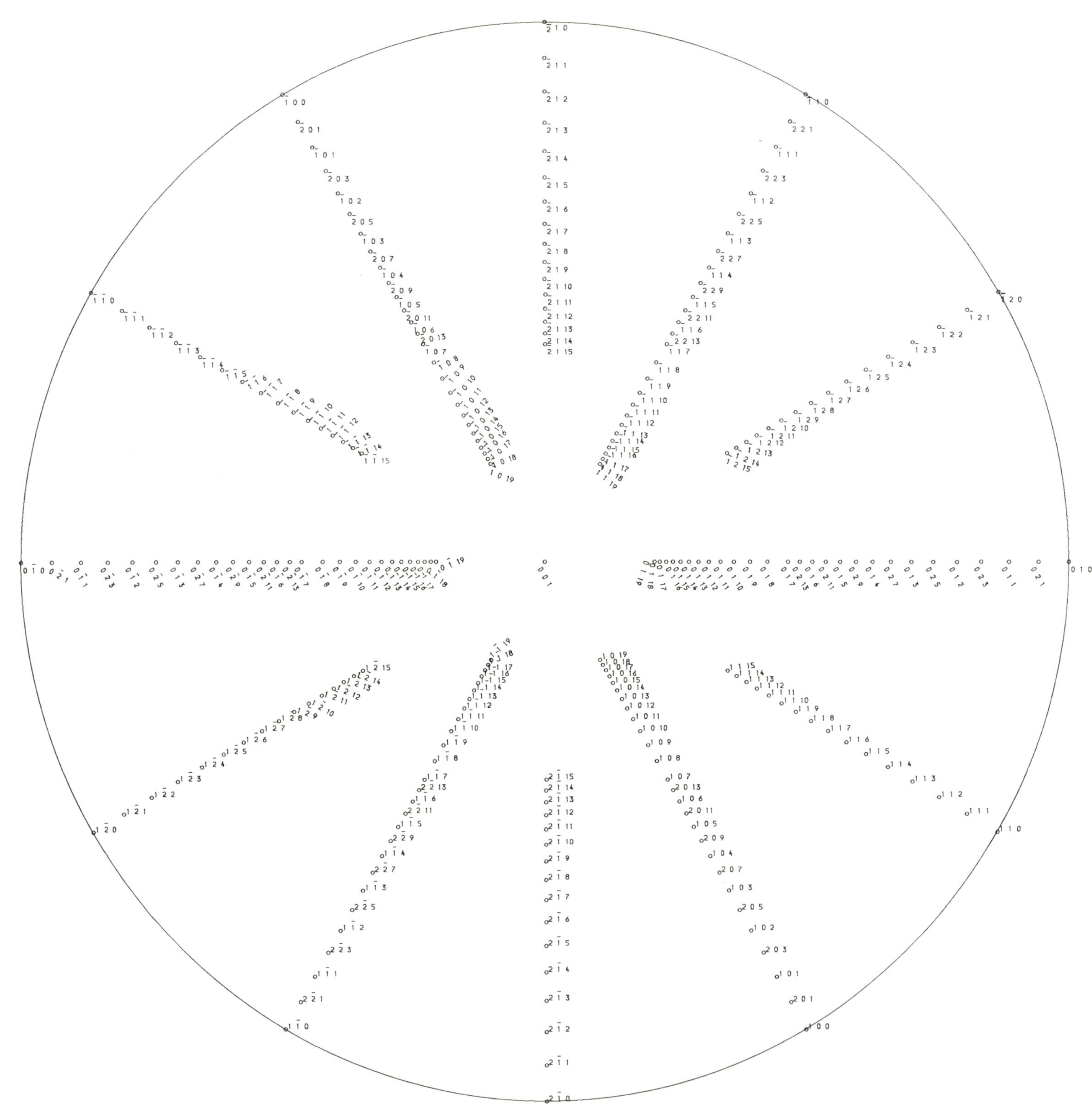

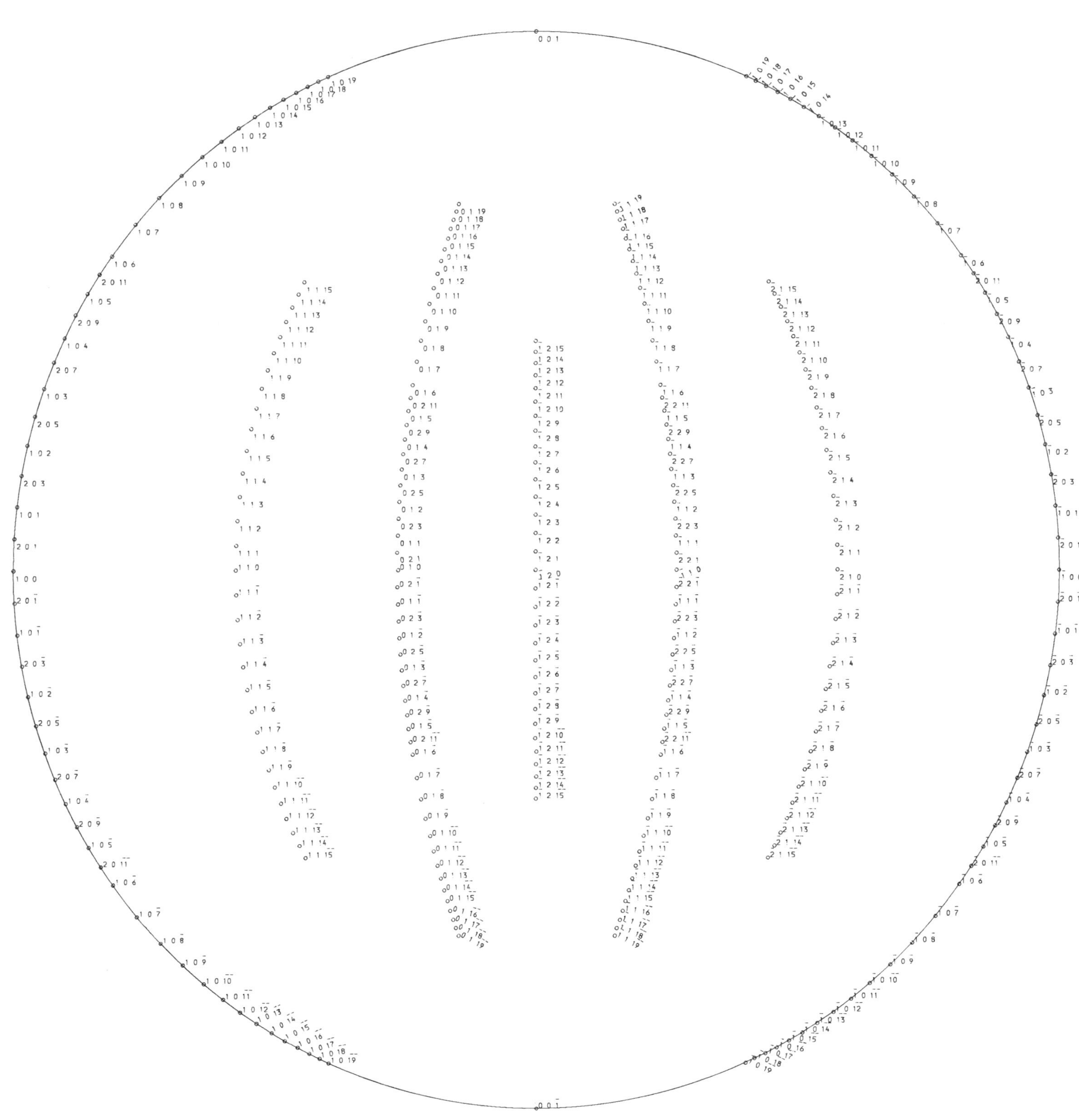

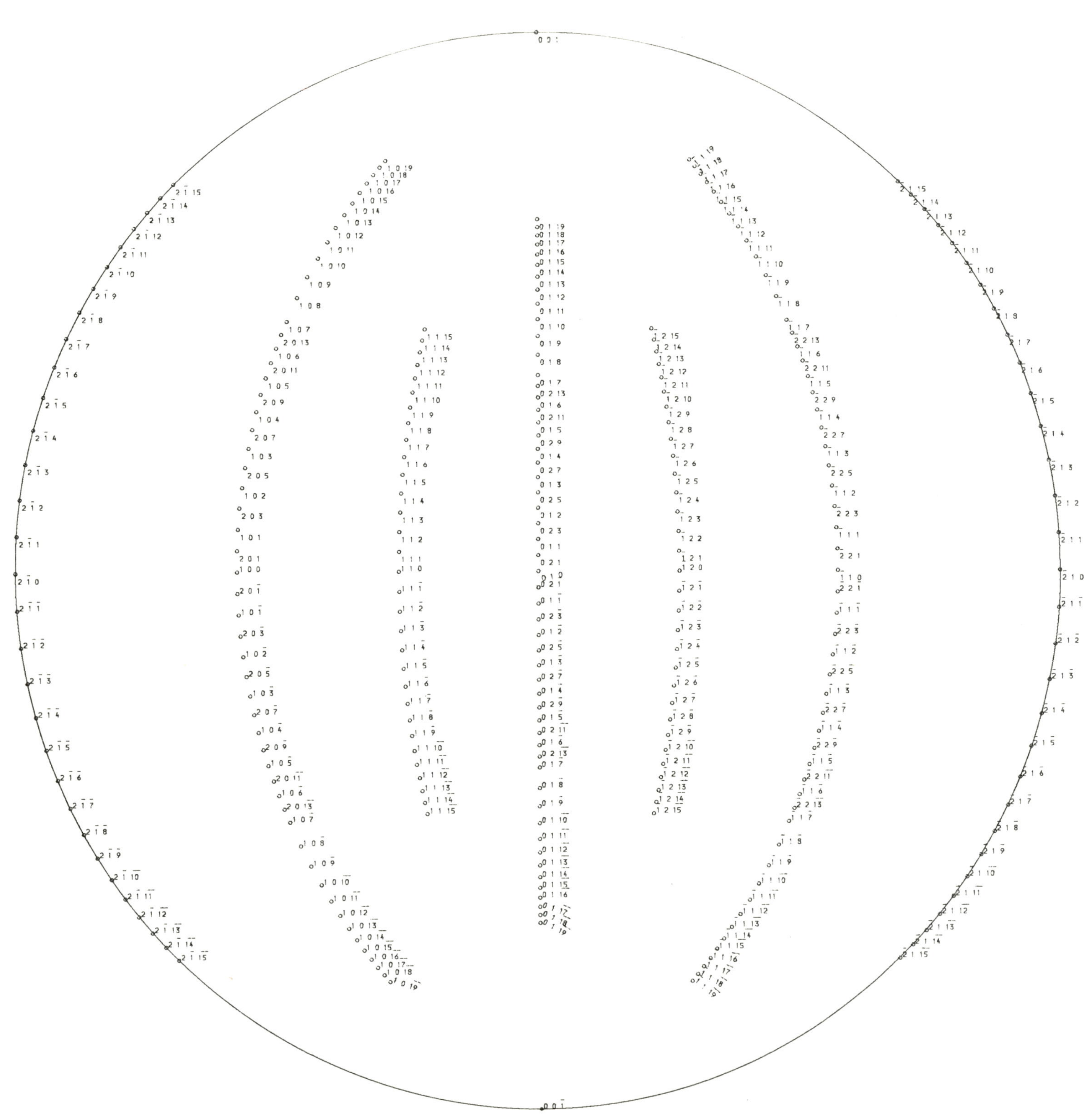

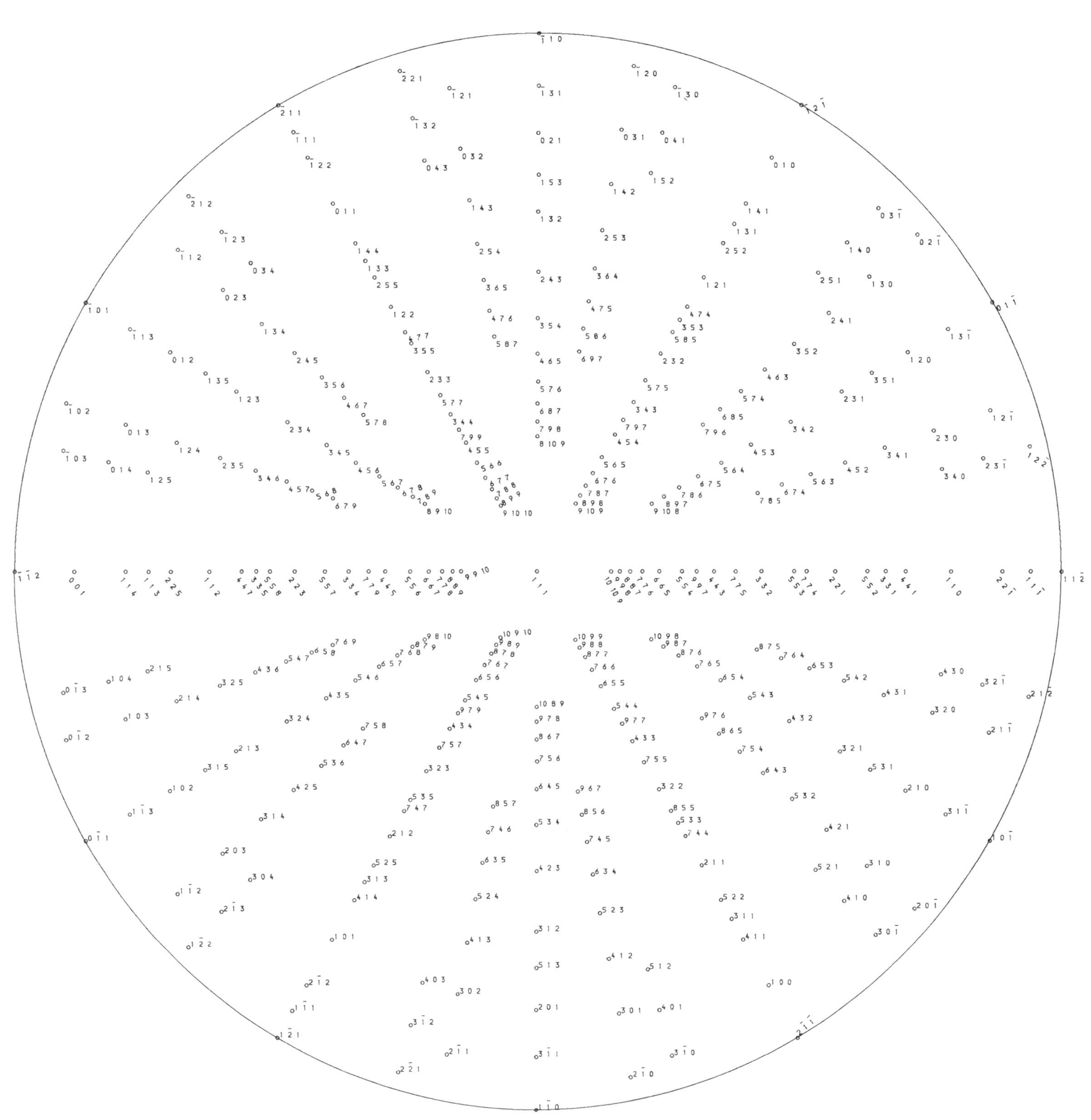

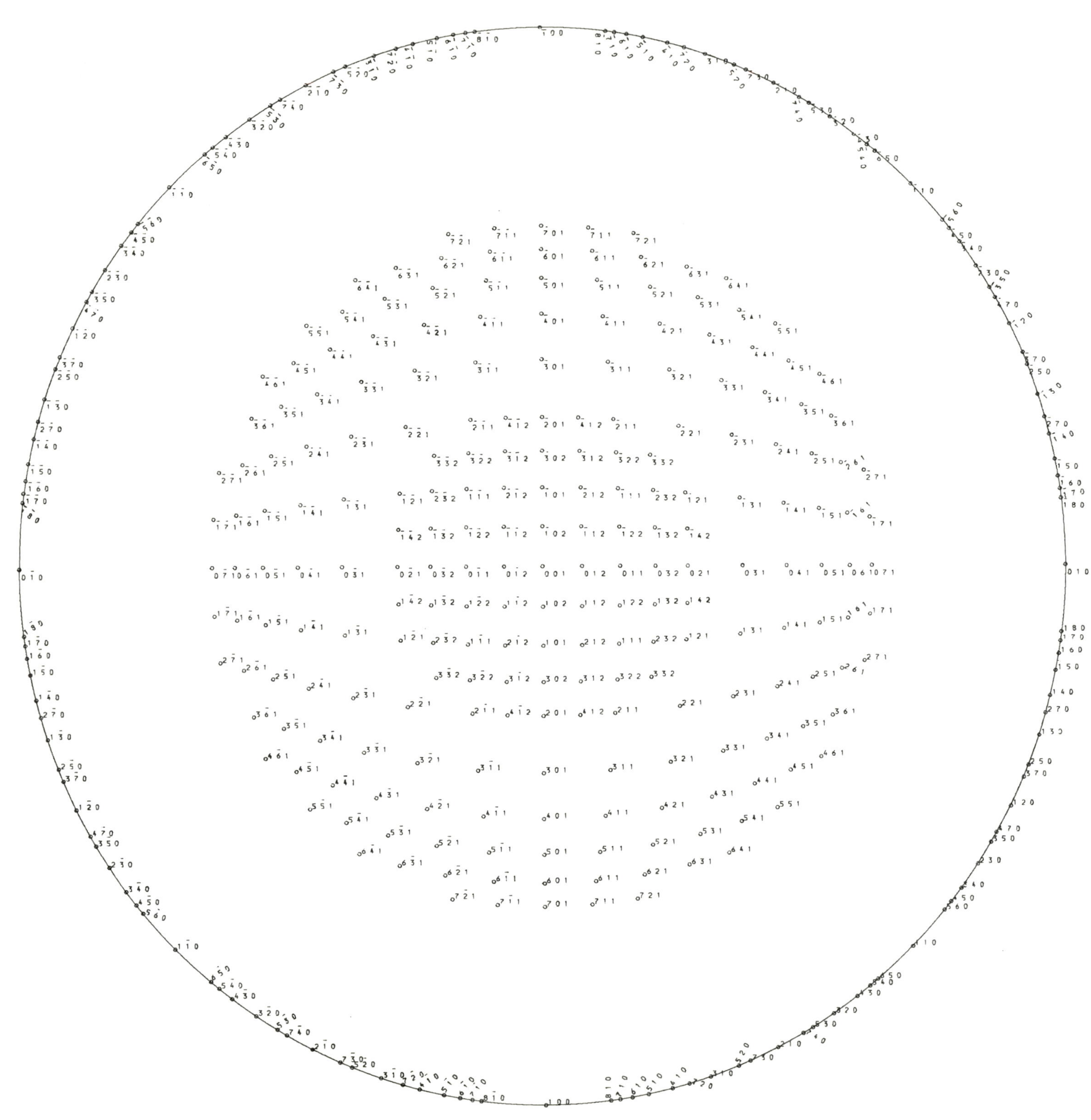

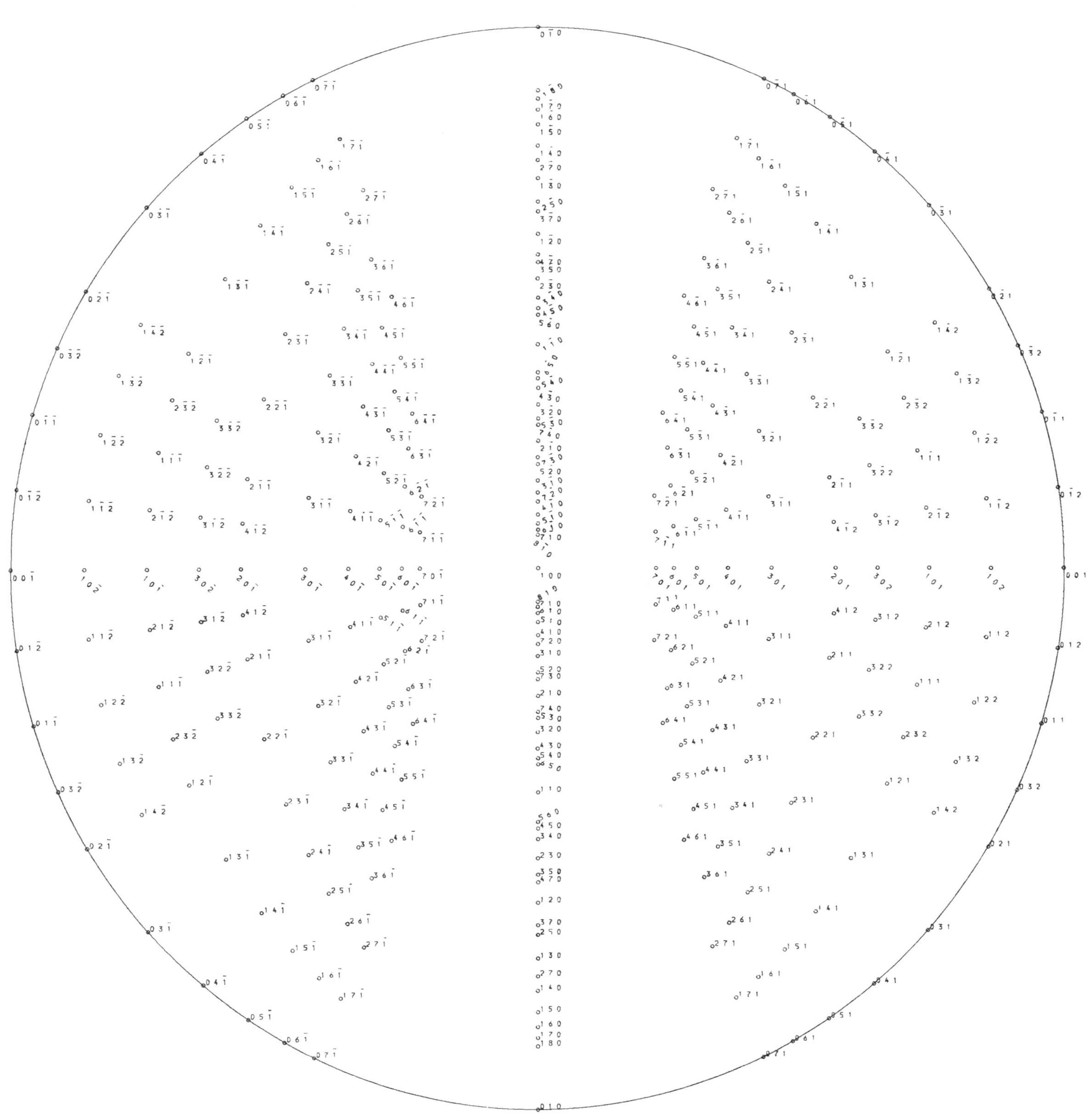

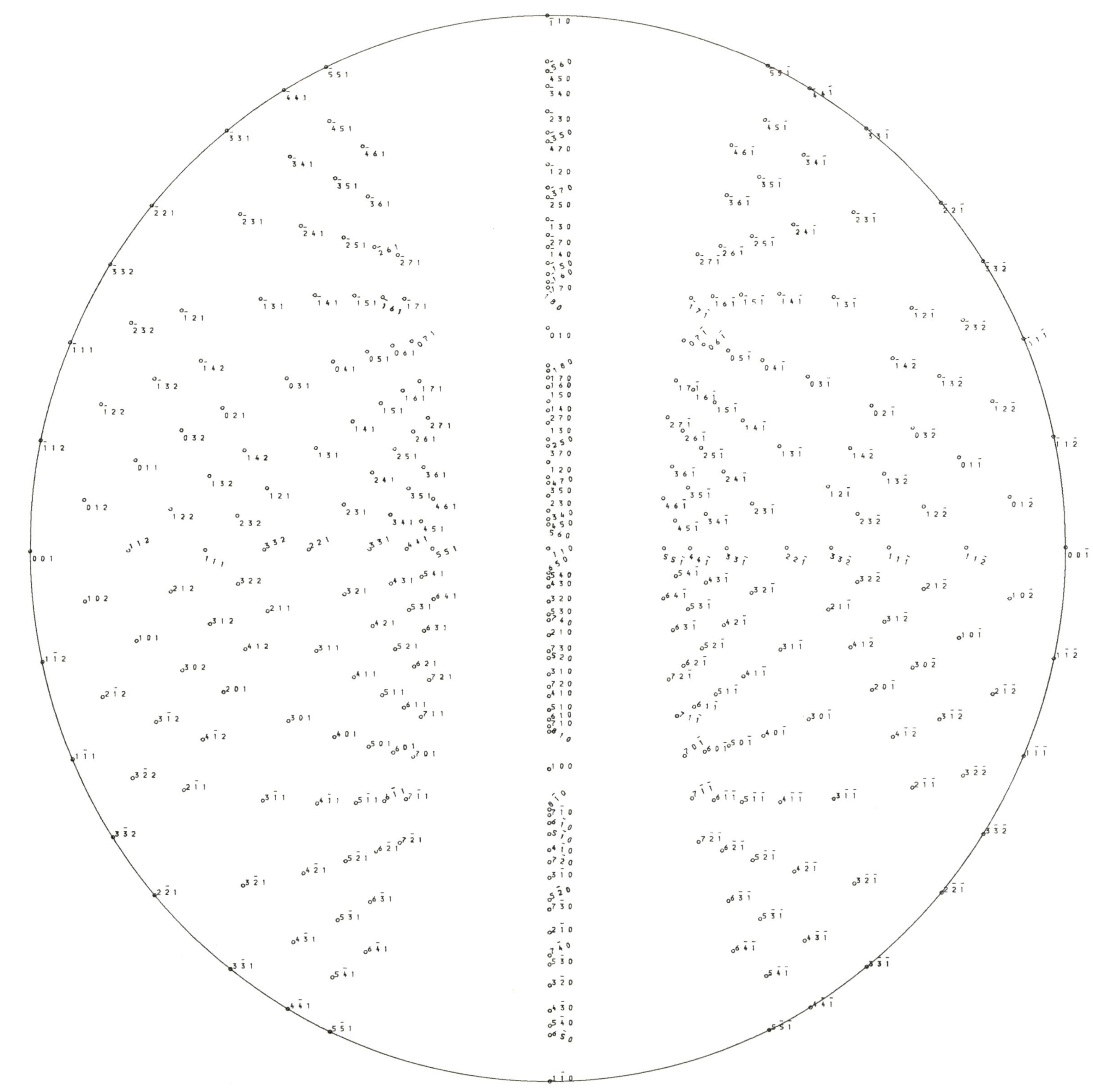

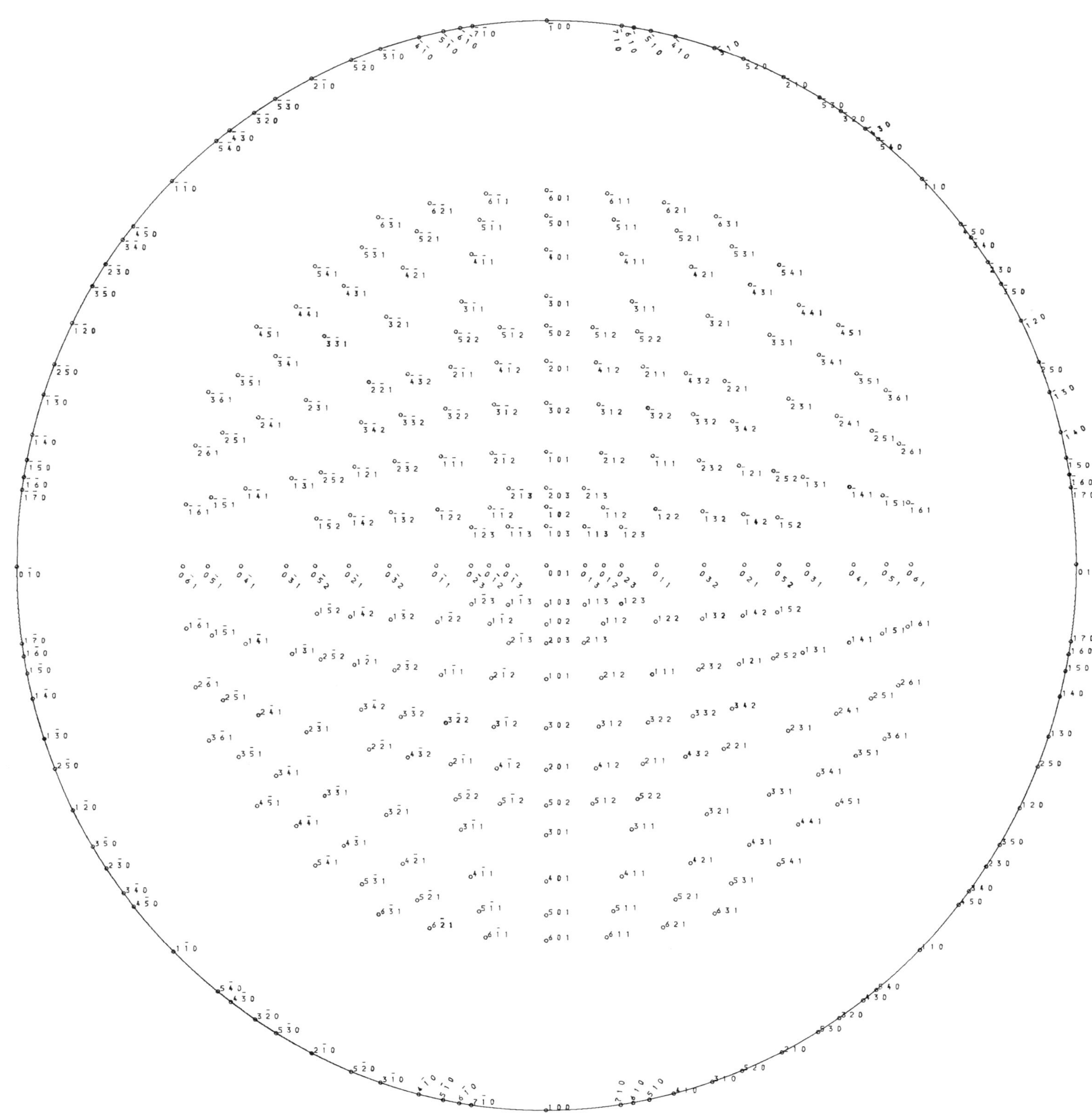

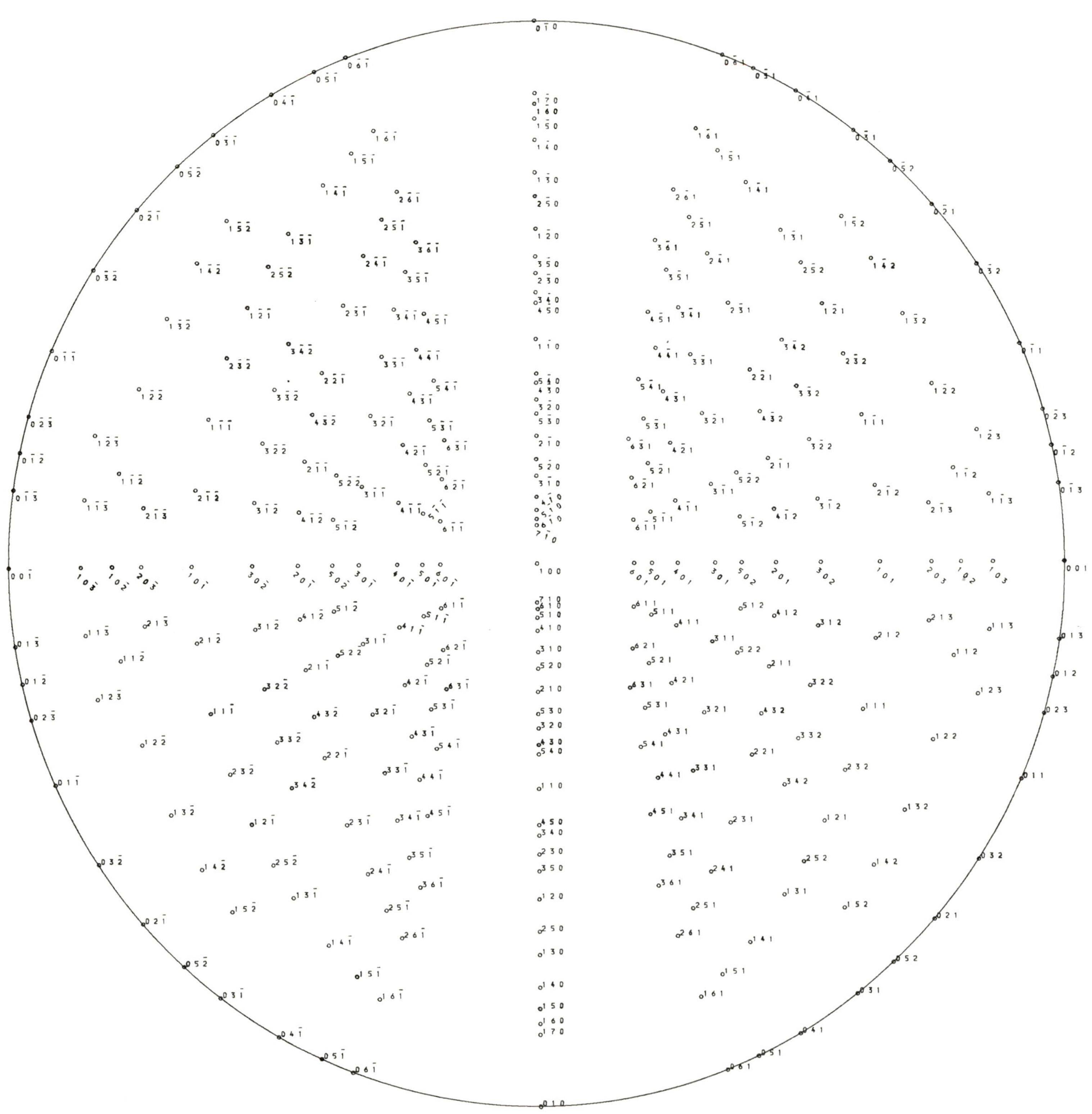

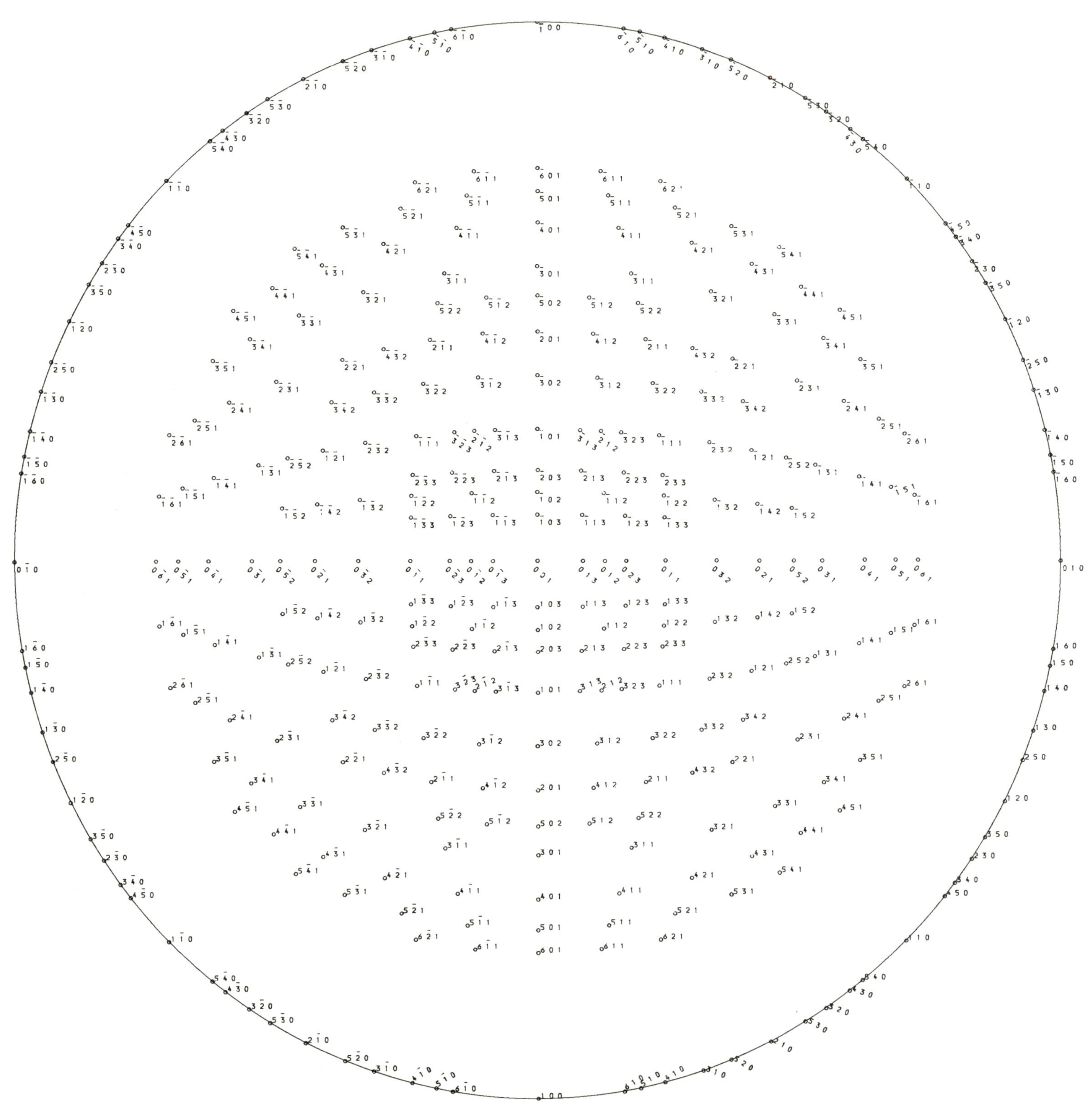

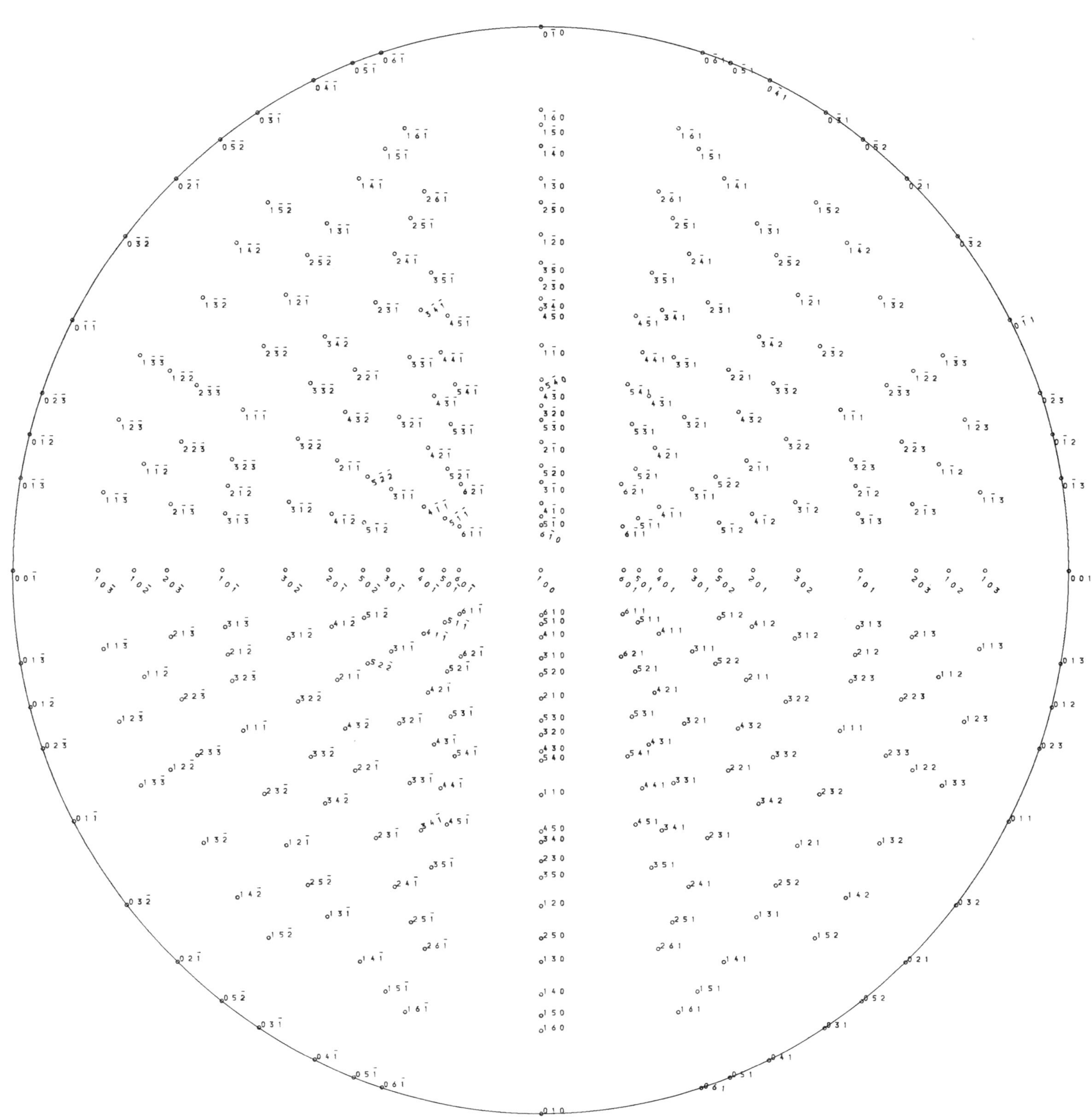

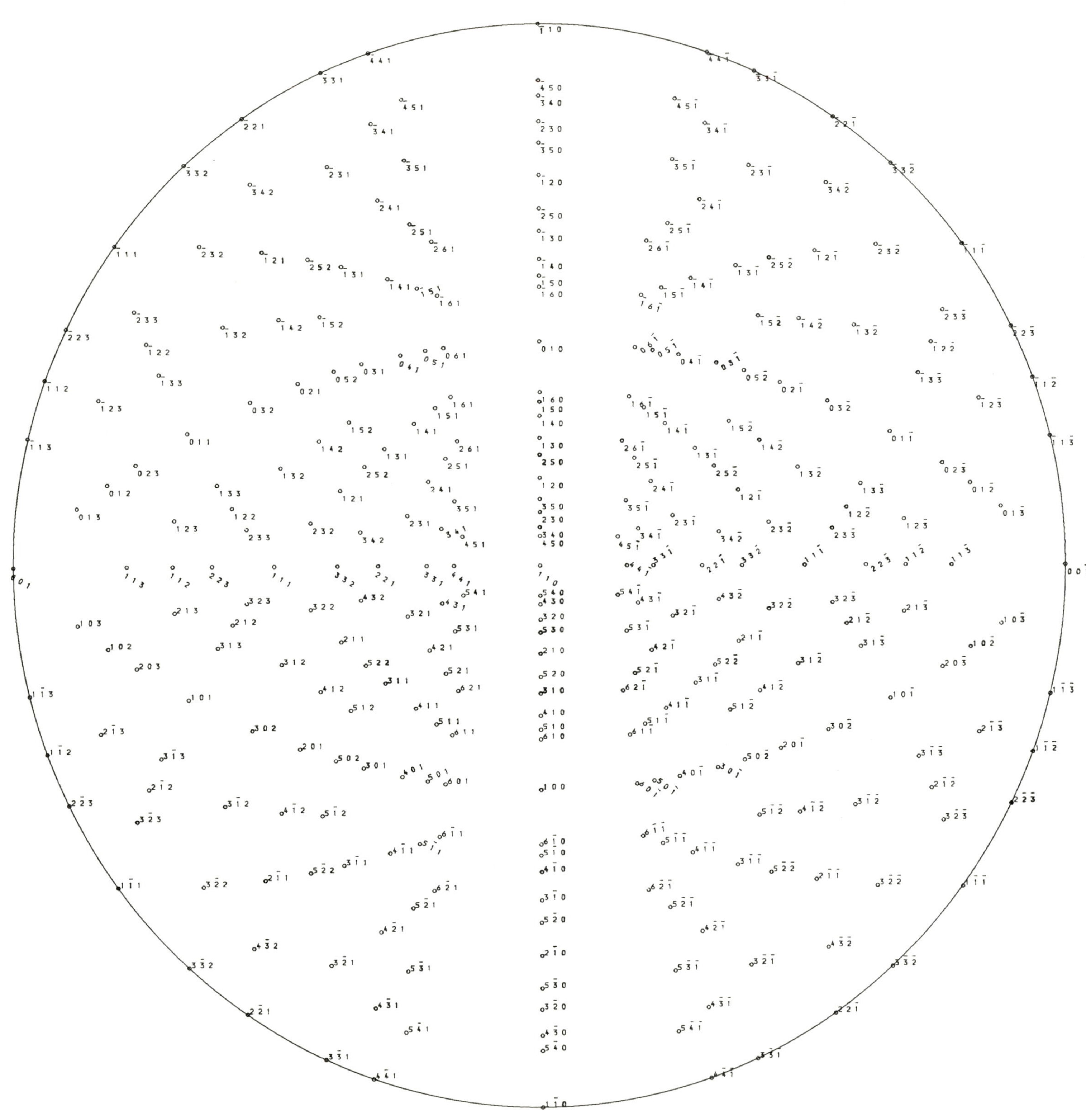

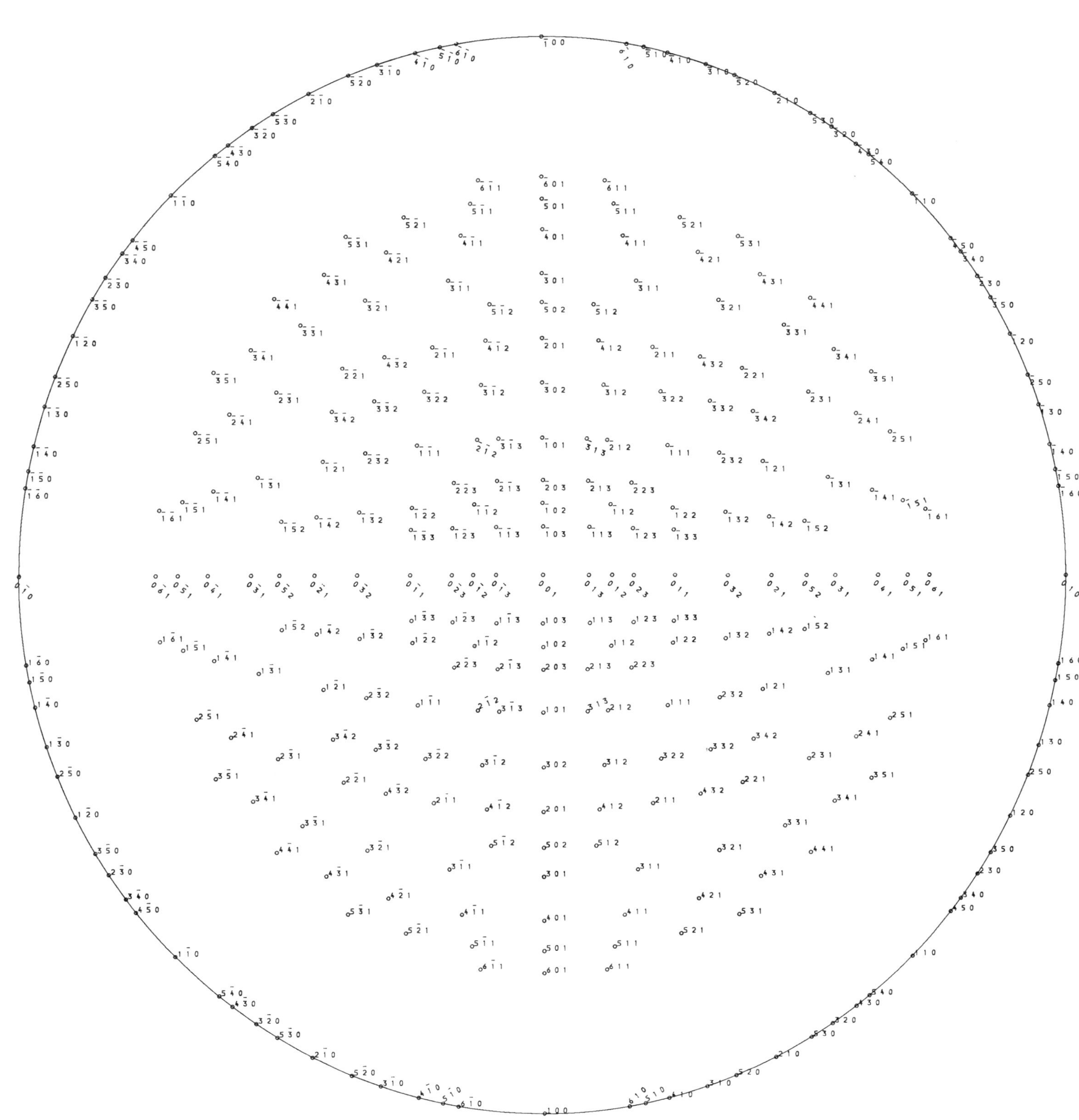

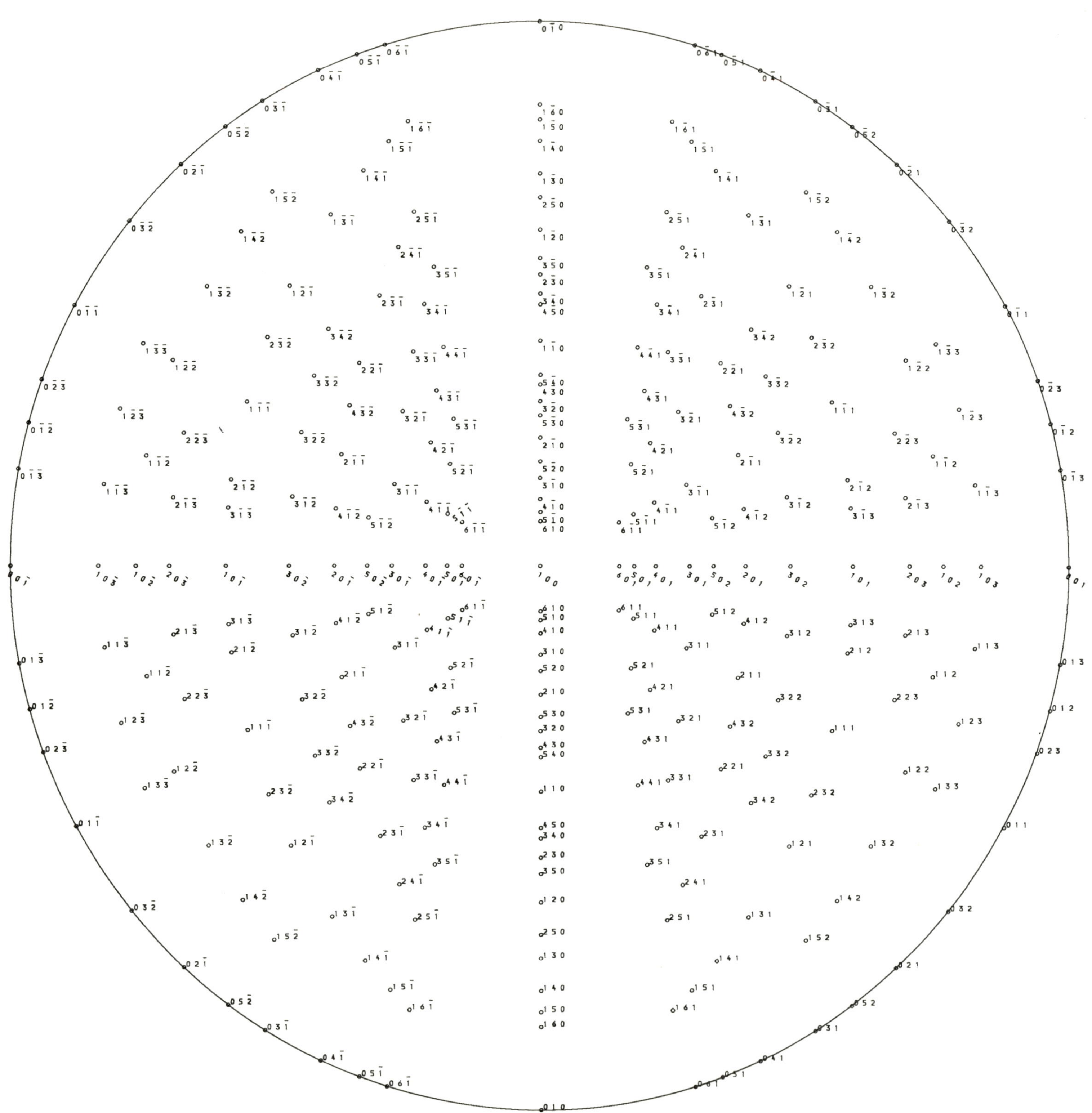

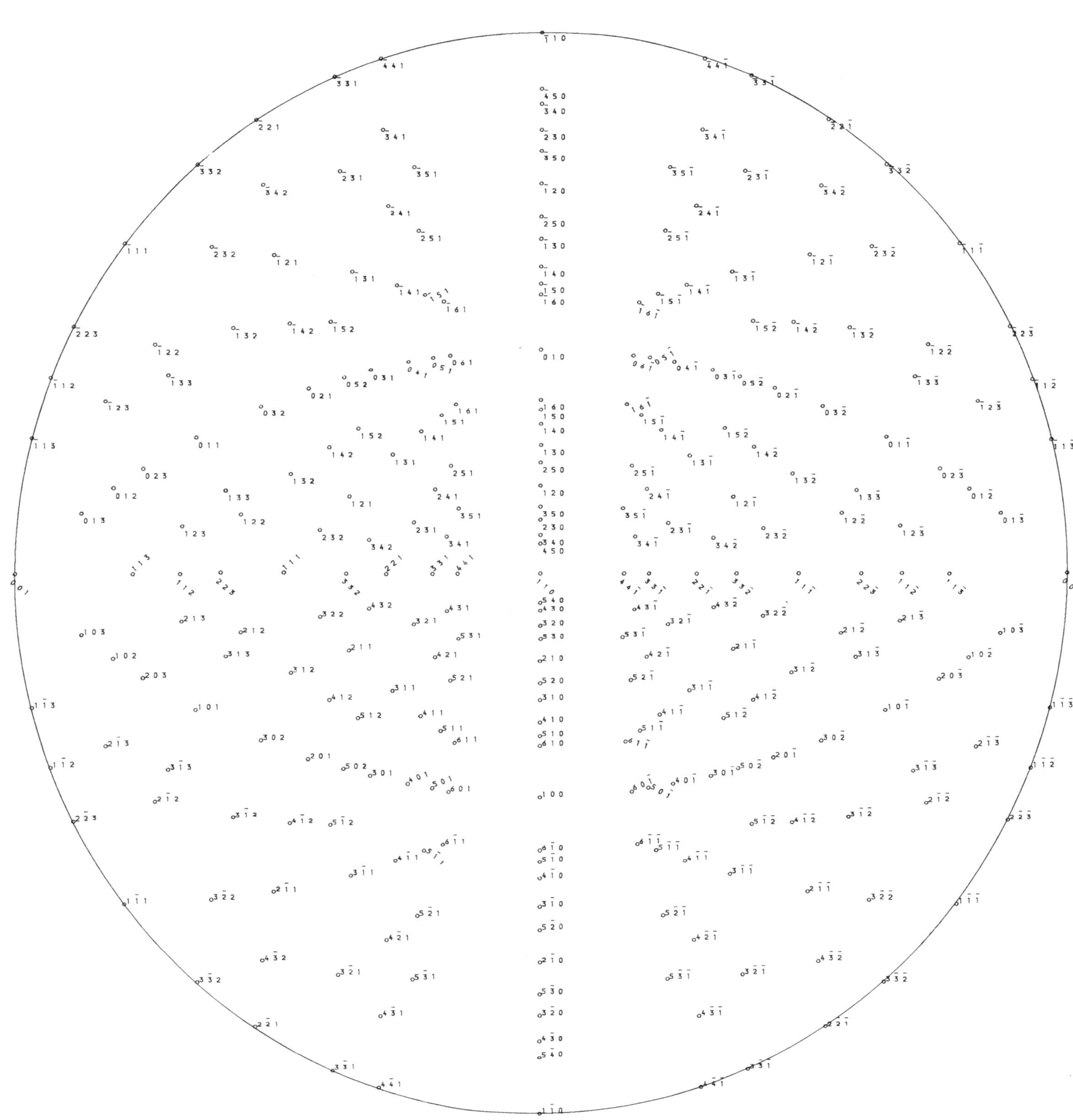

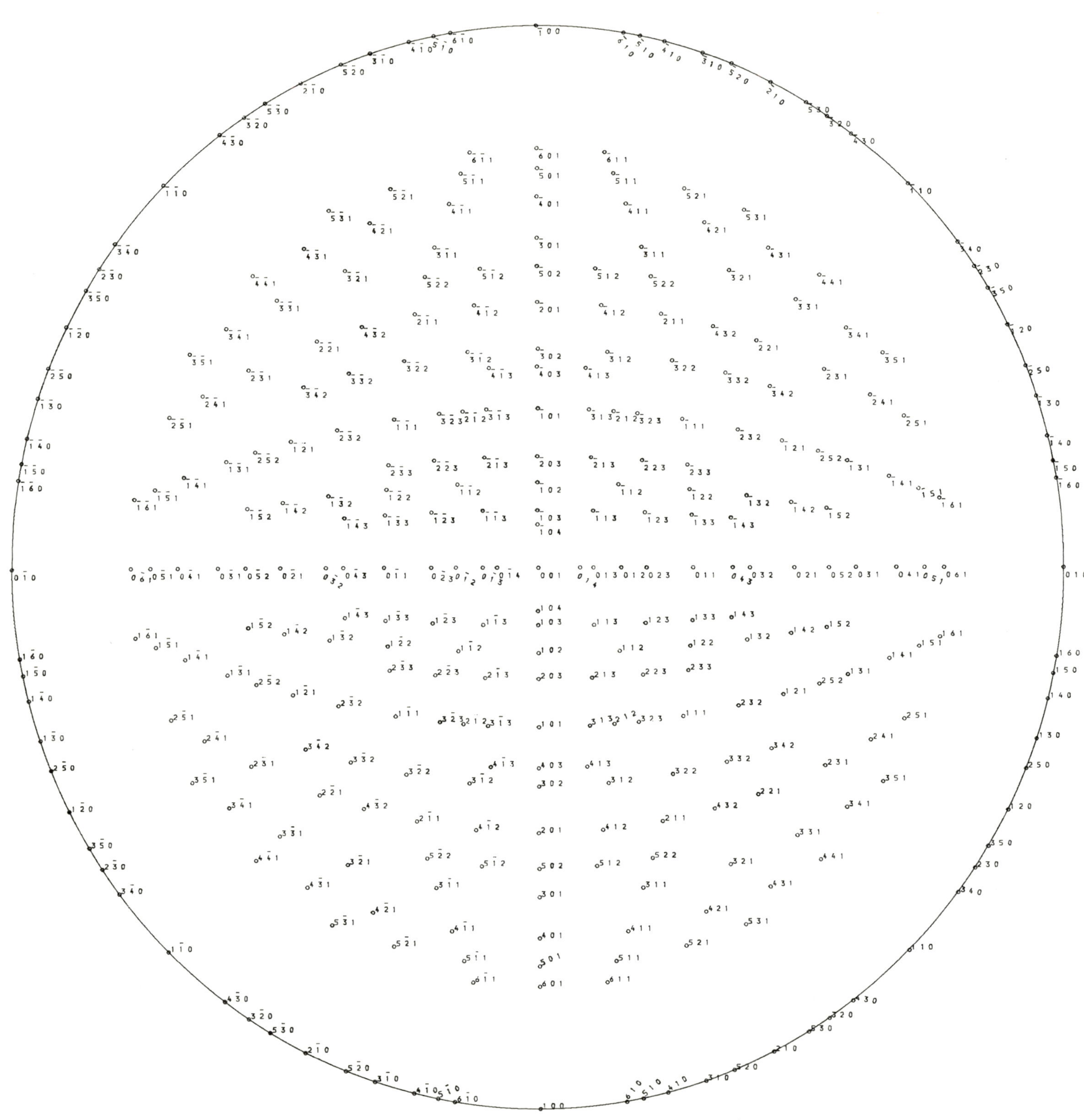

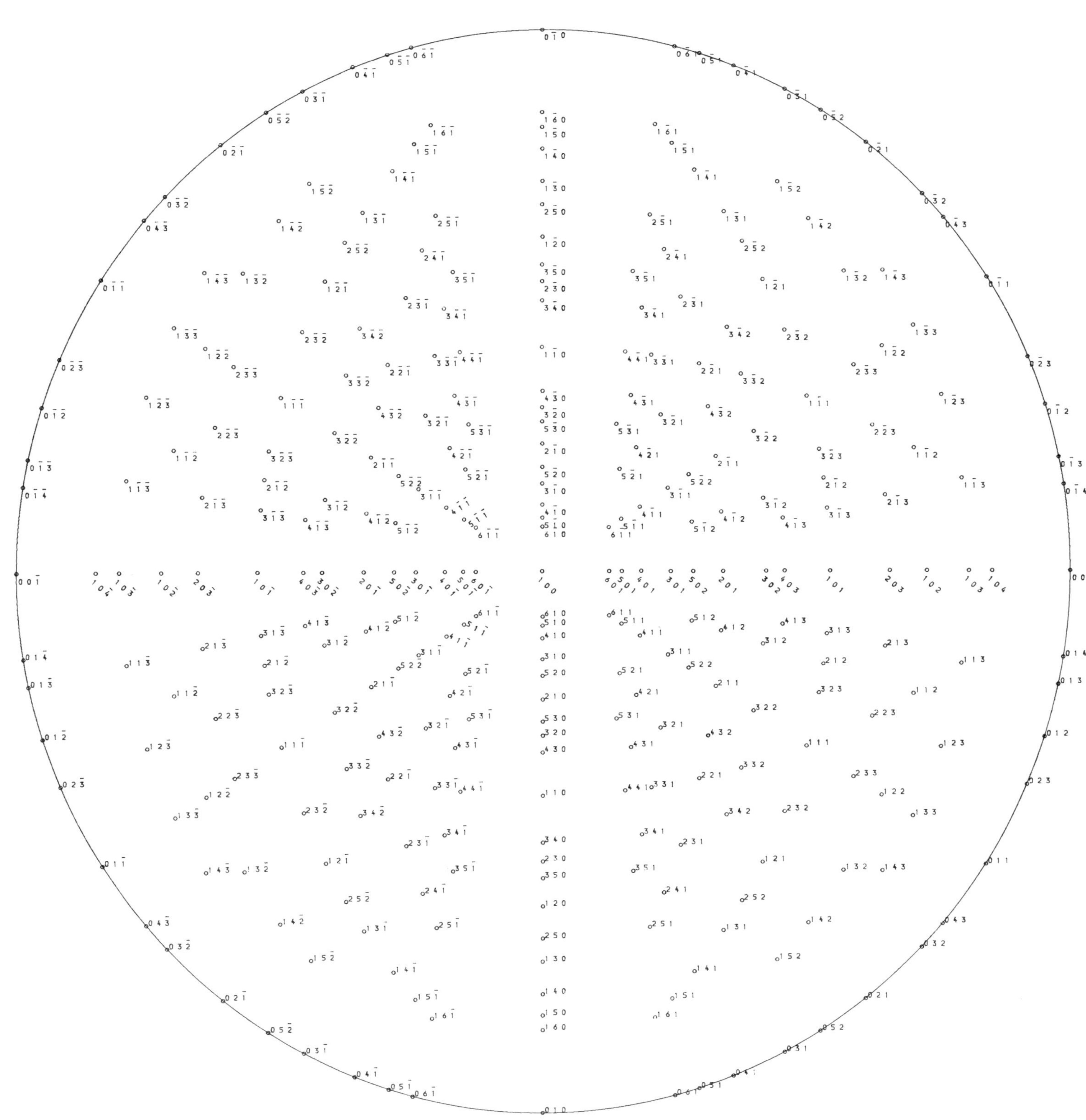

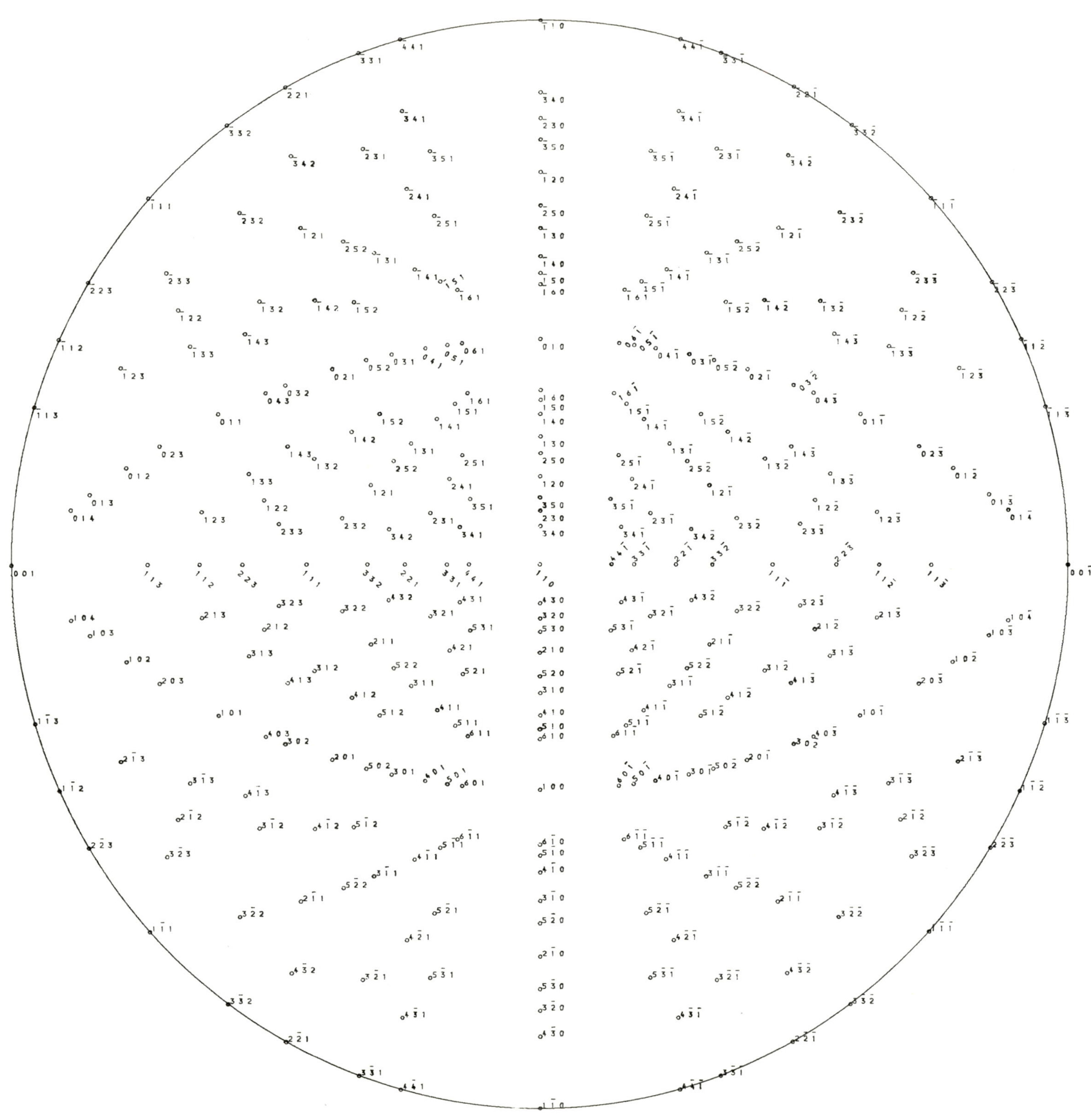

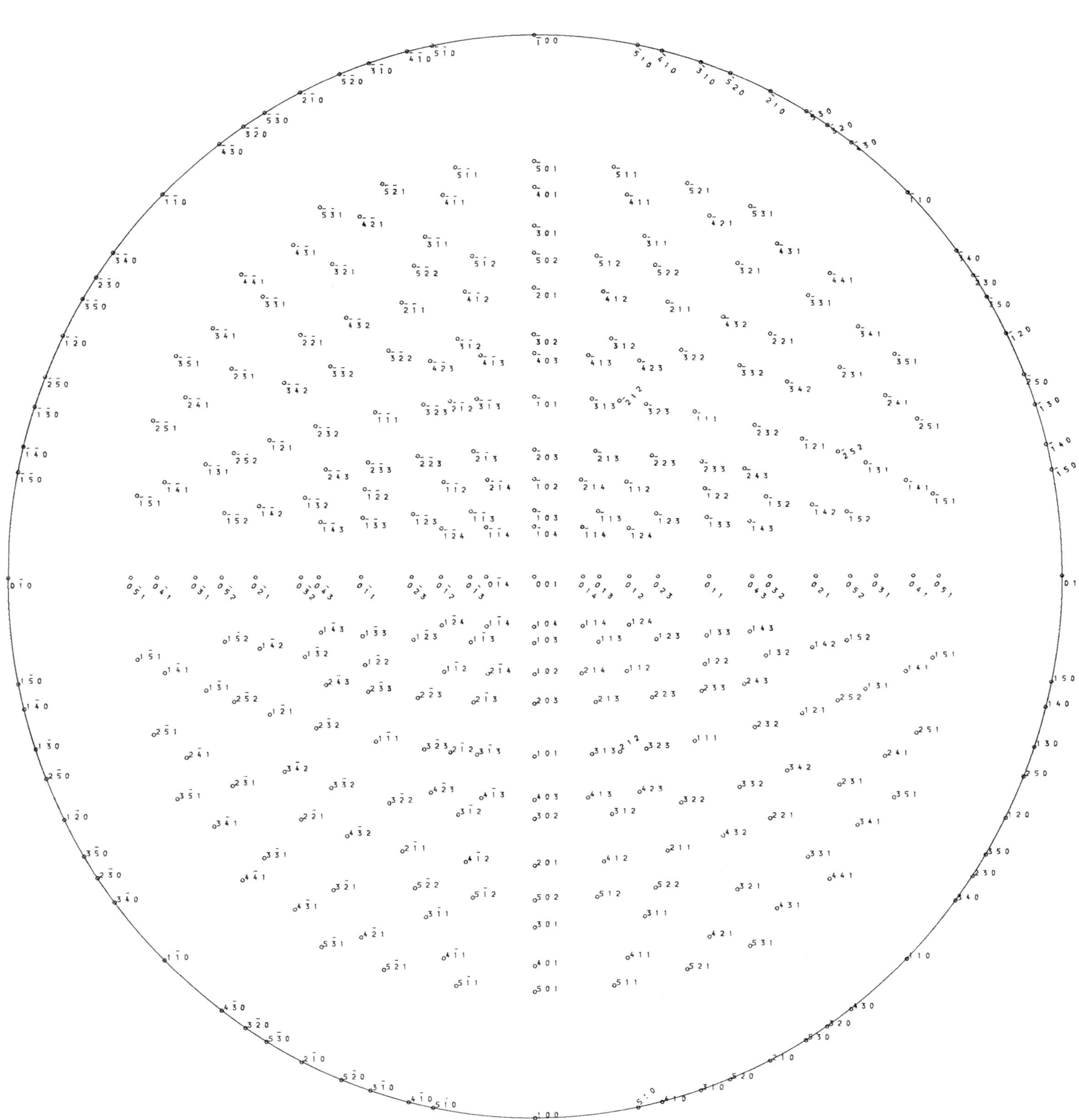

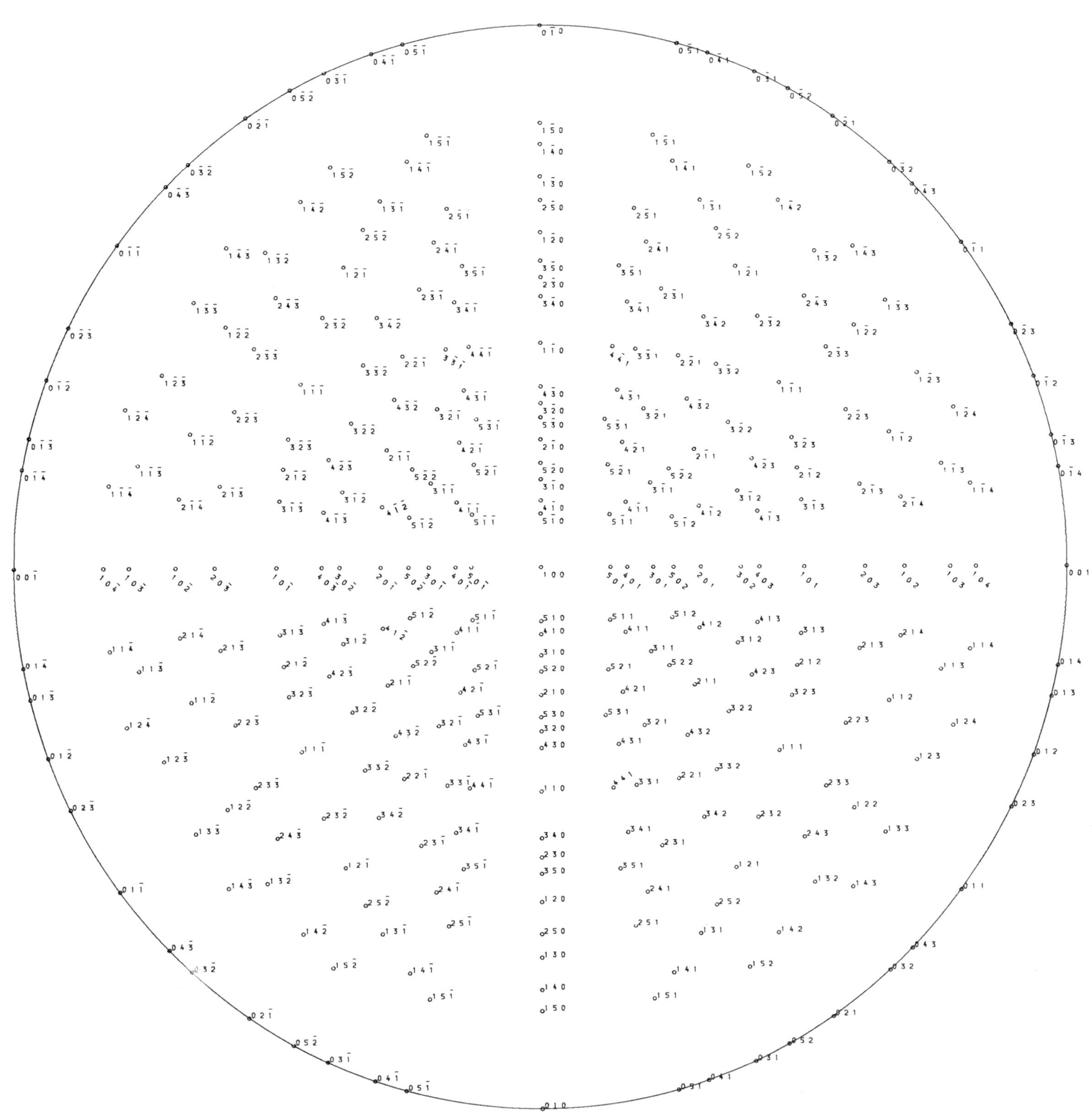

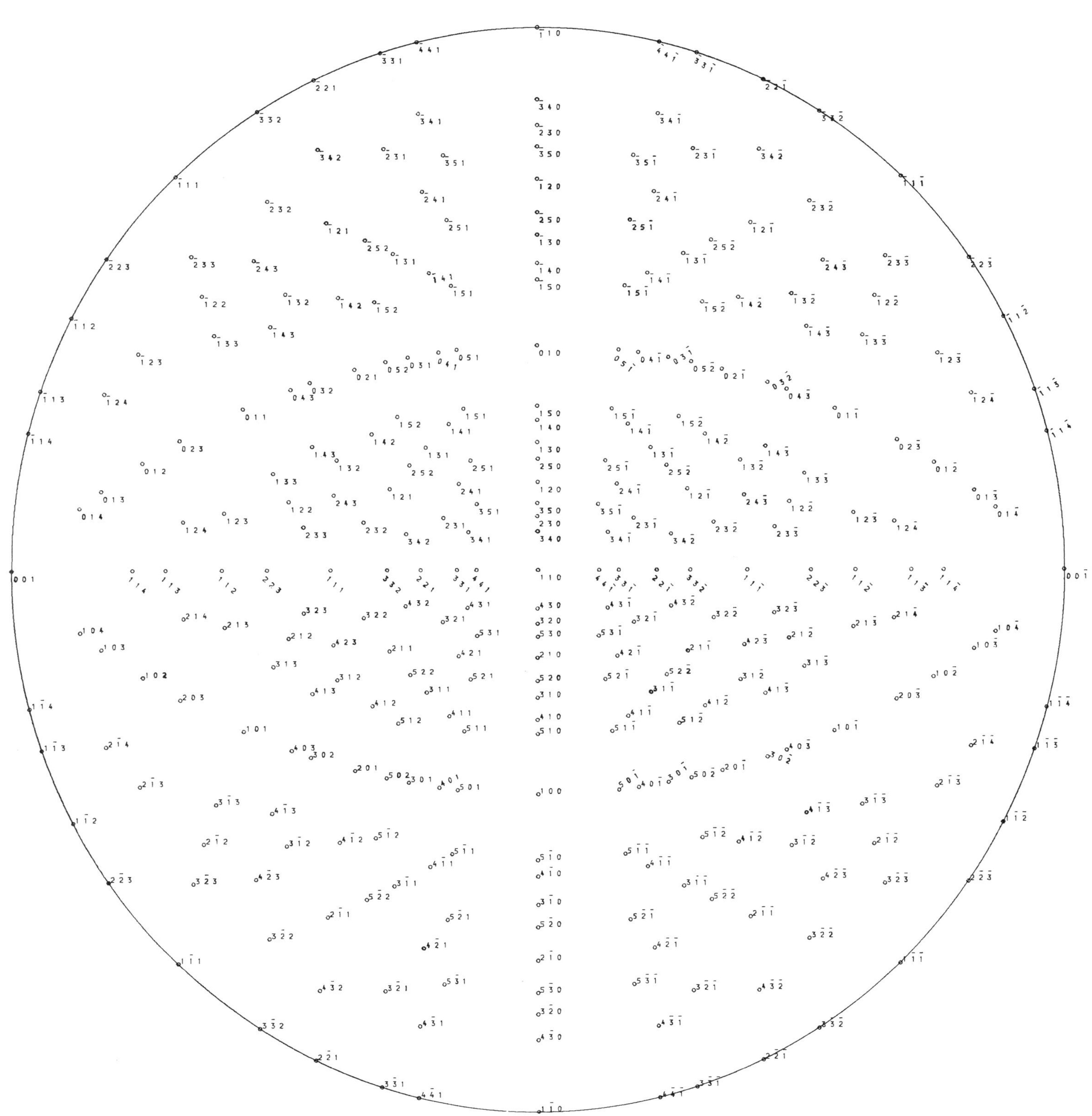

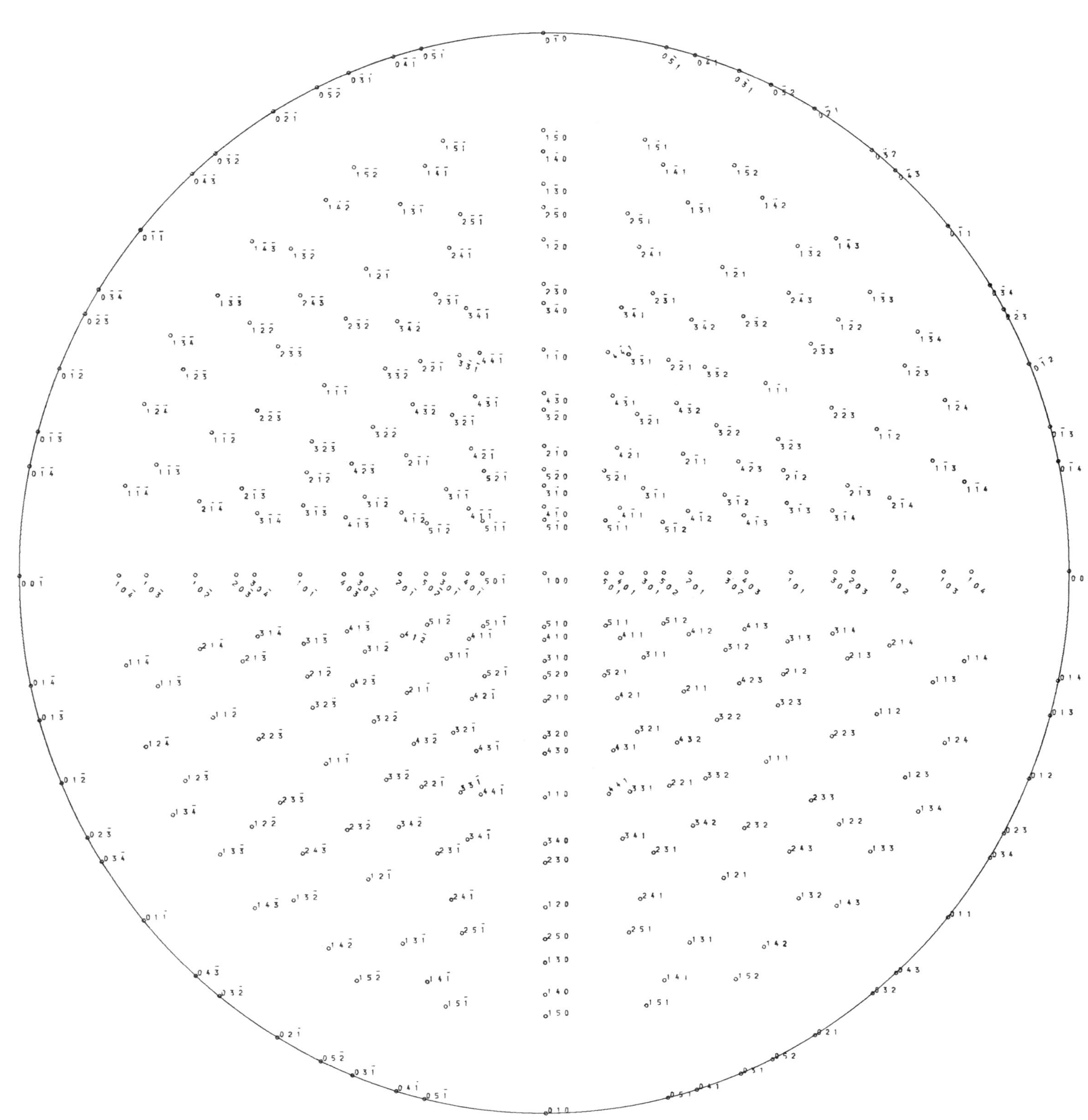

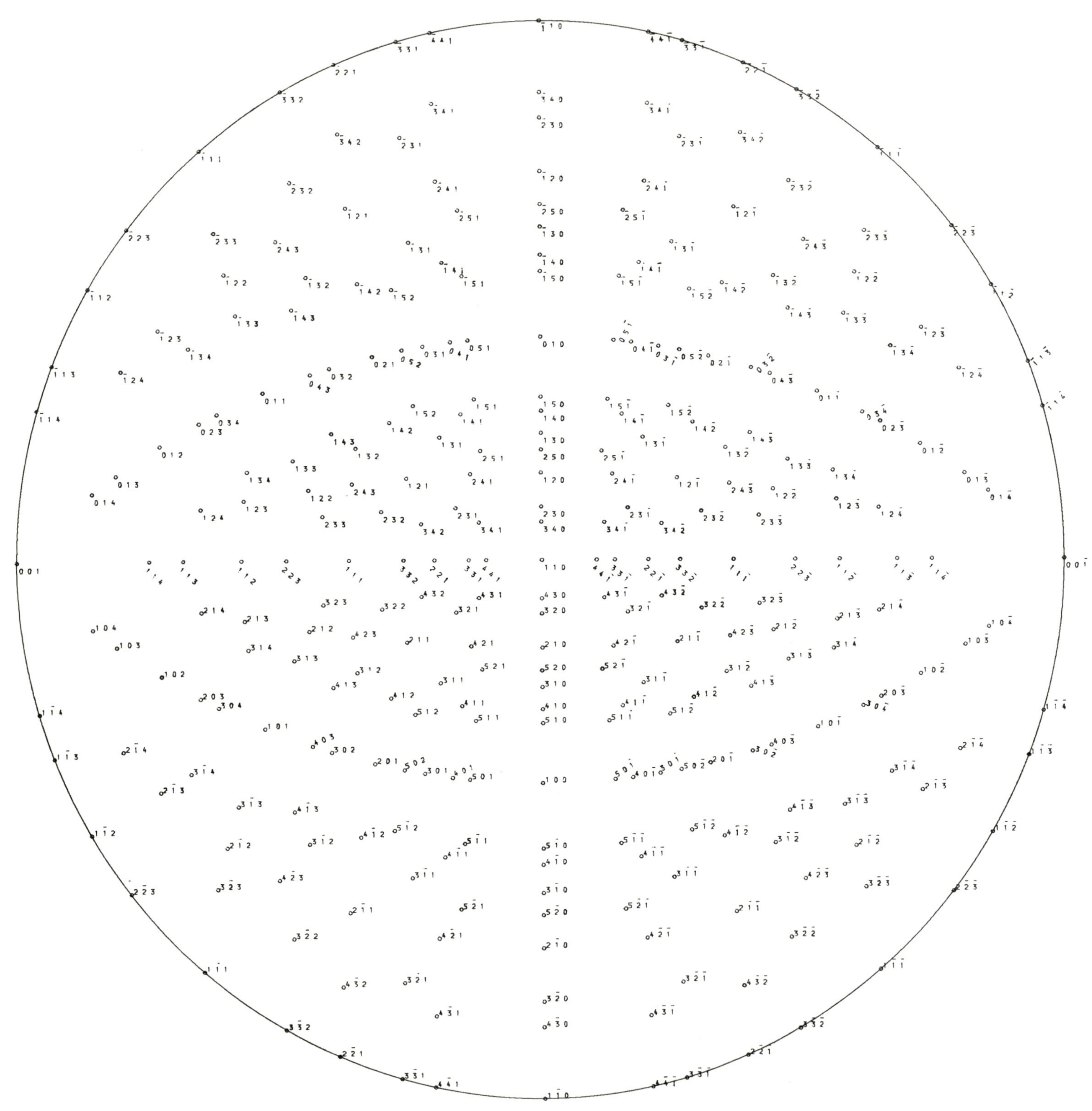

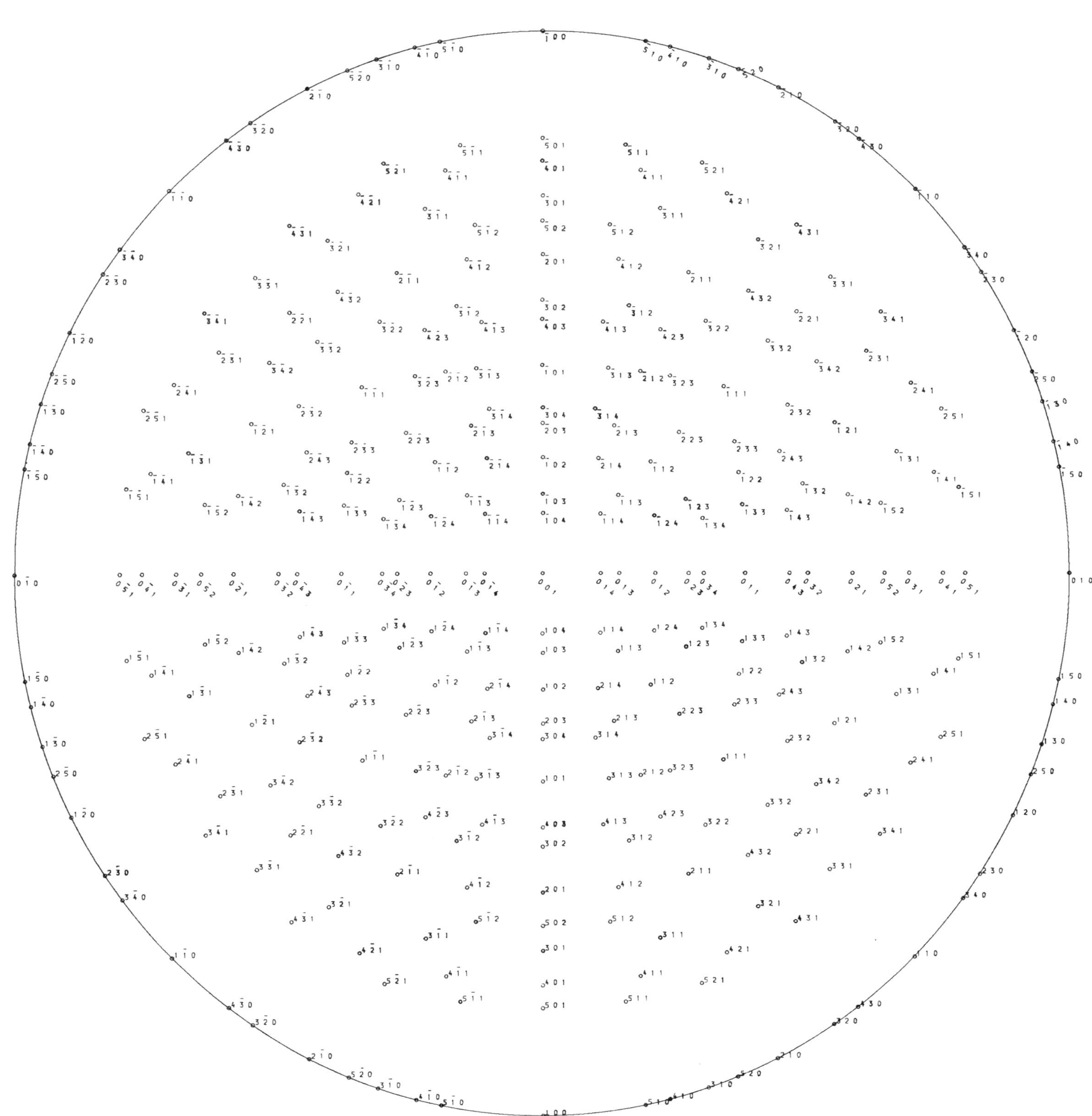

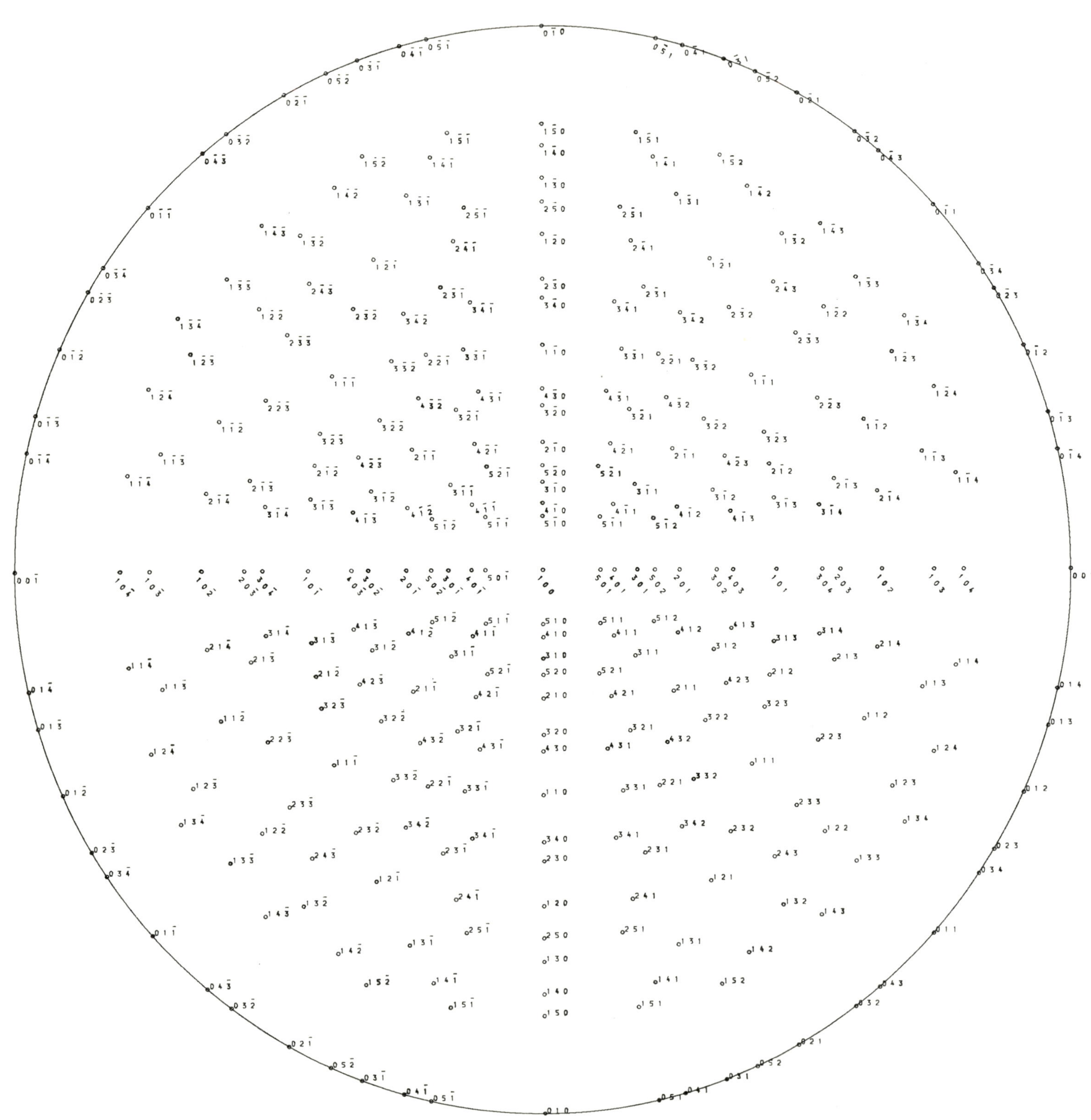

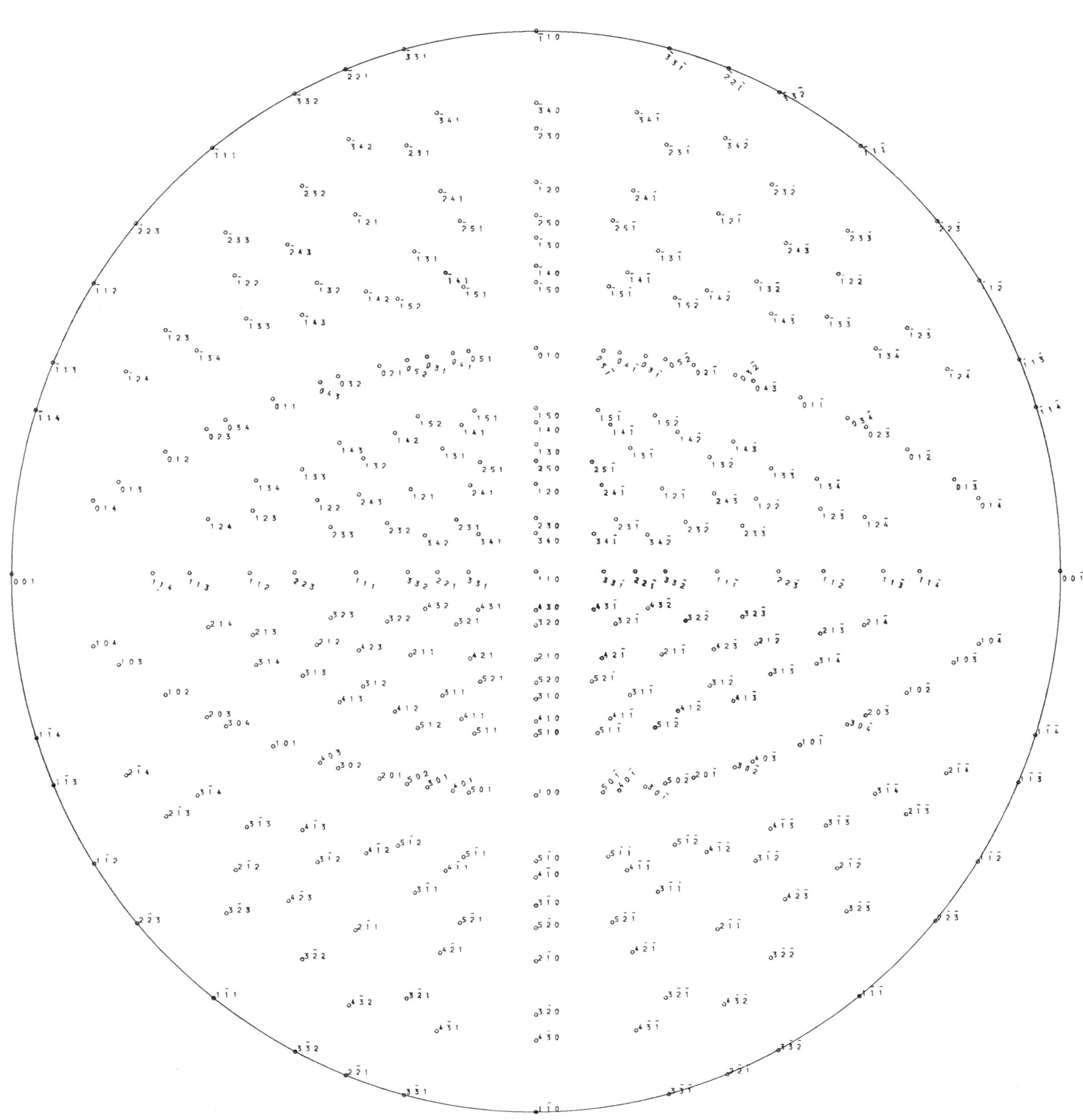

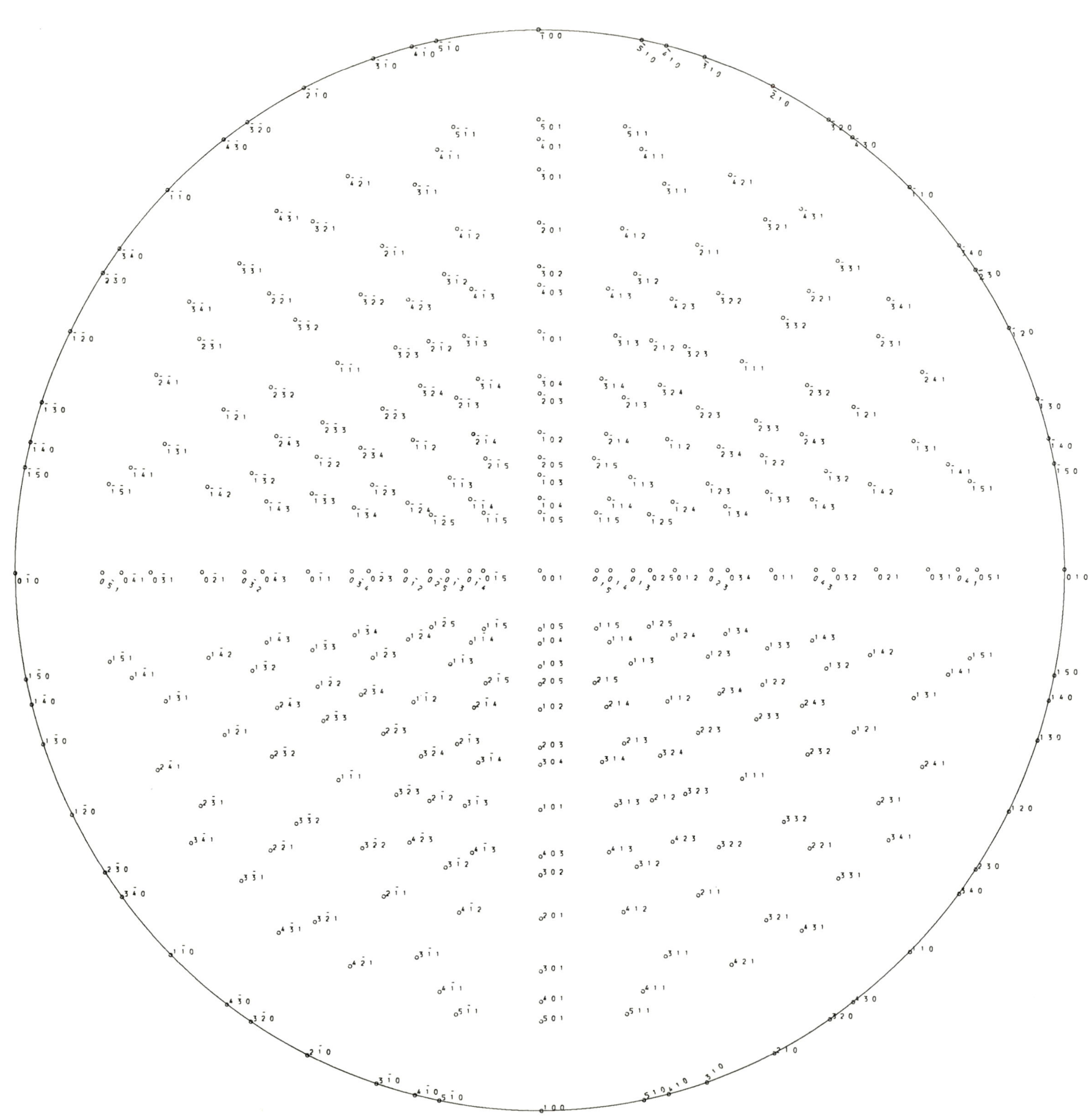

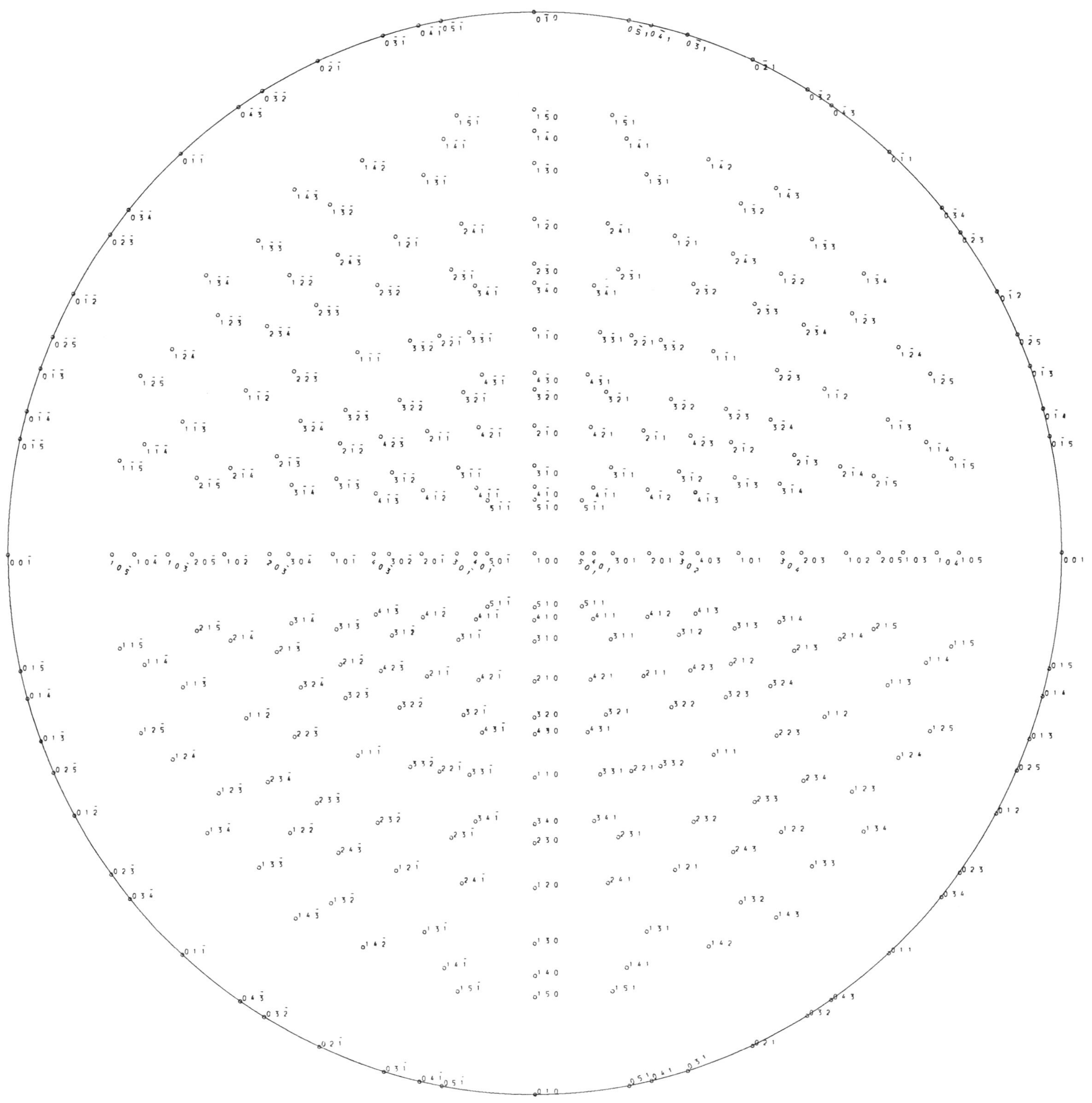

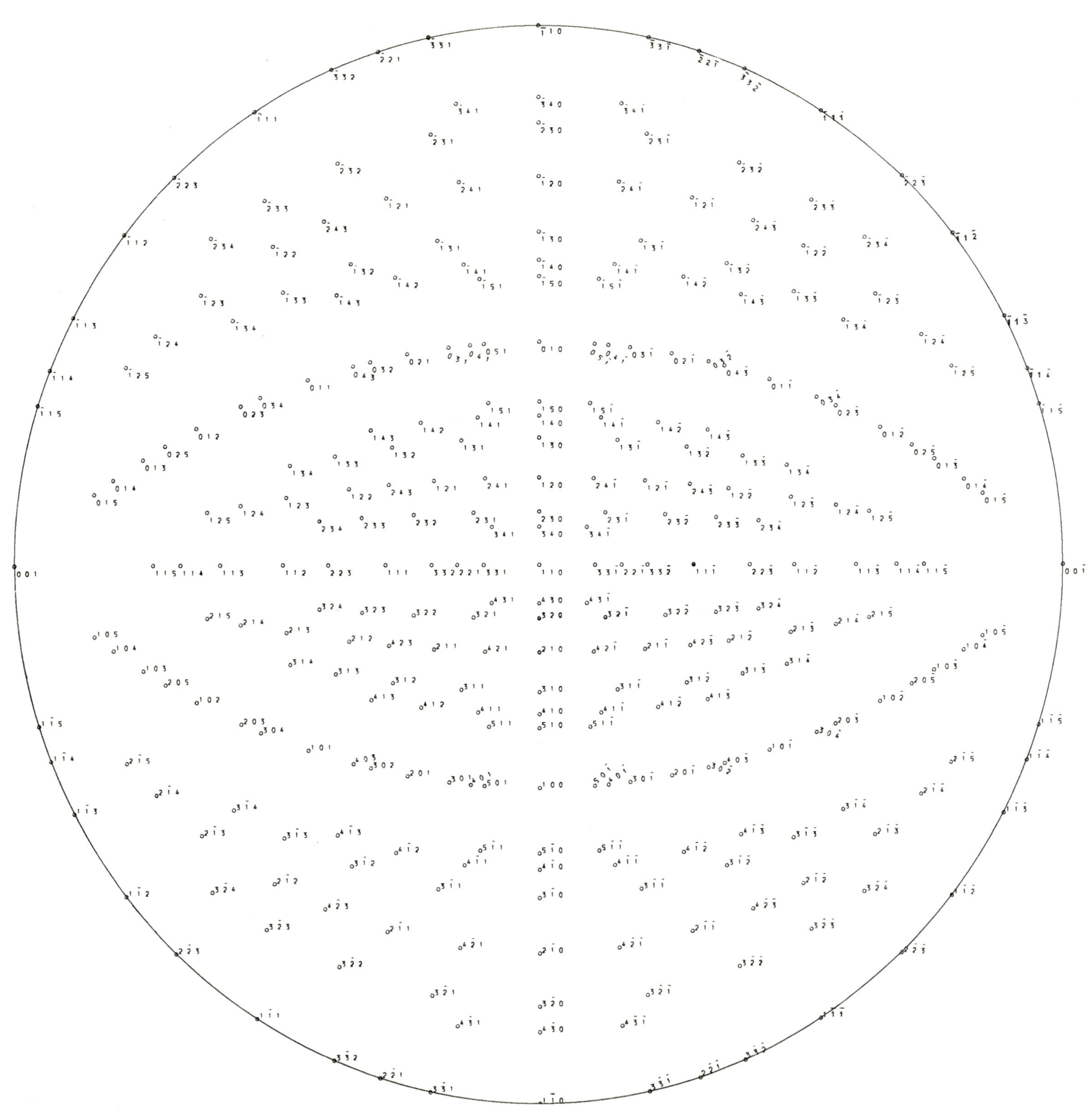

Beta spodumene $LiAlSi_2O_6$ Axial ratio 1.2144 001 axis

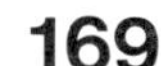

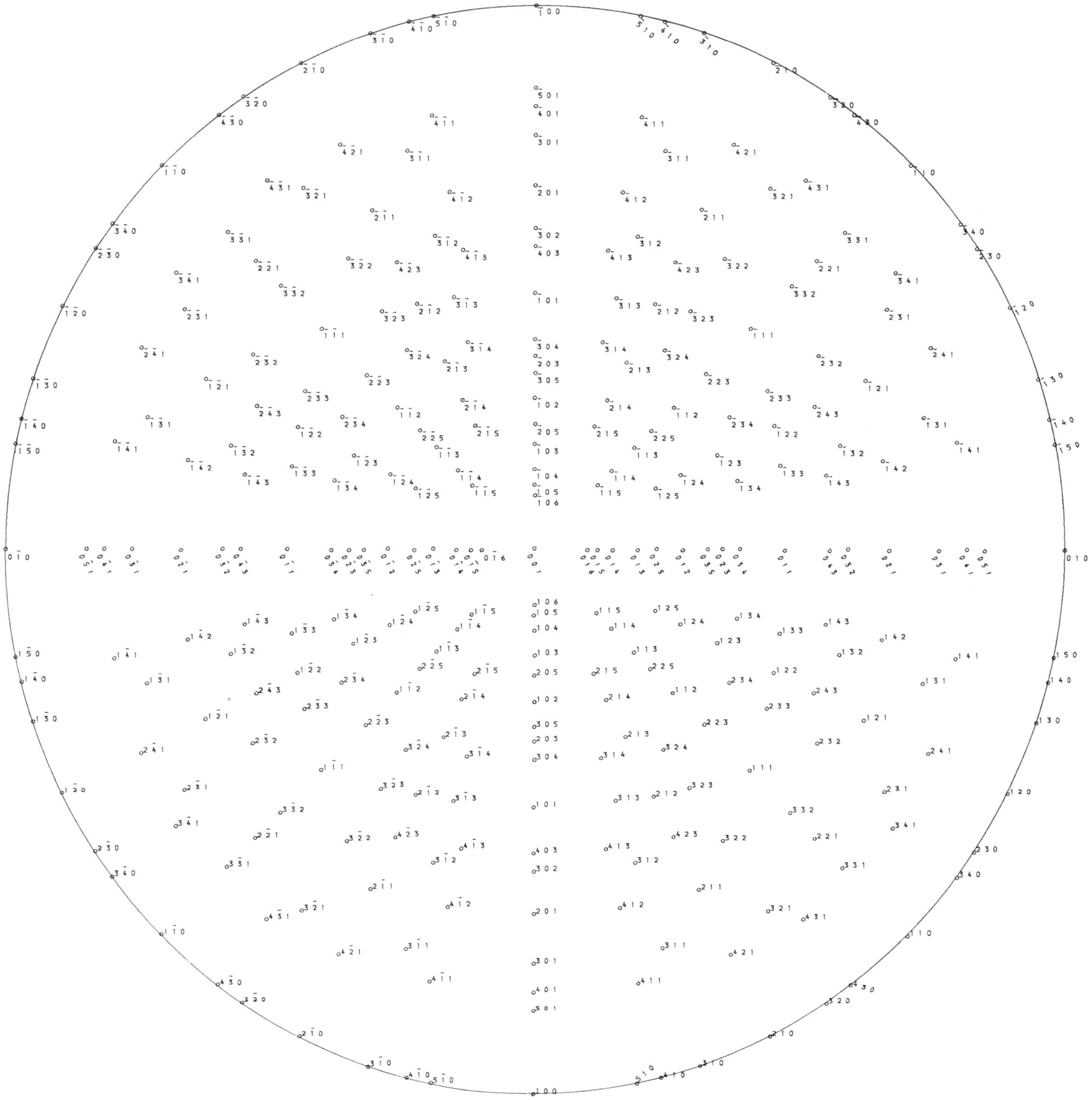

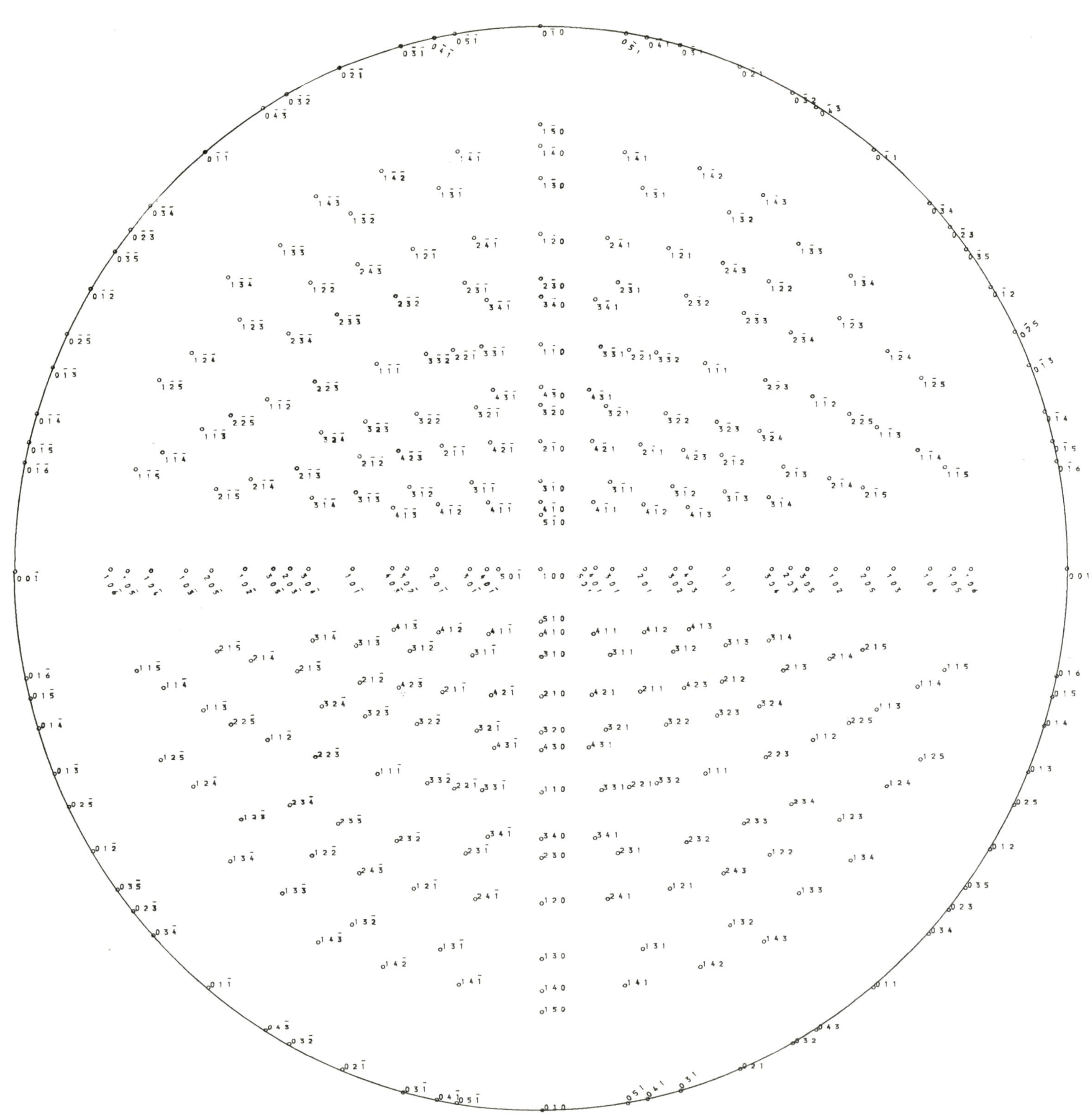

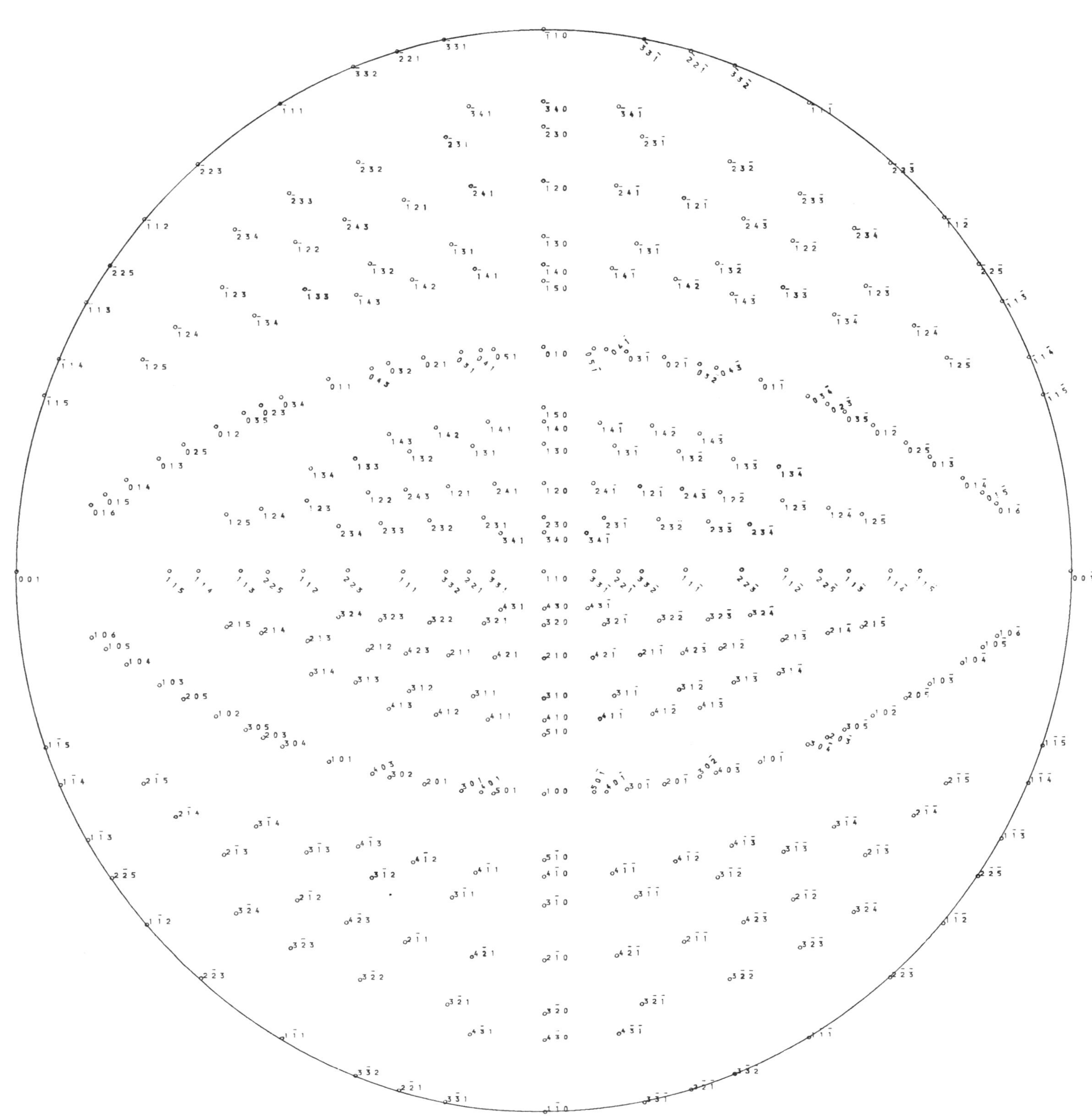

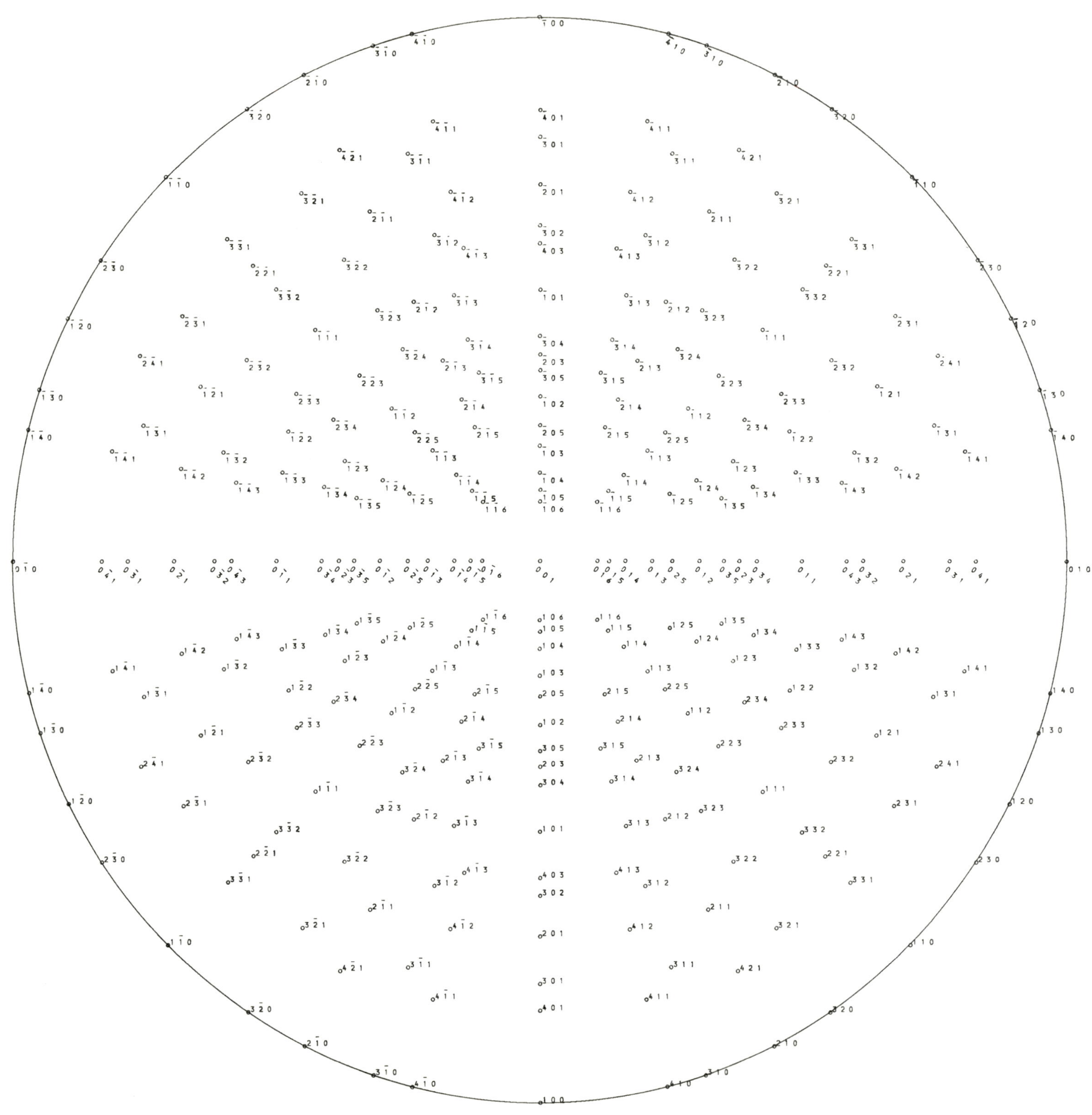

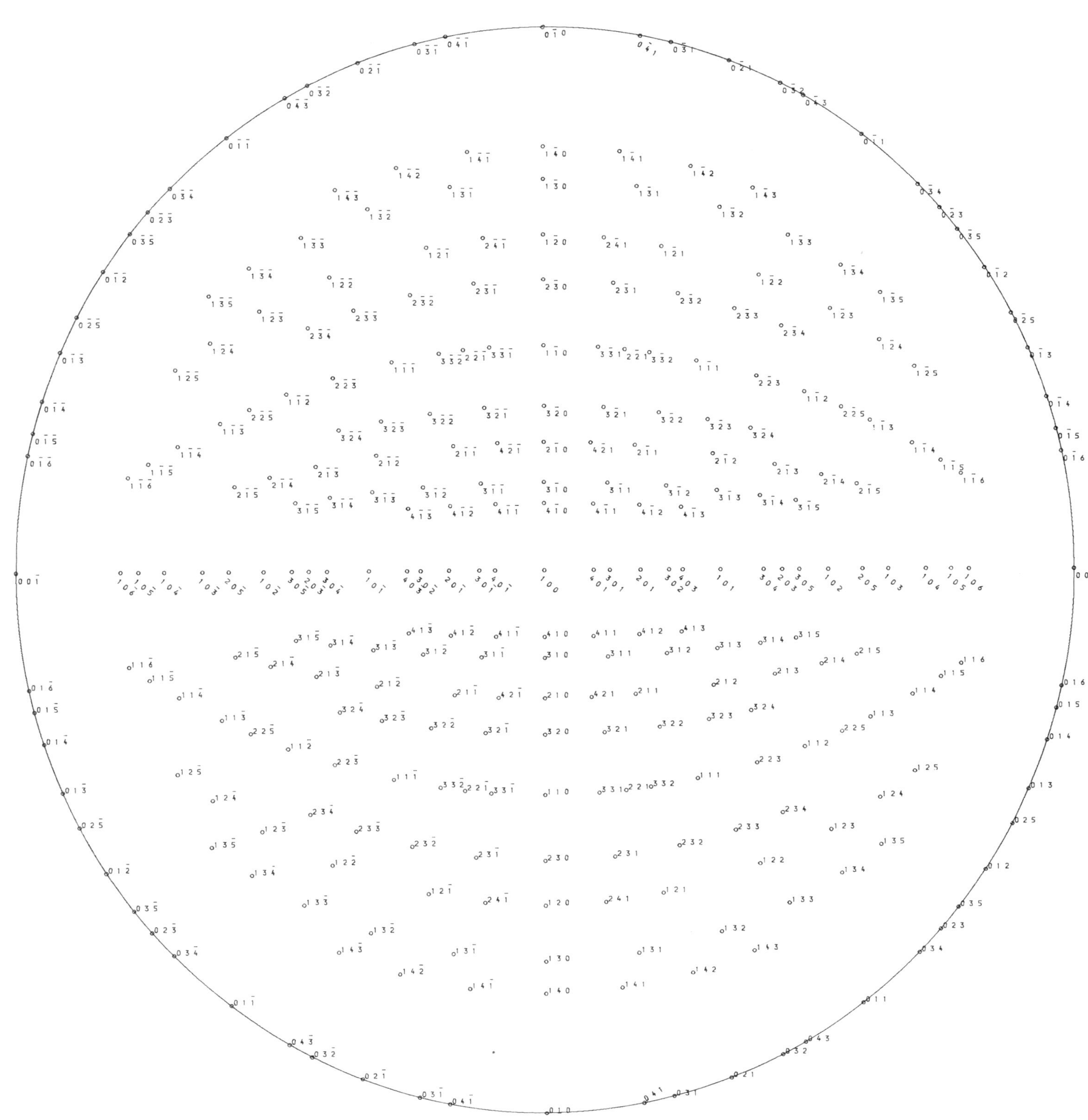

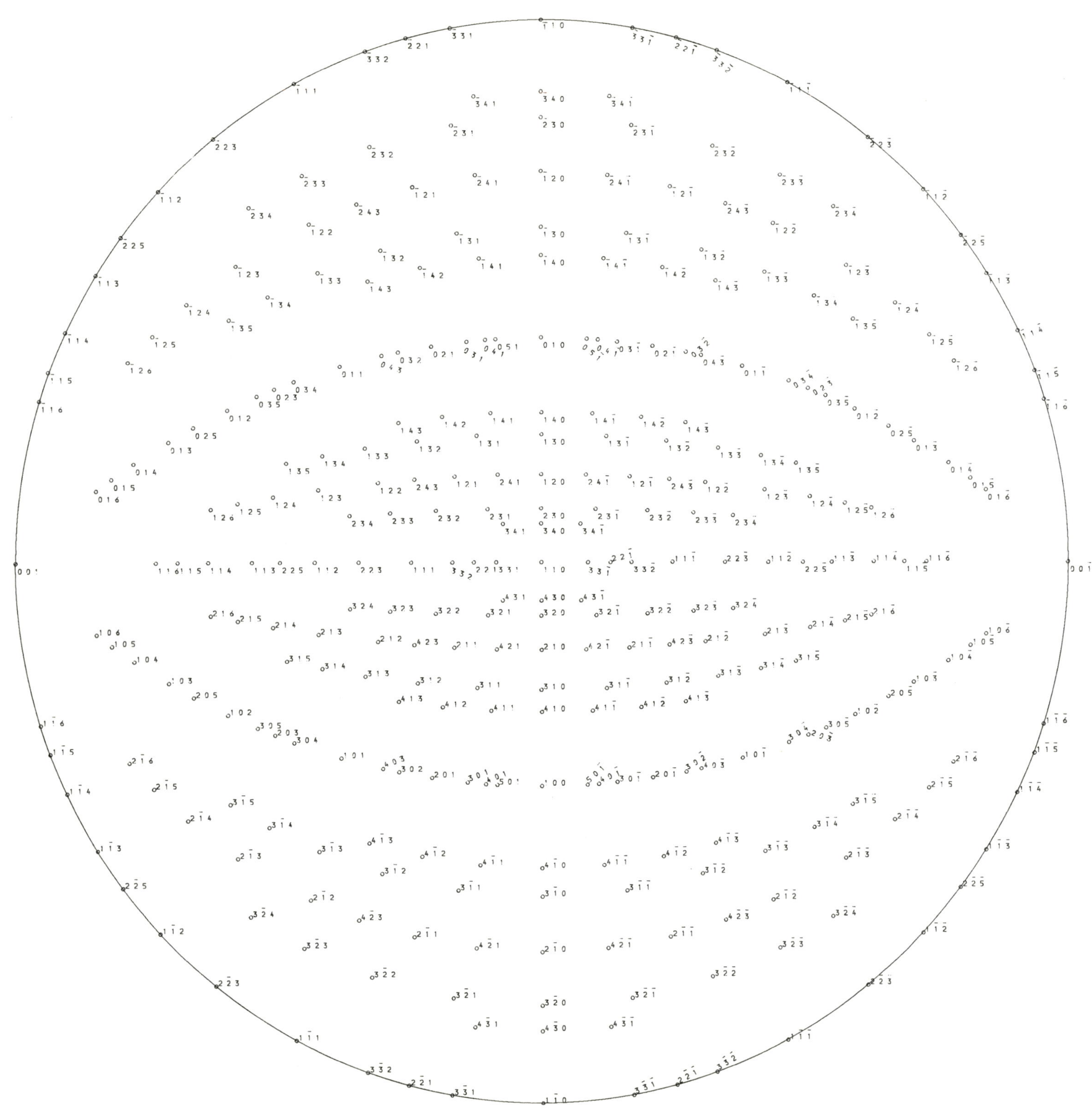

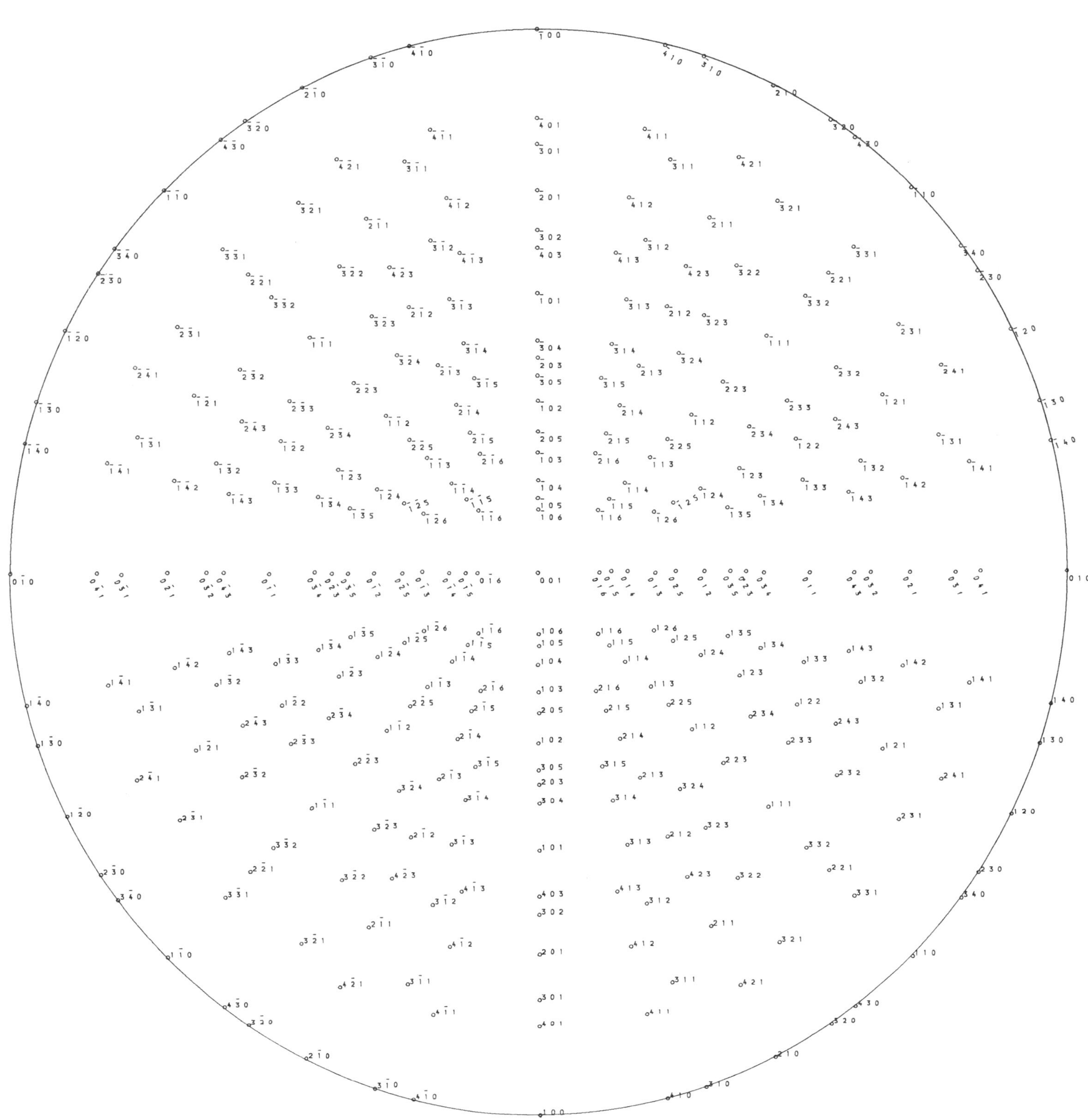

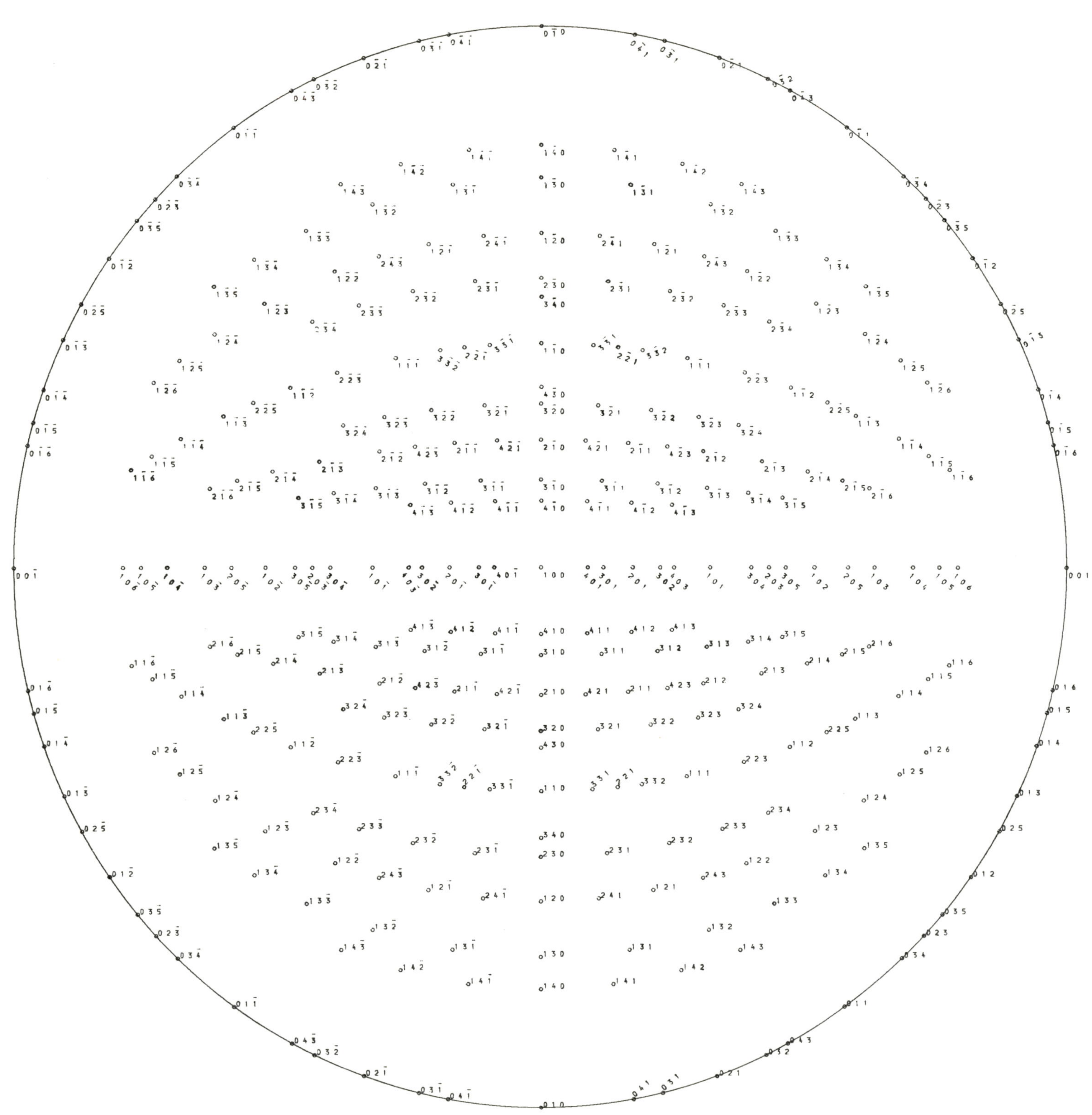

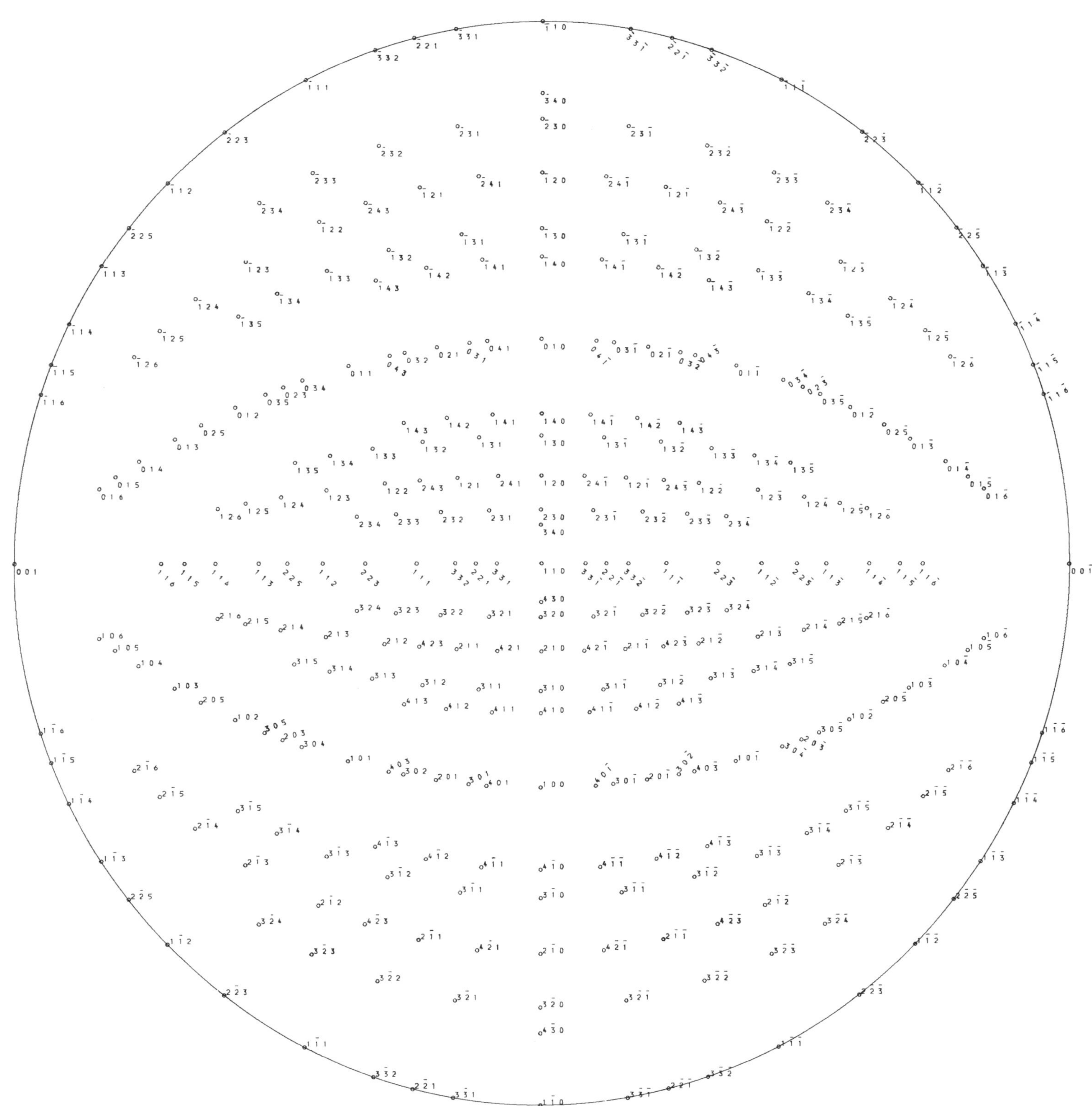

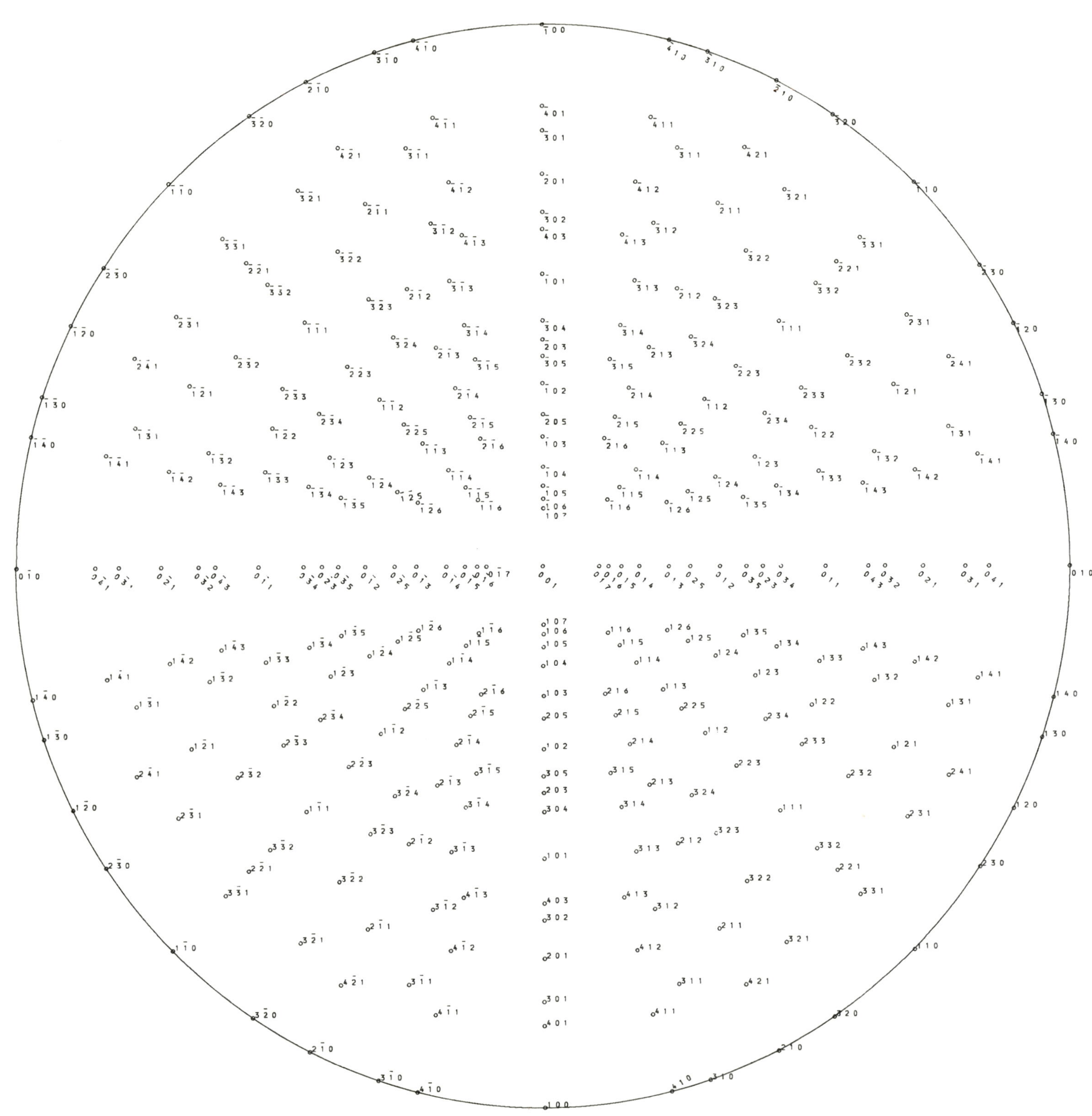

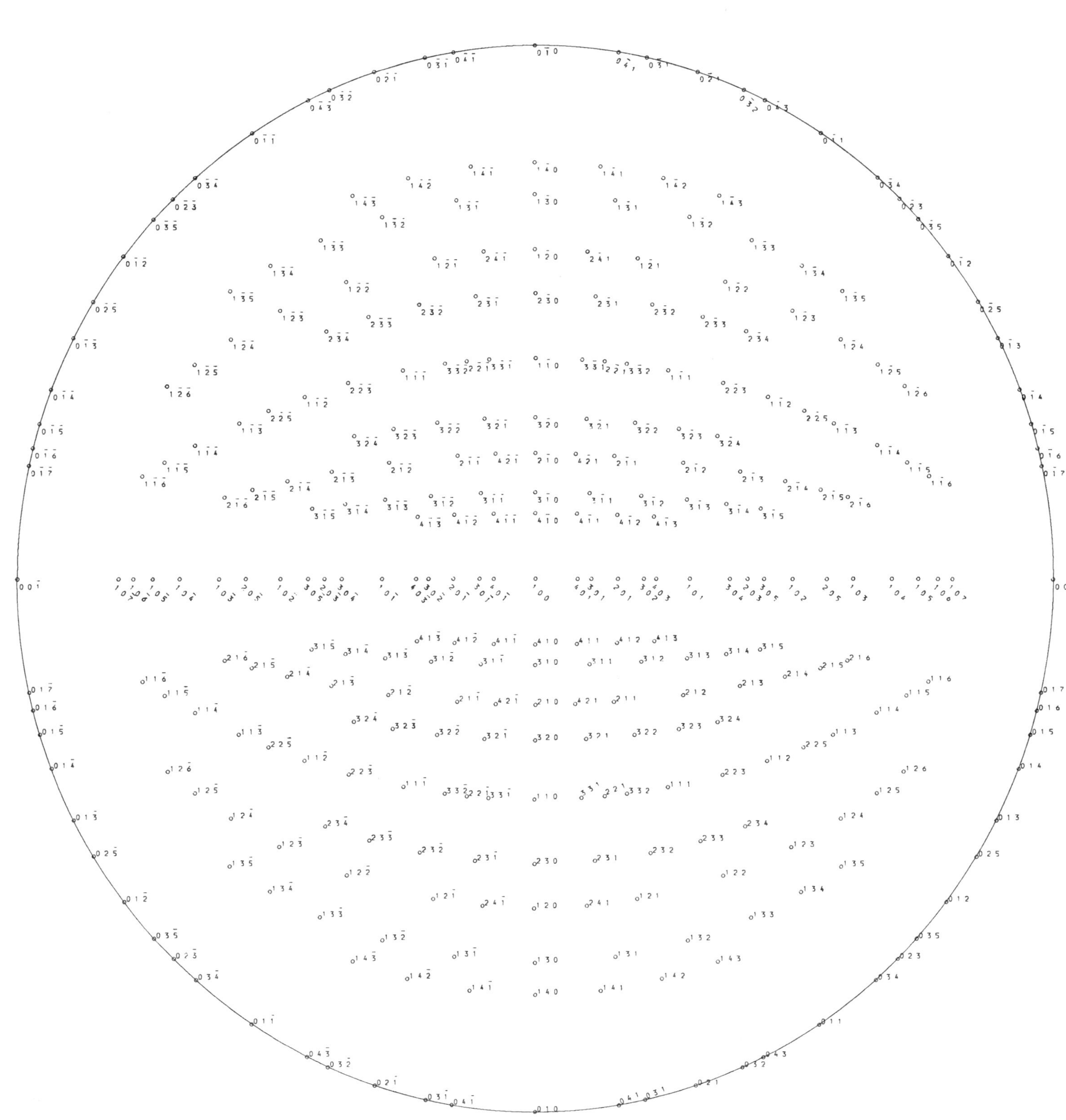

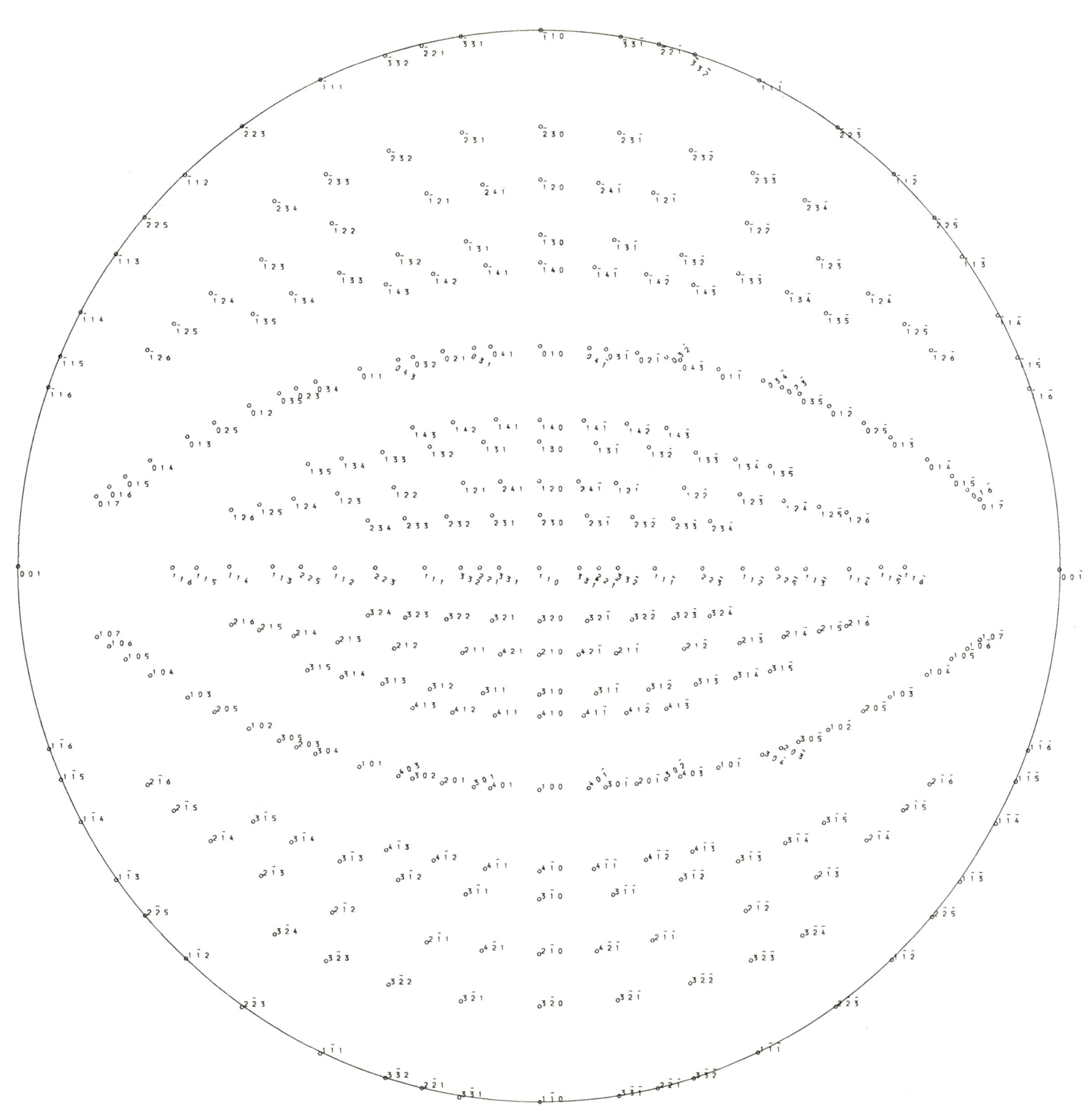

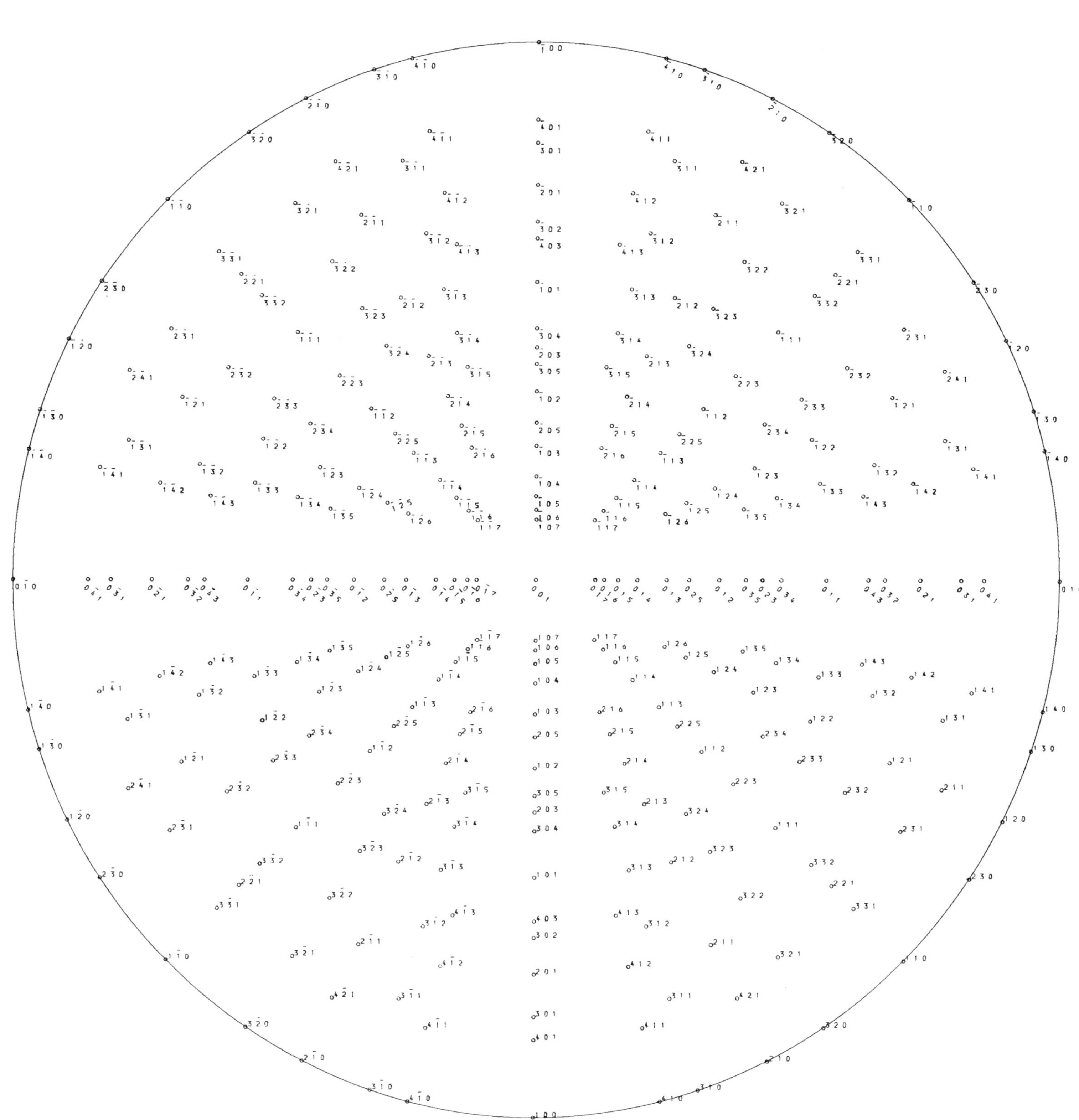

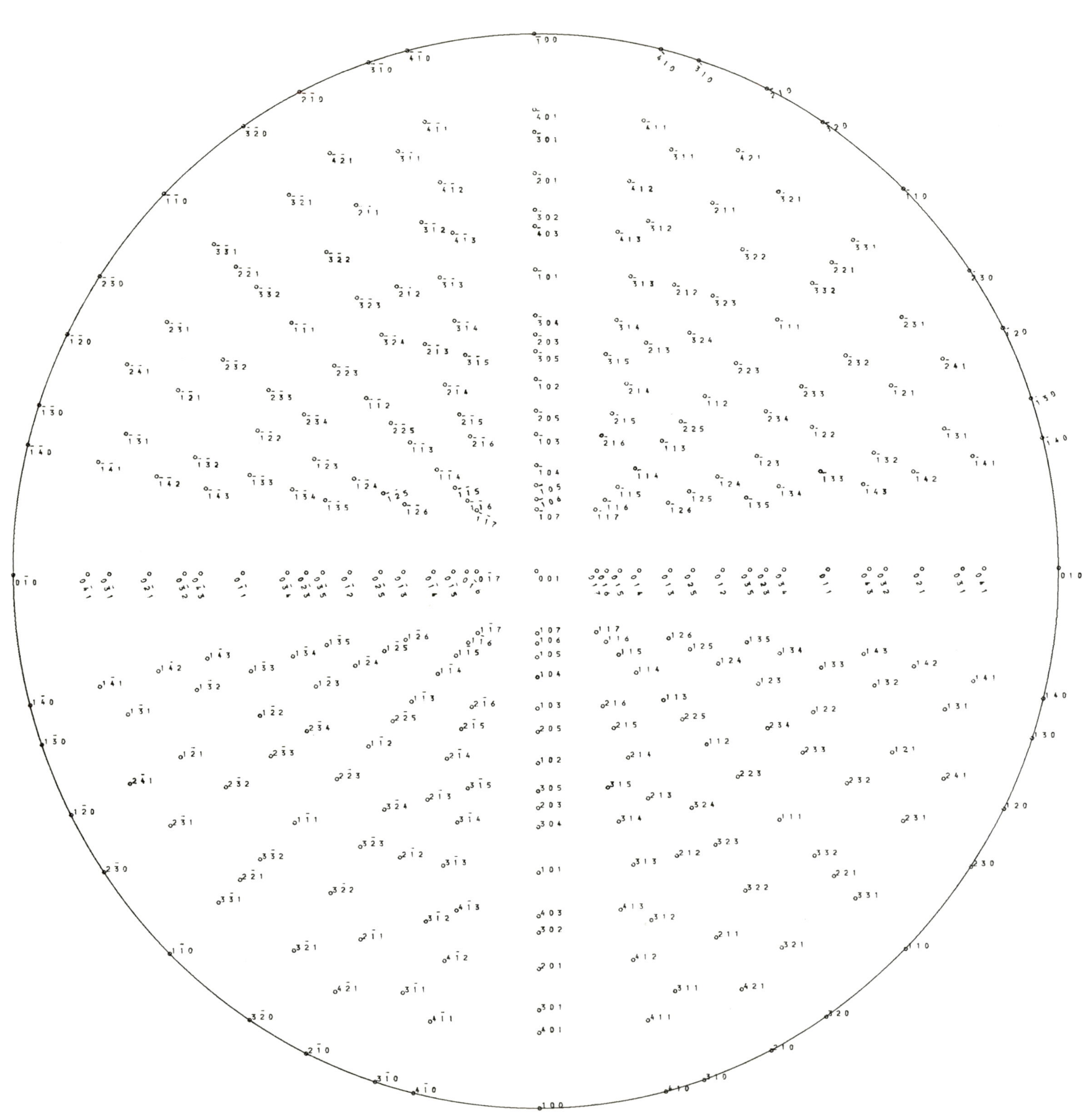

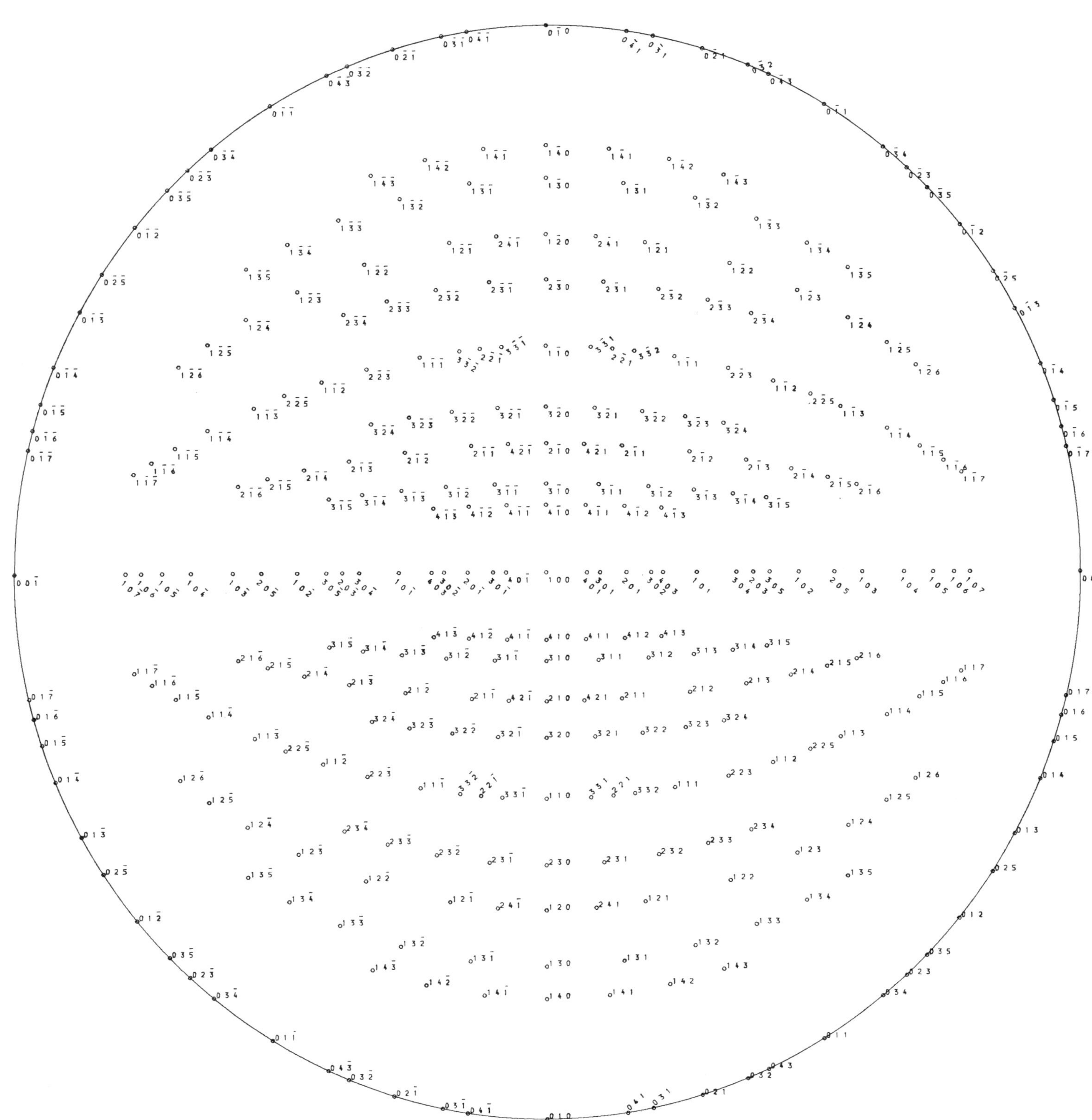

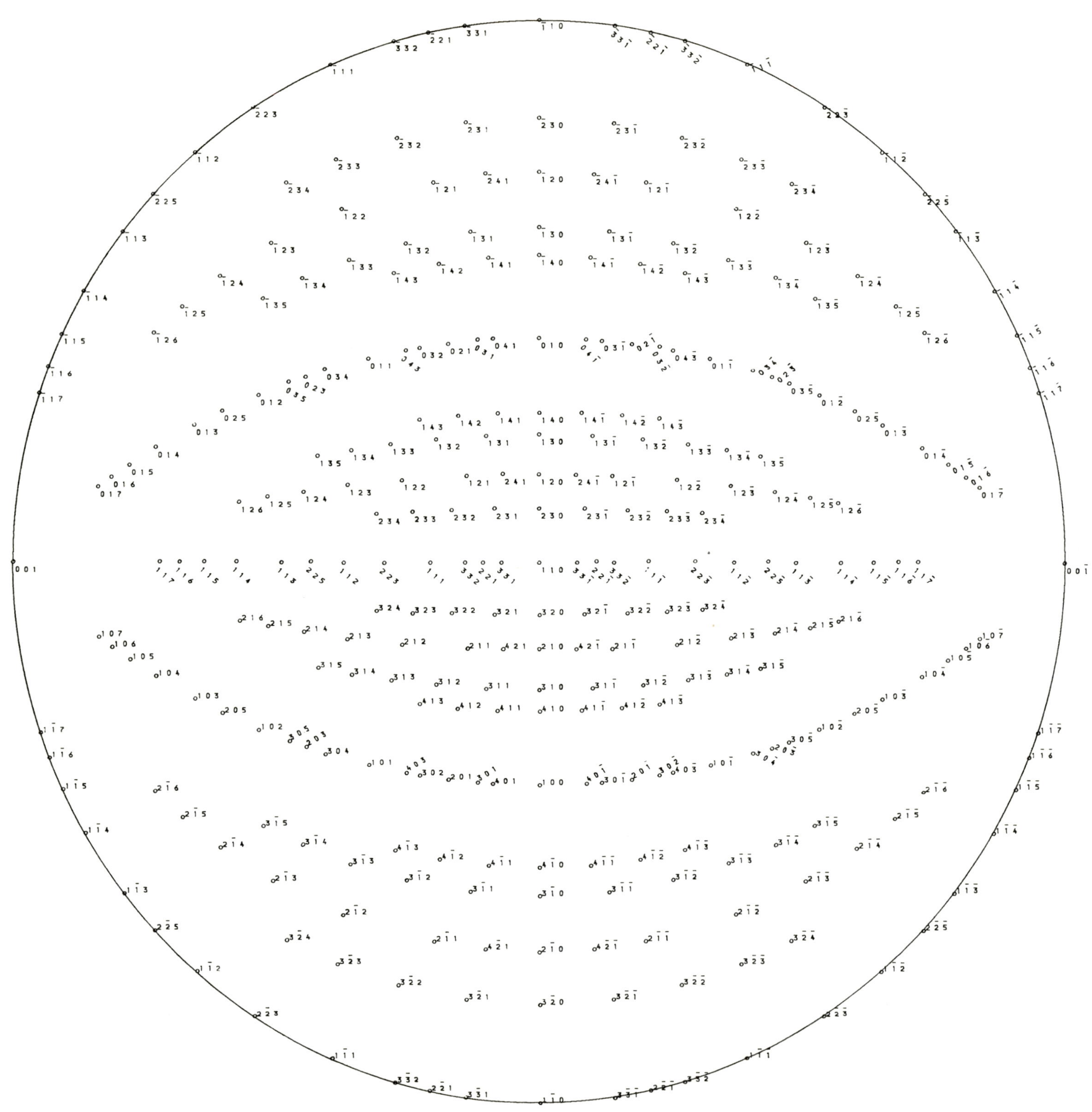

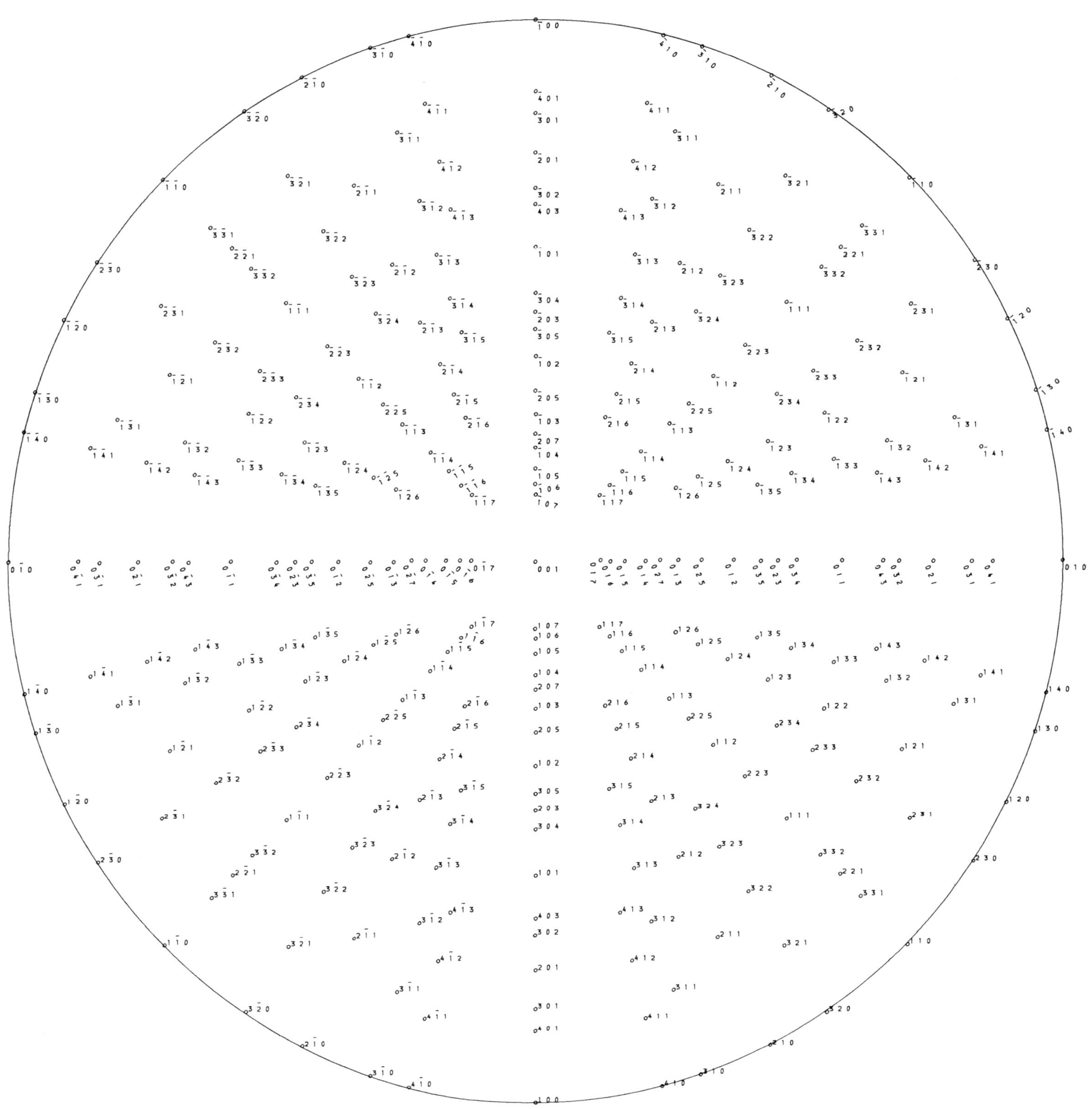

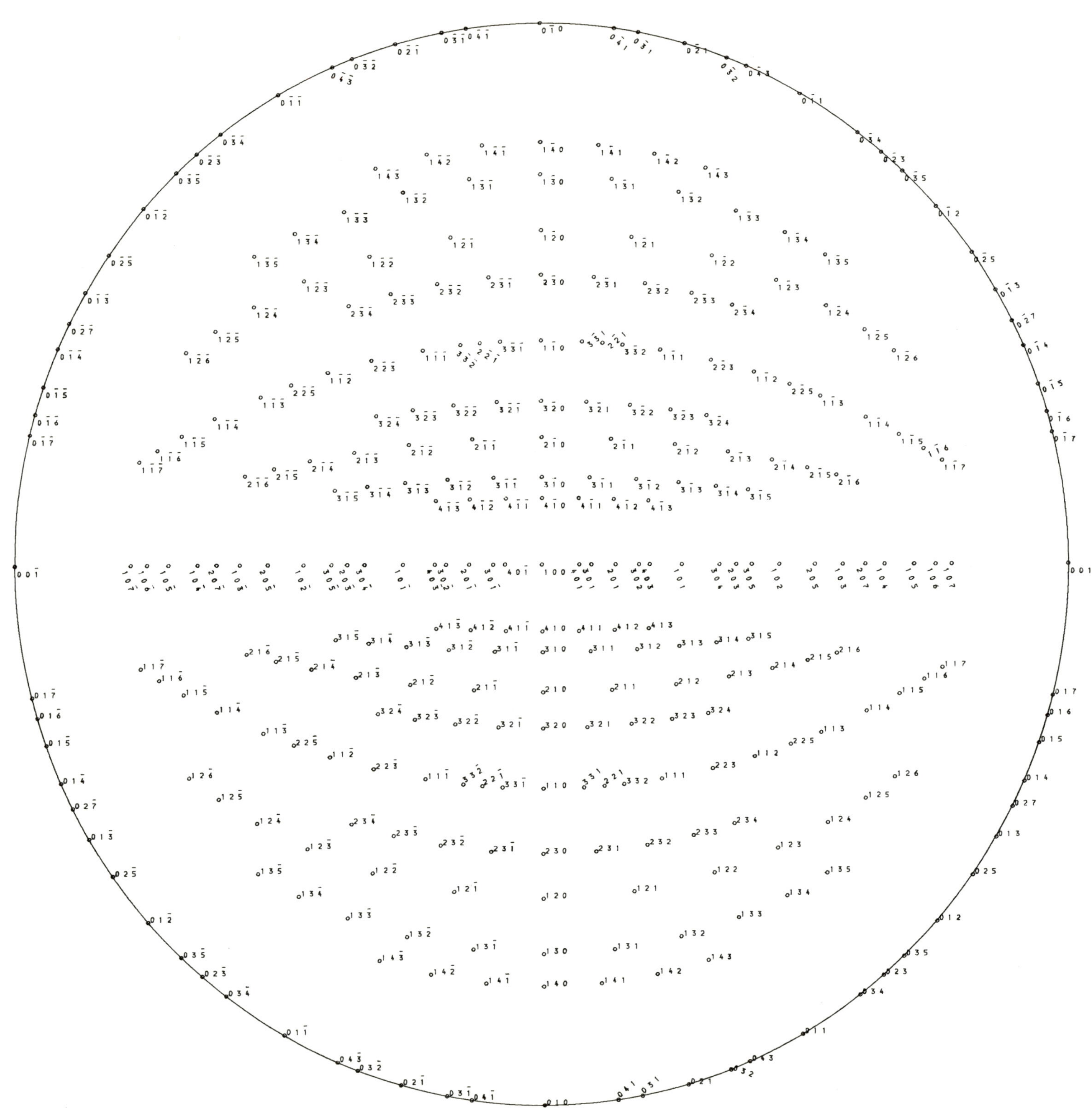

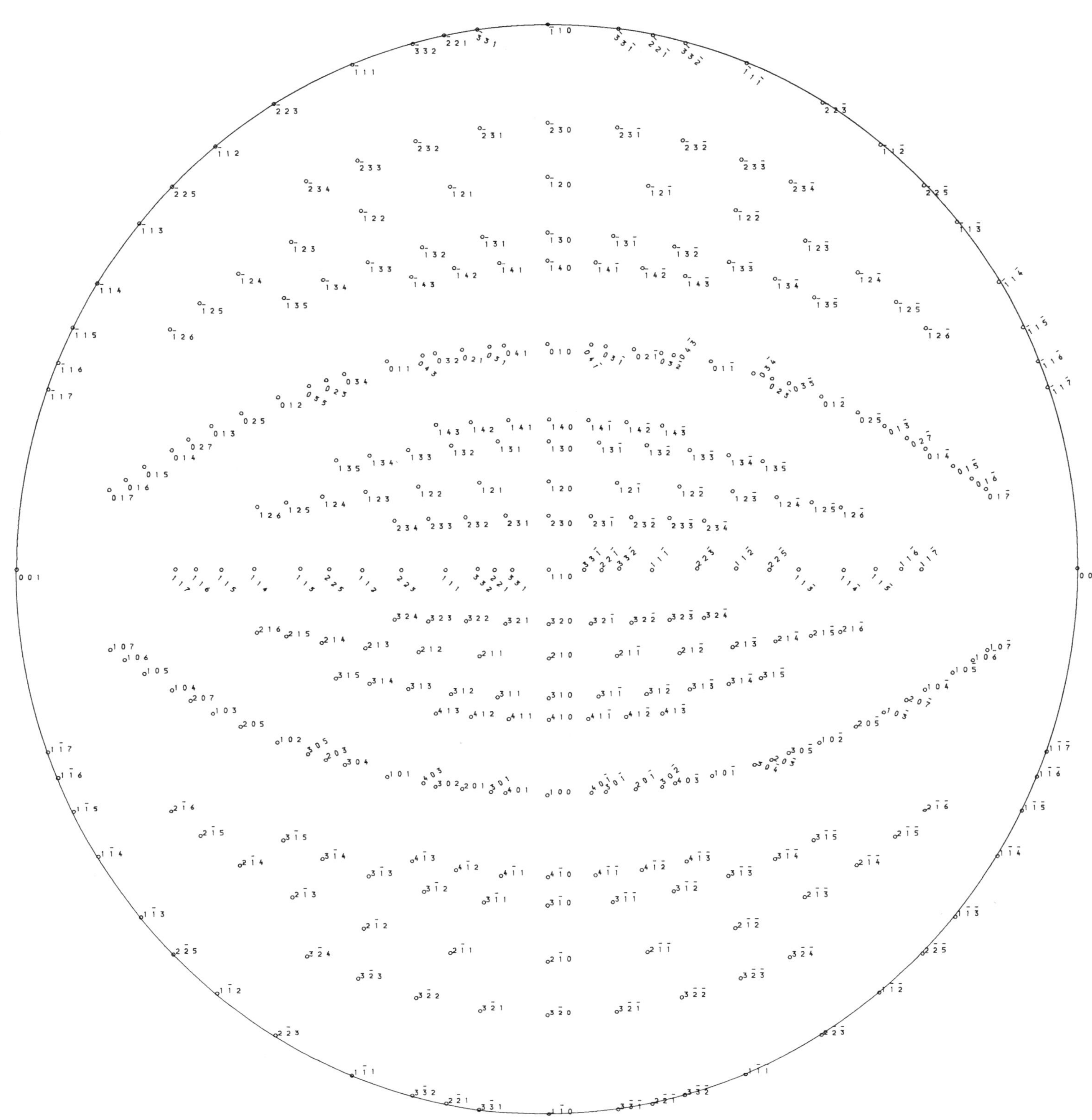

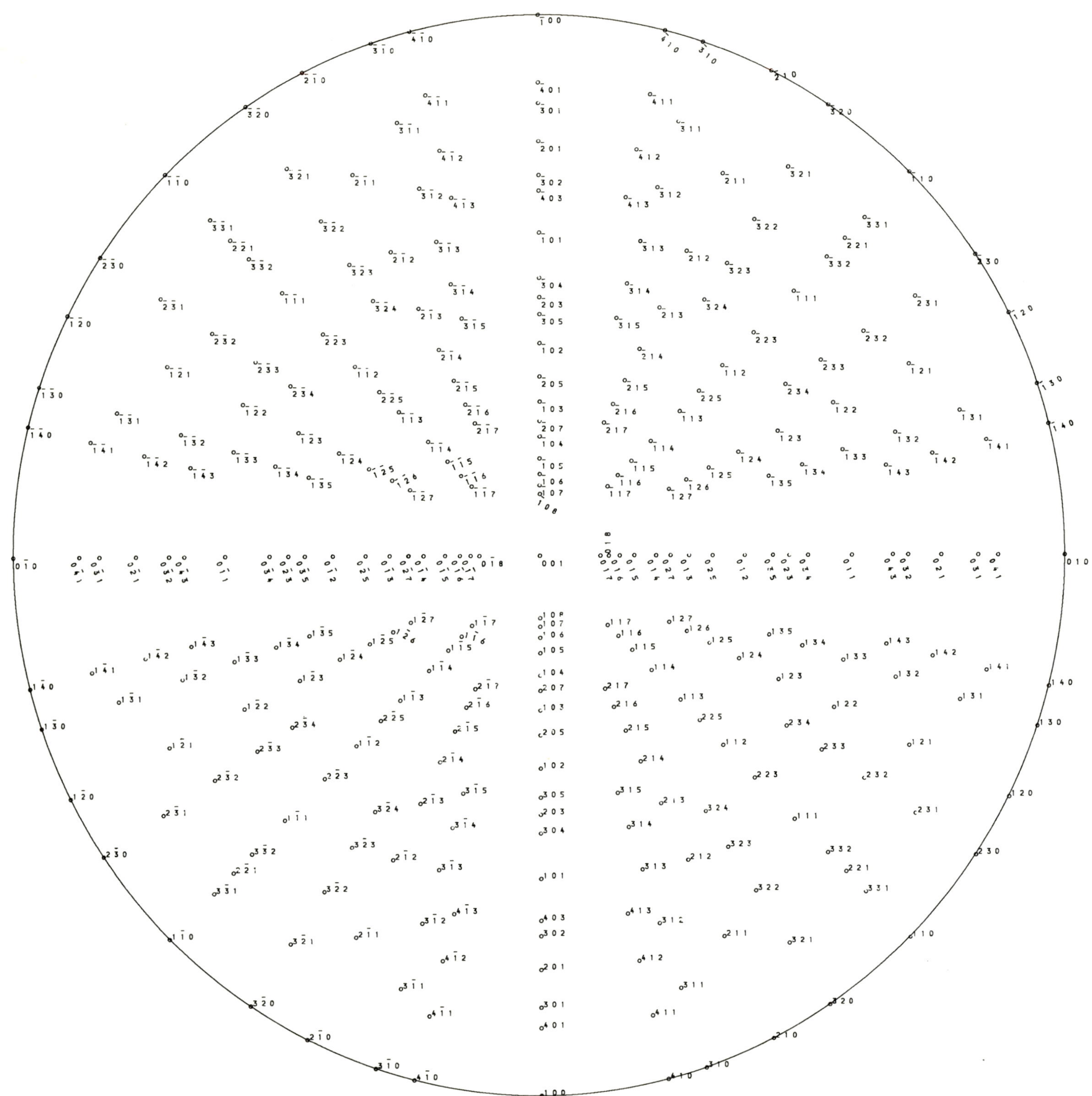

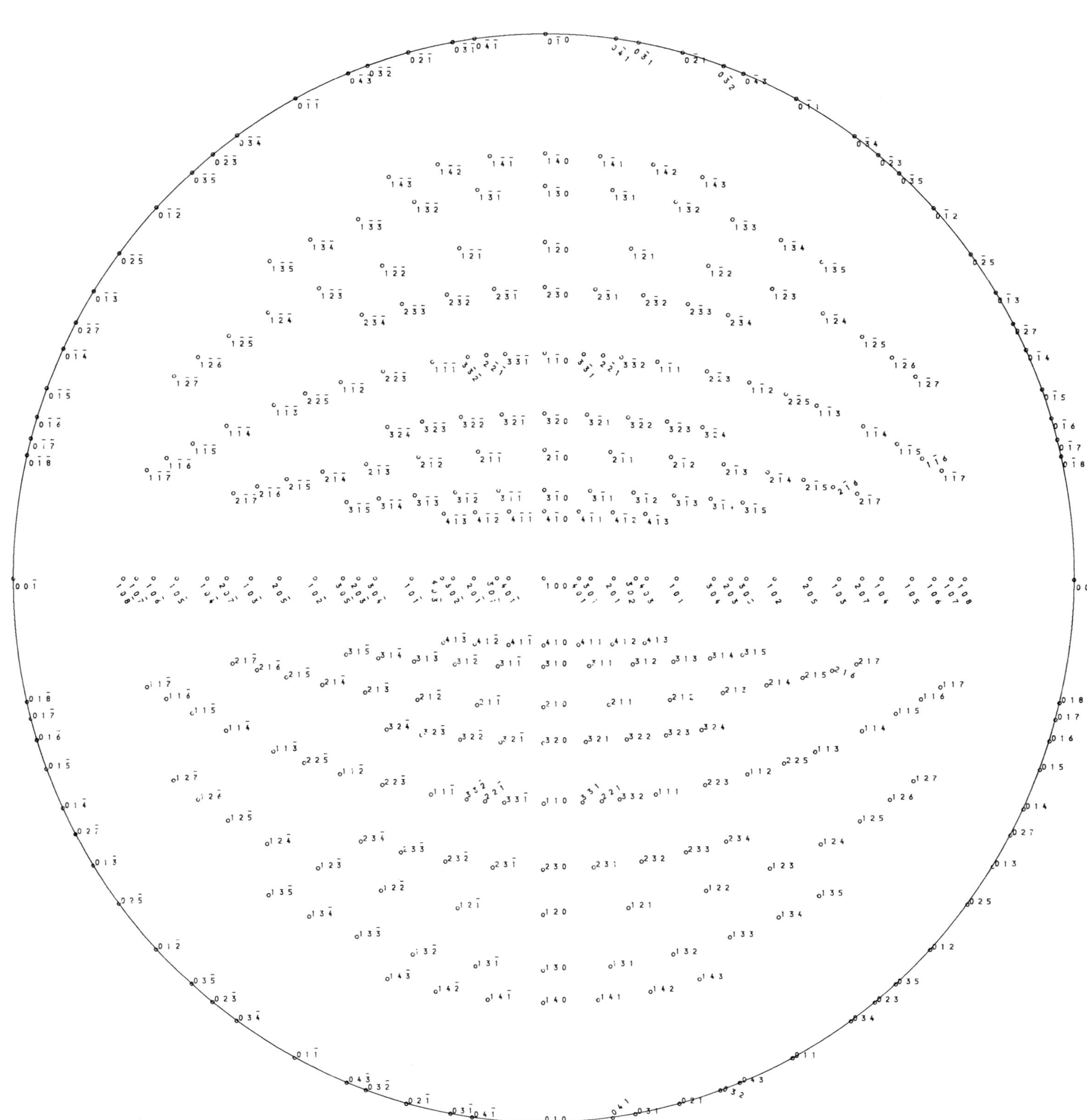

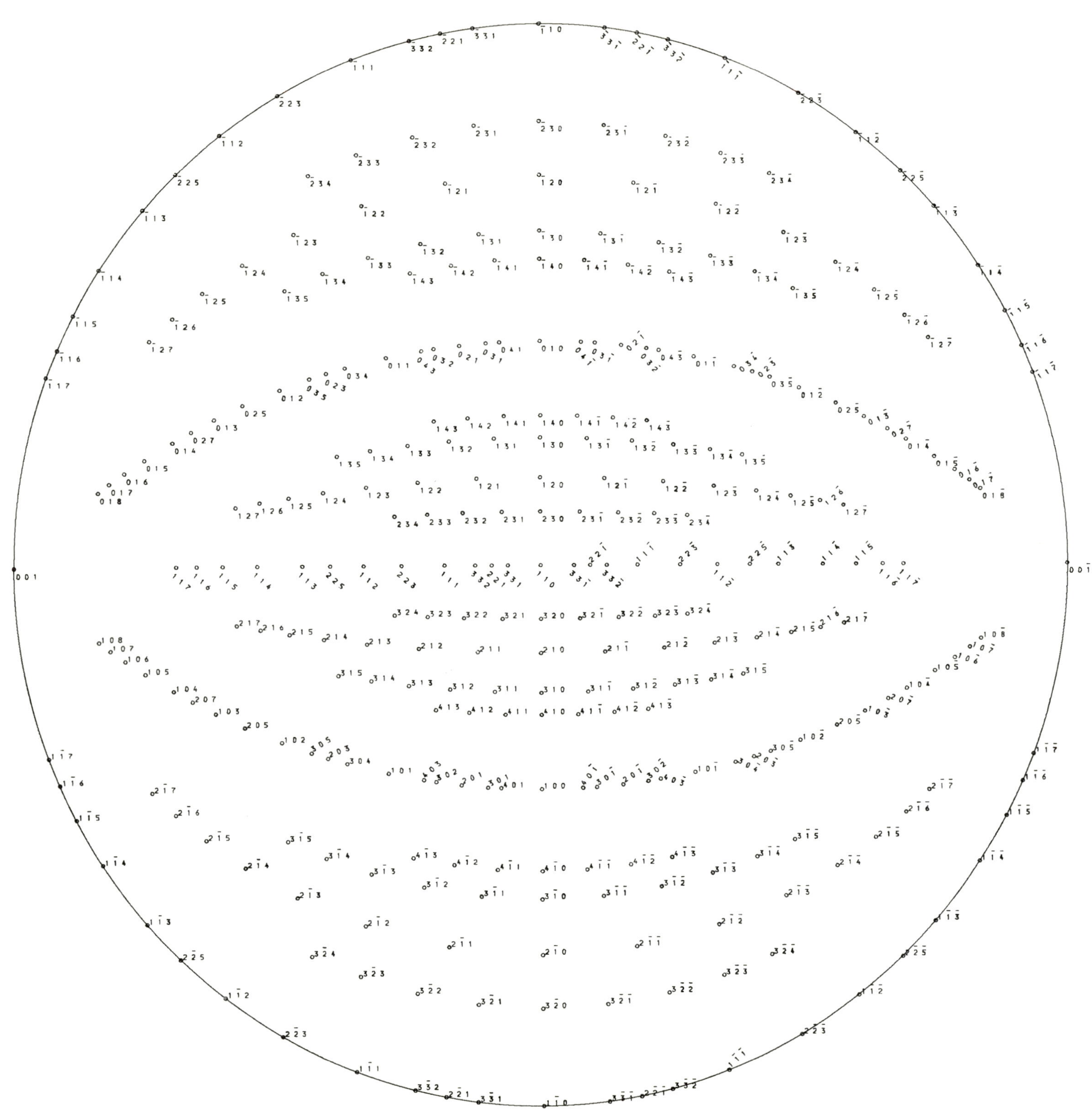

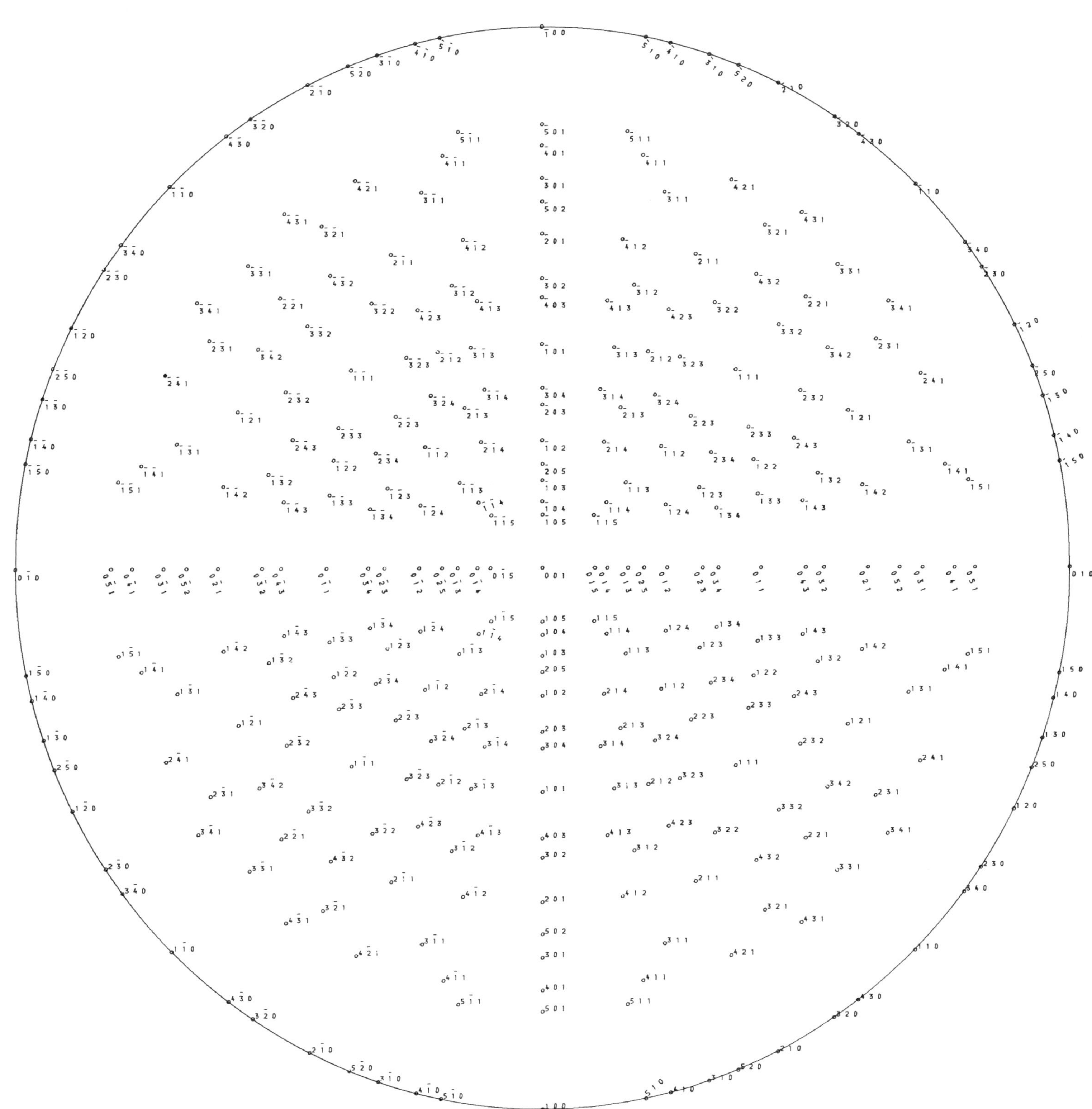

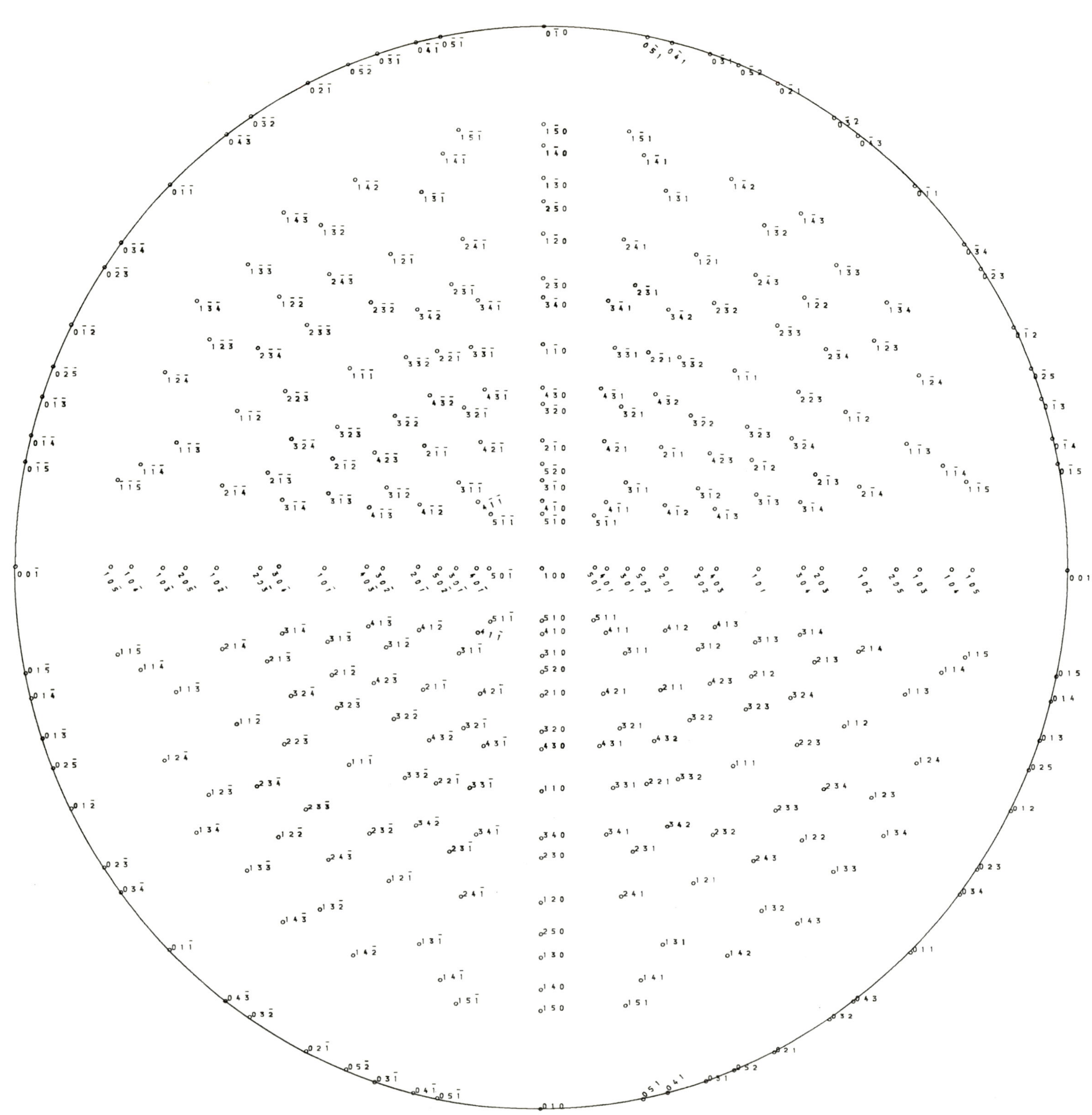

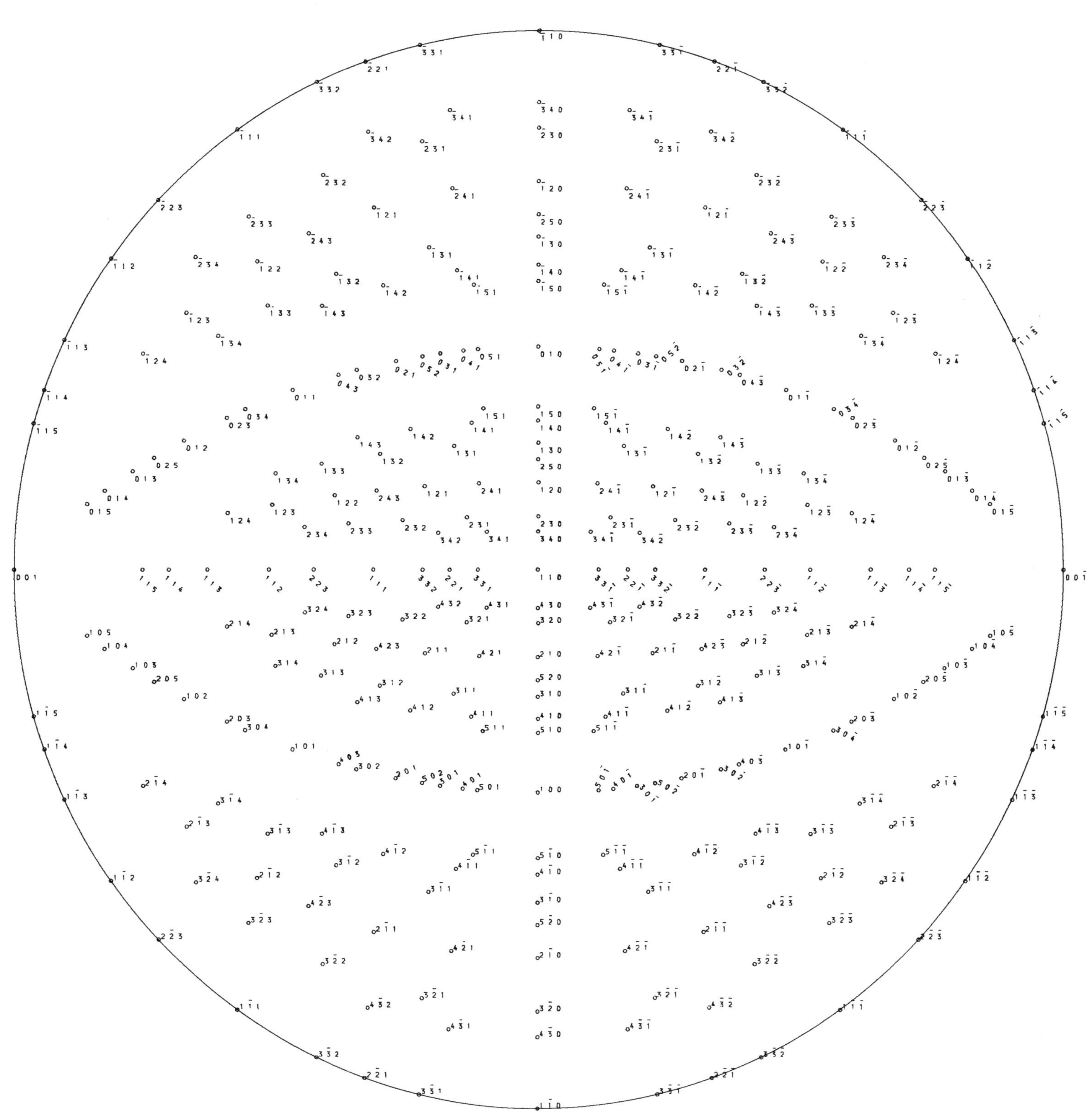

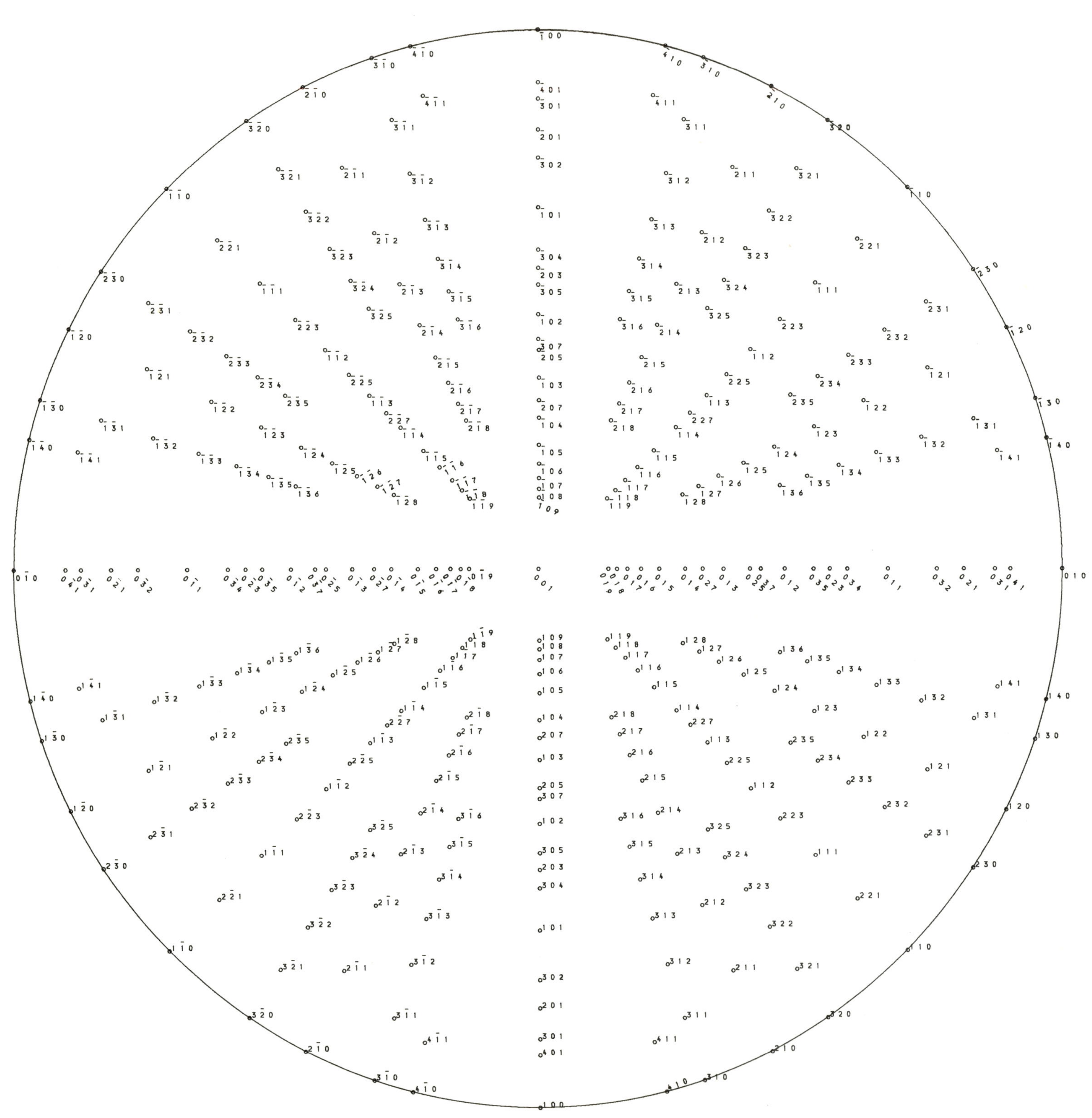

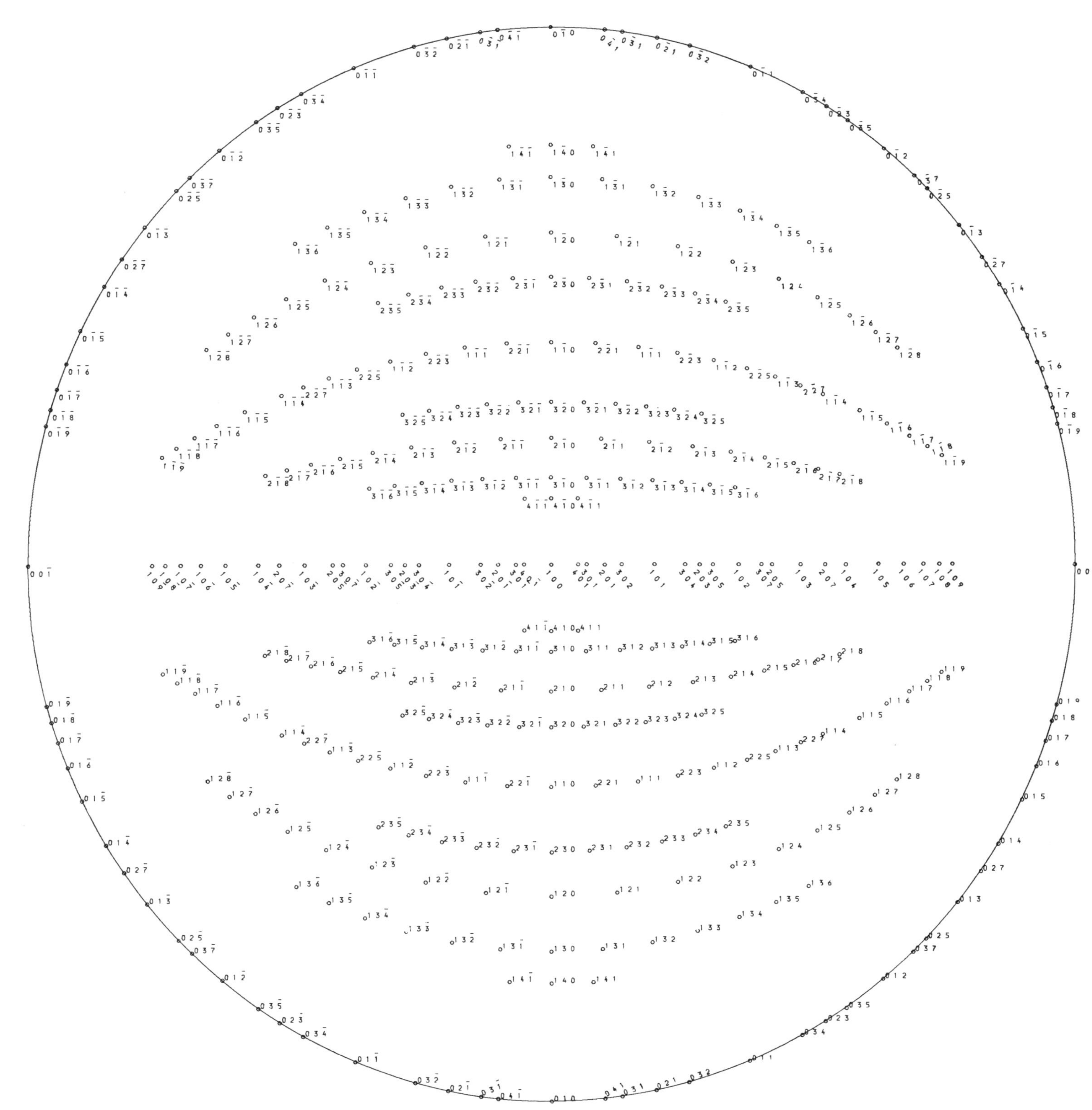

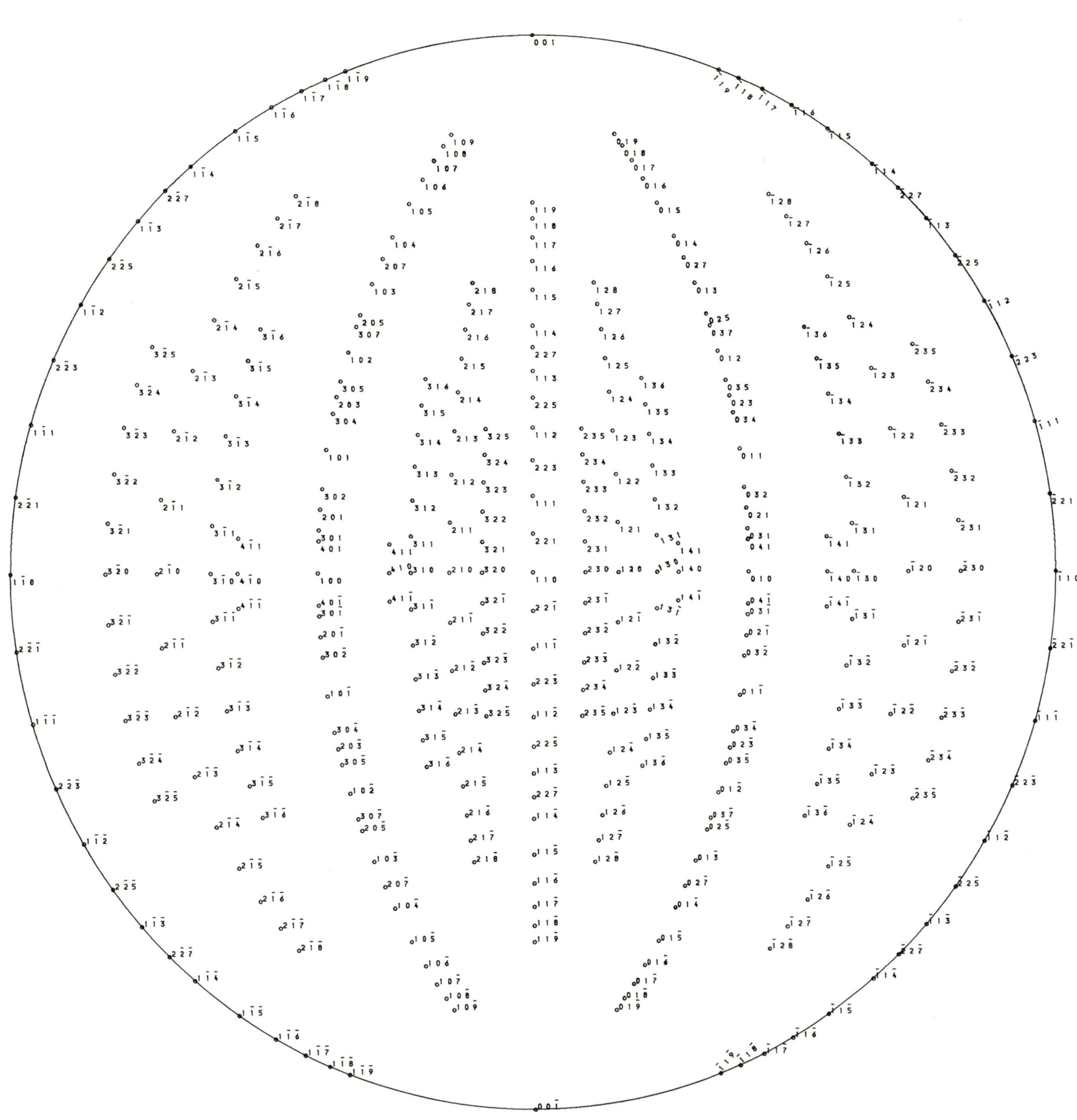

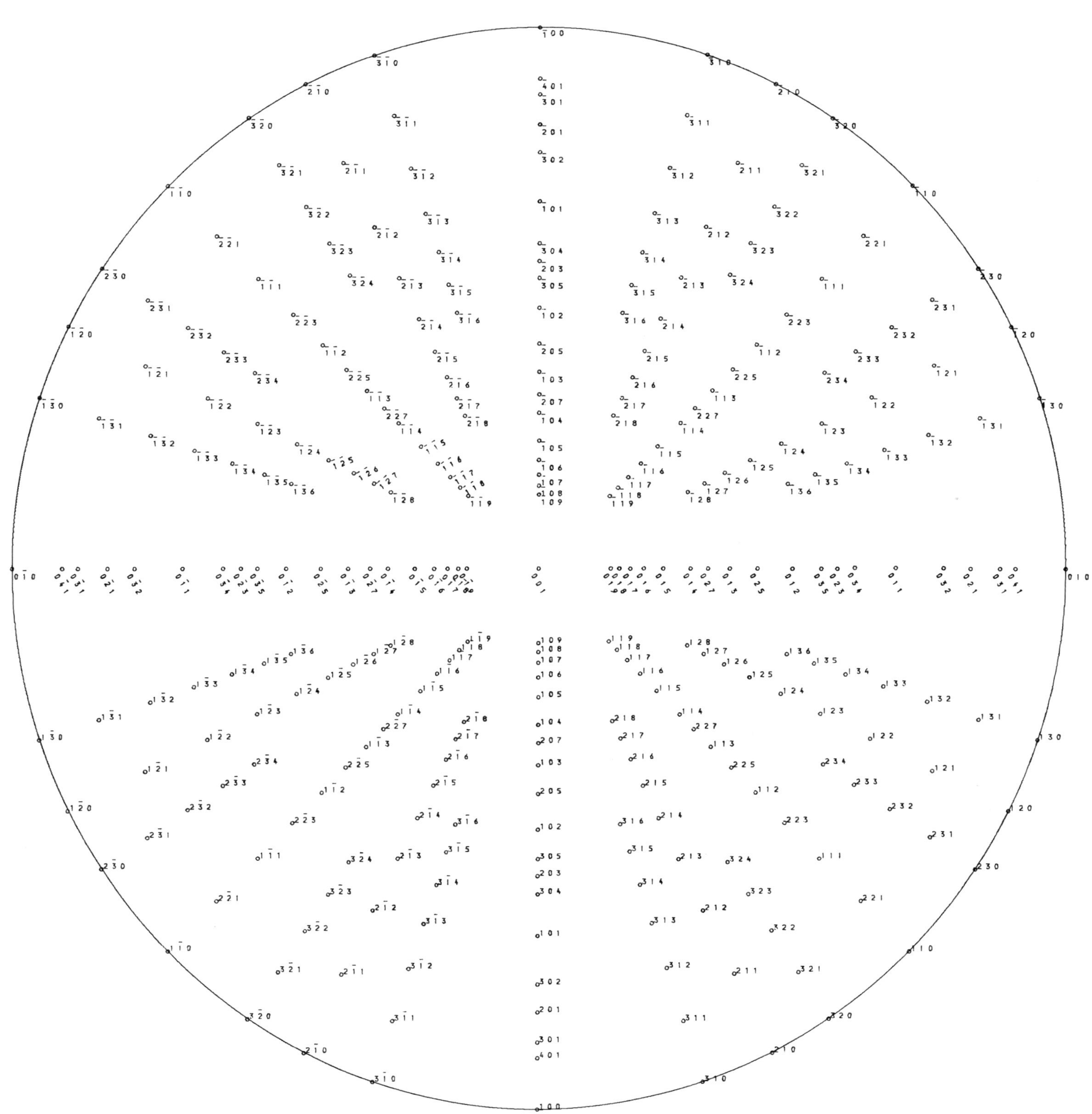

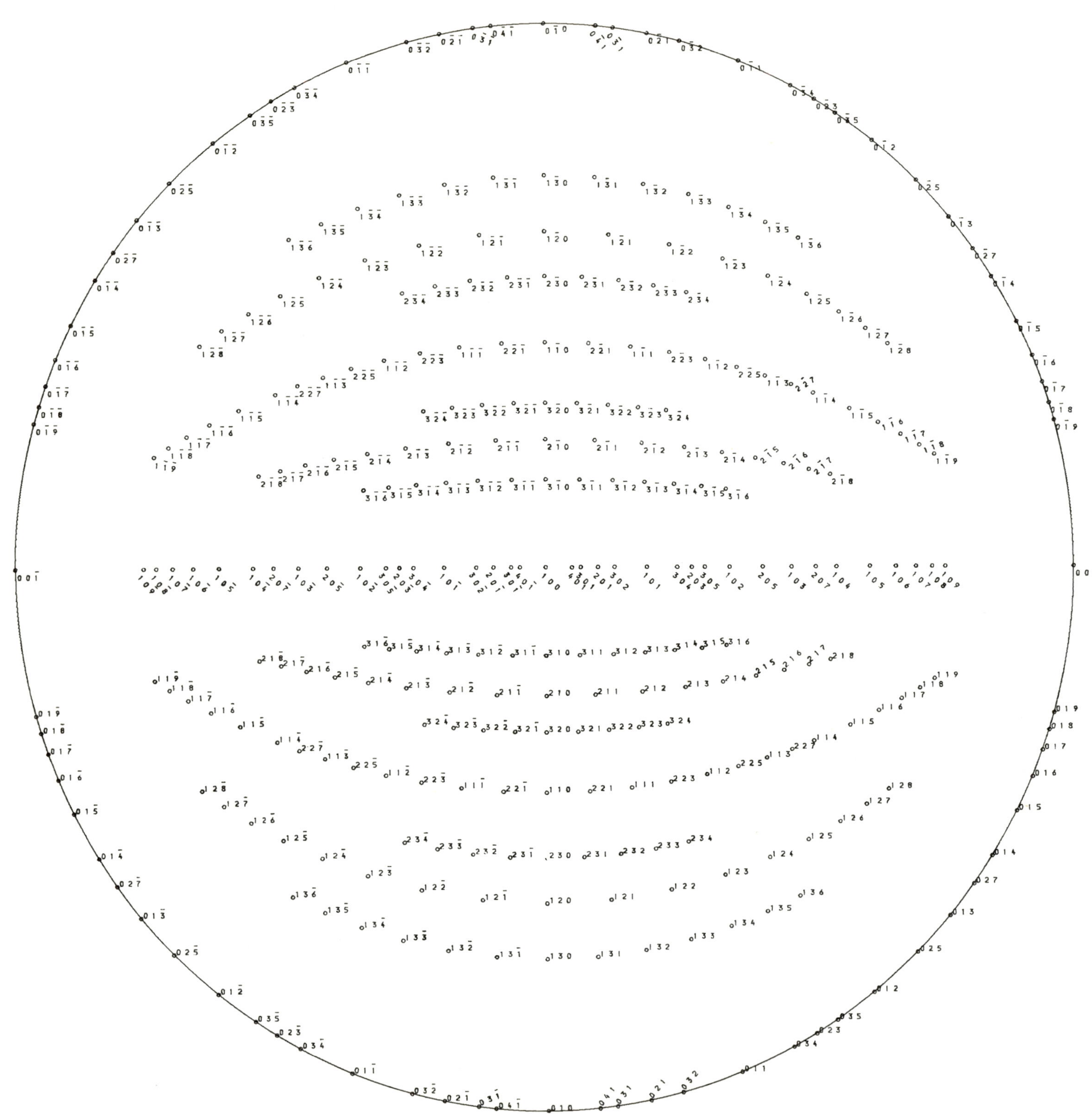

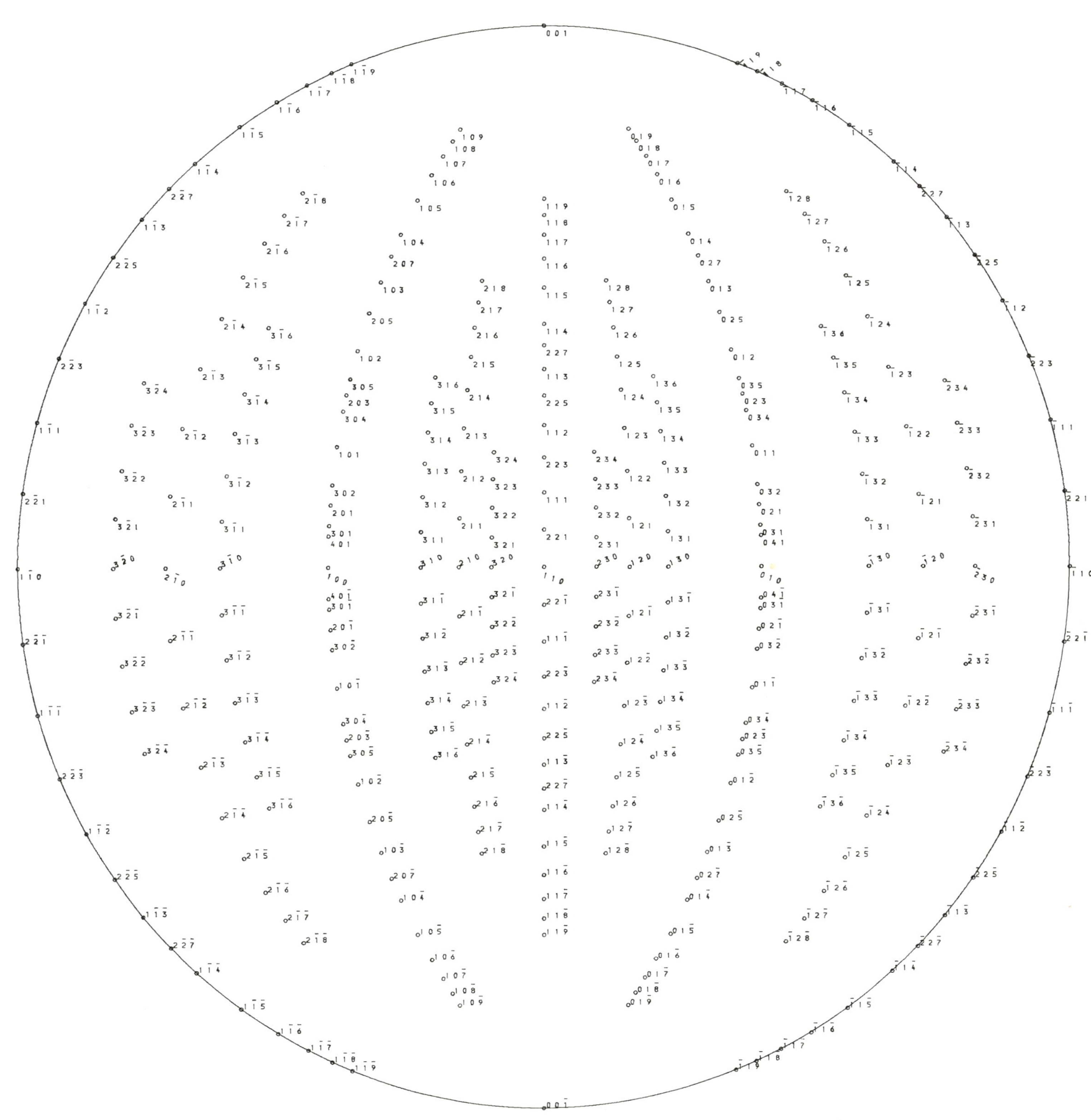

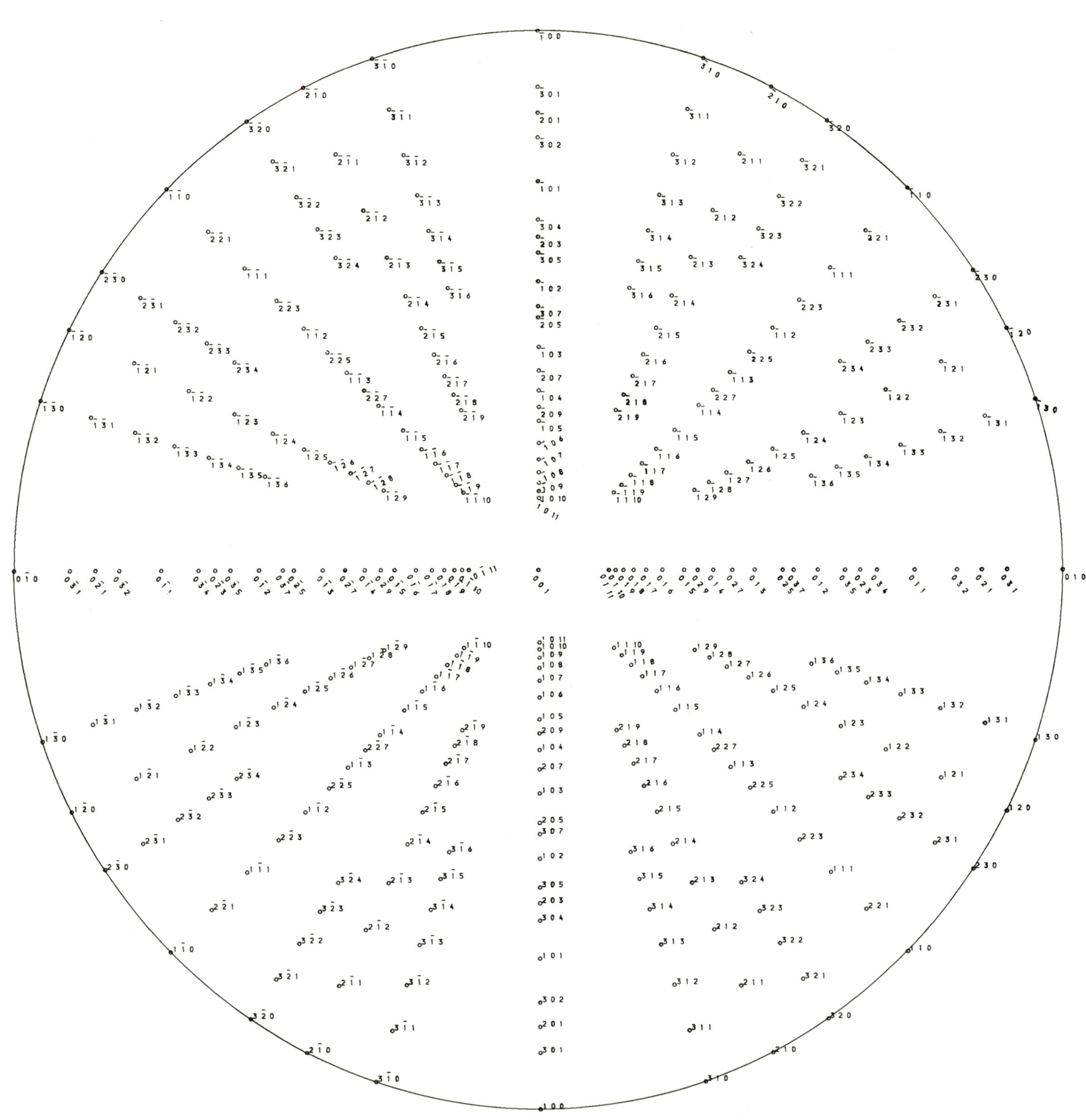

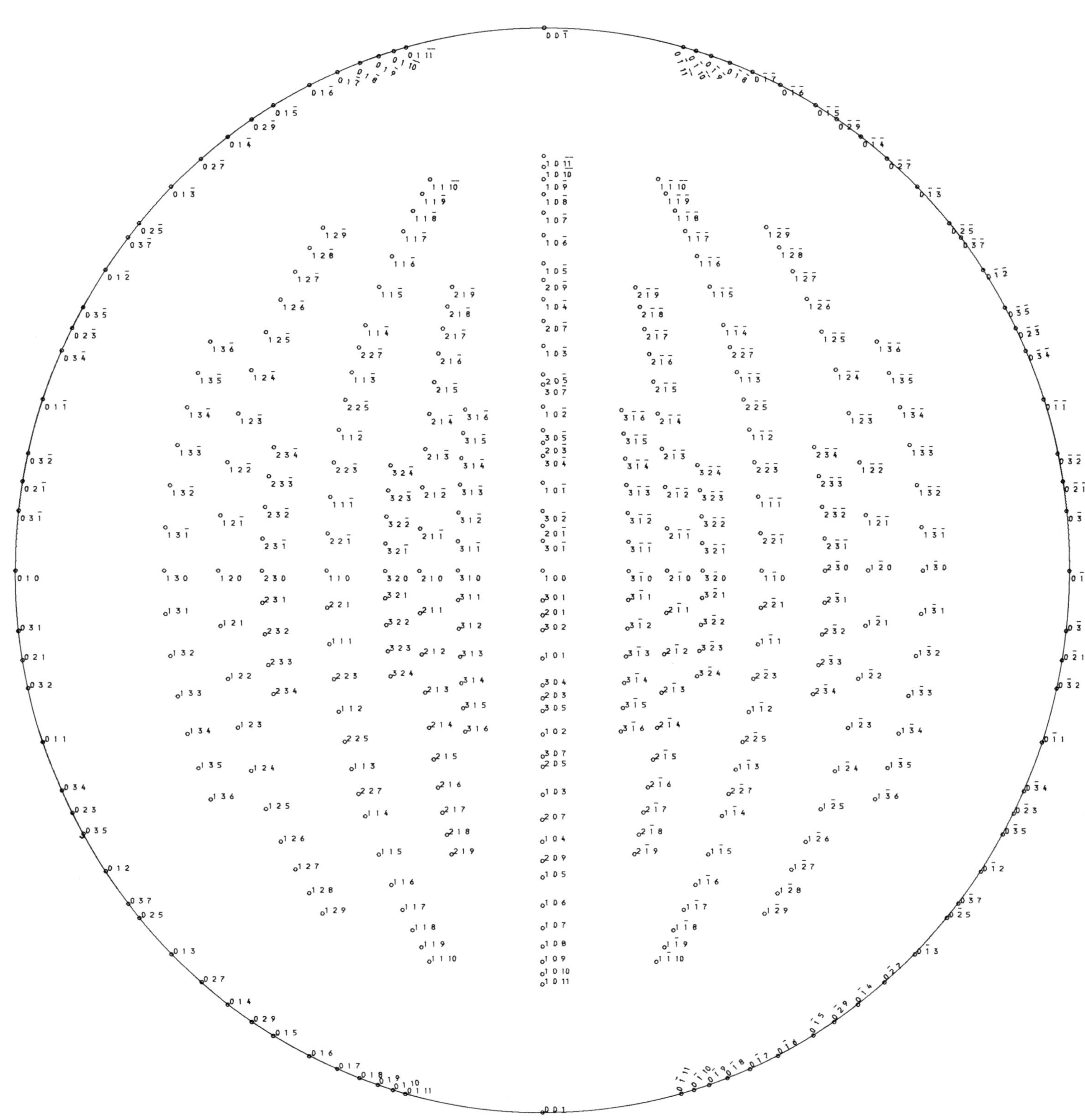

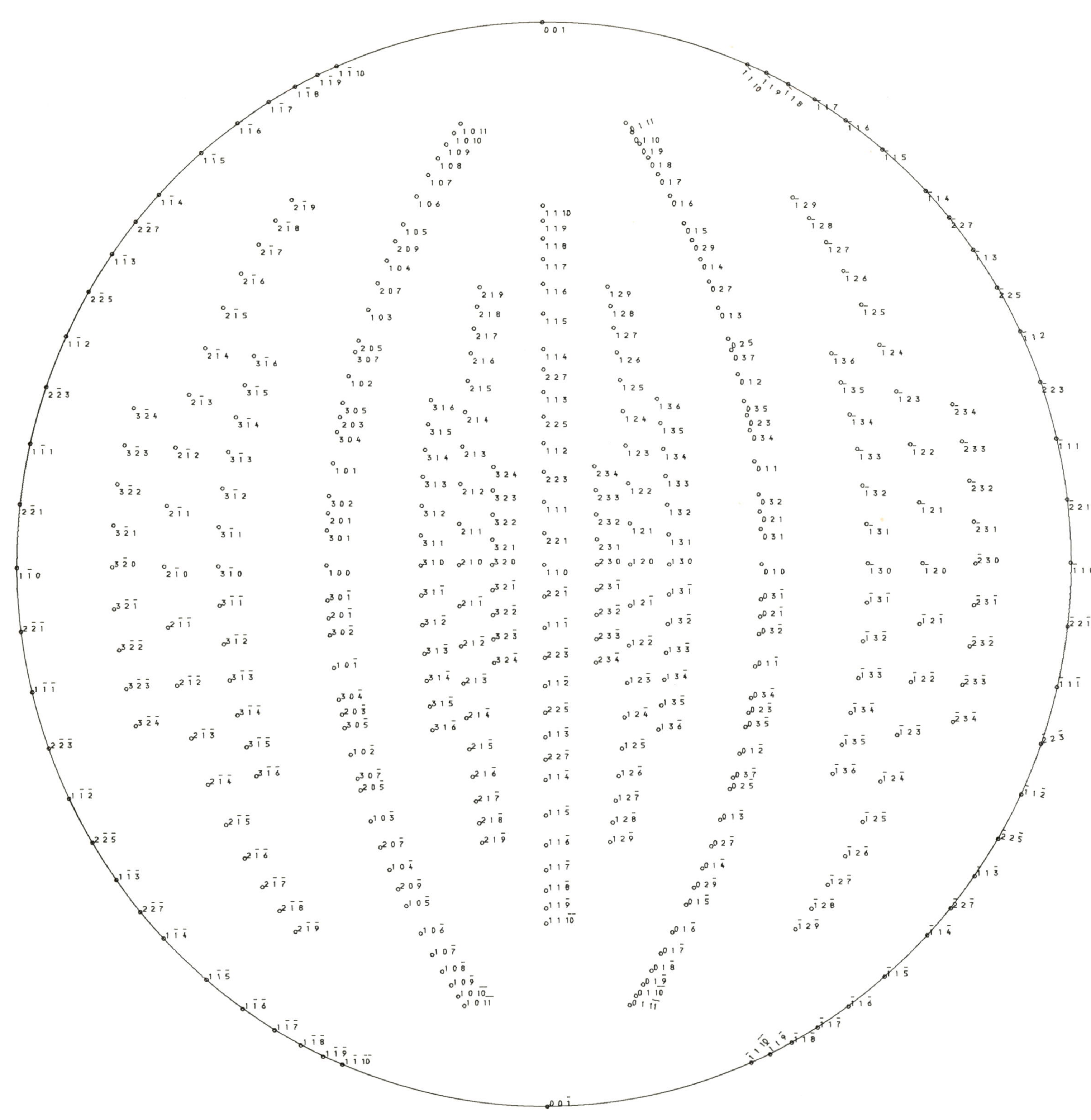

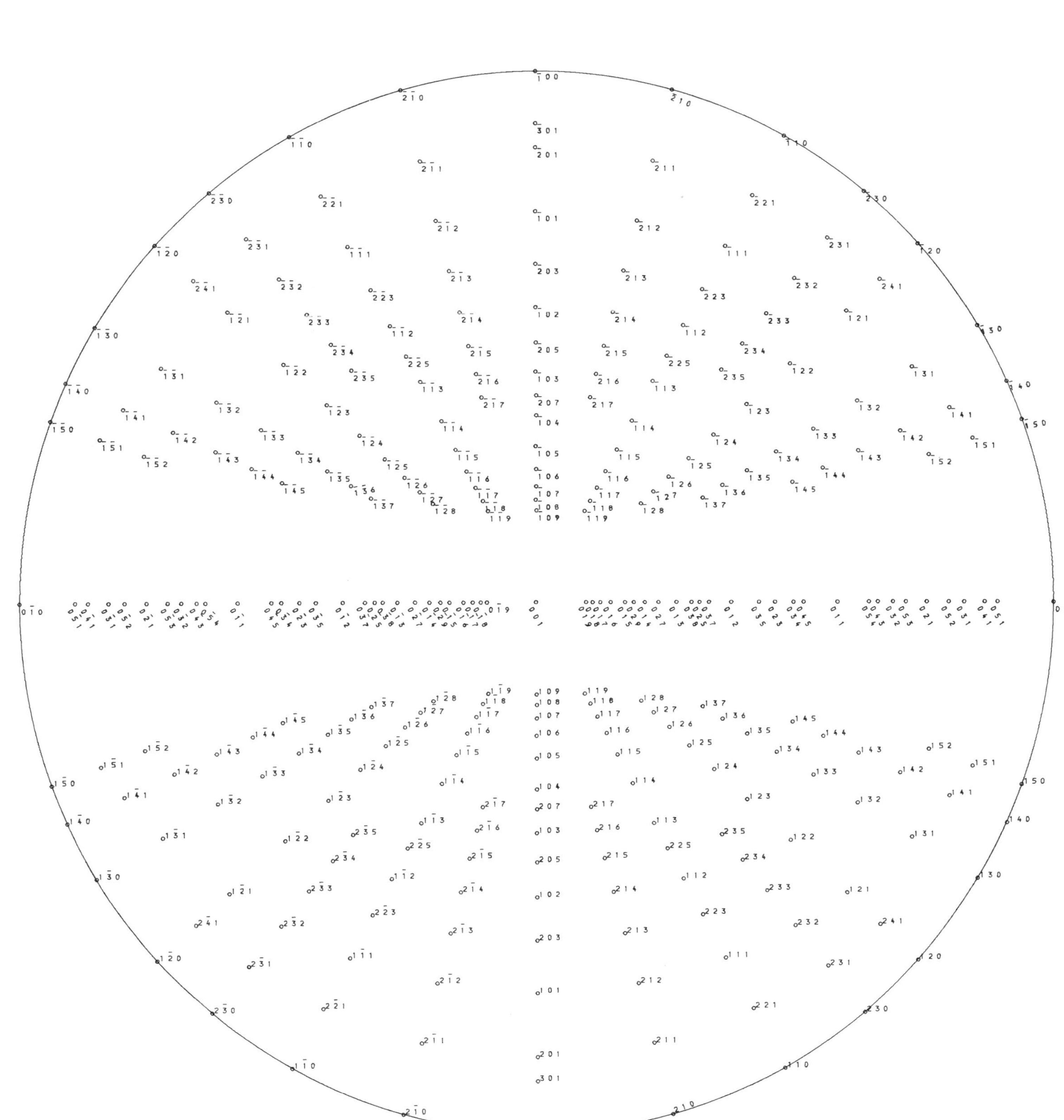

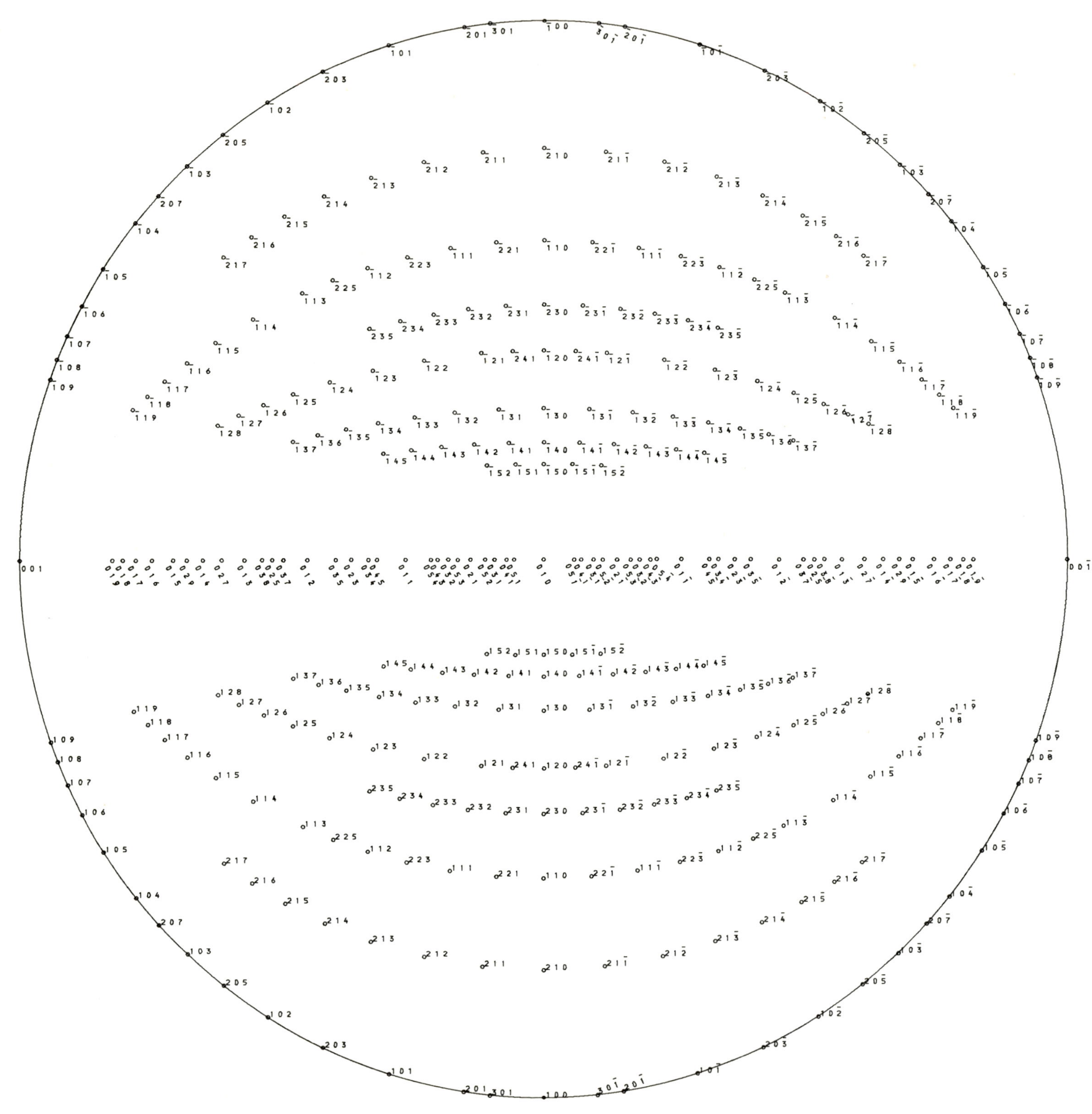

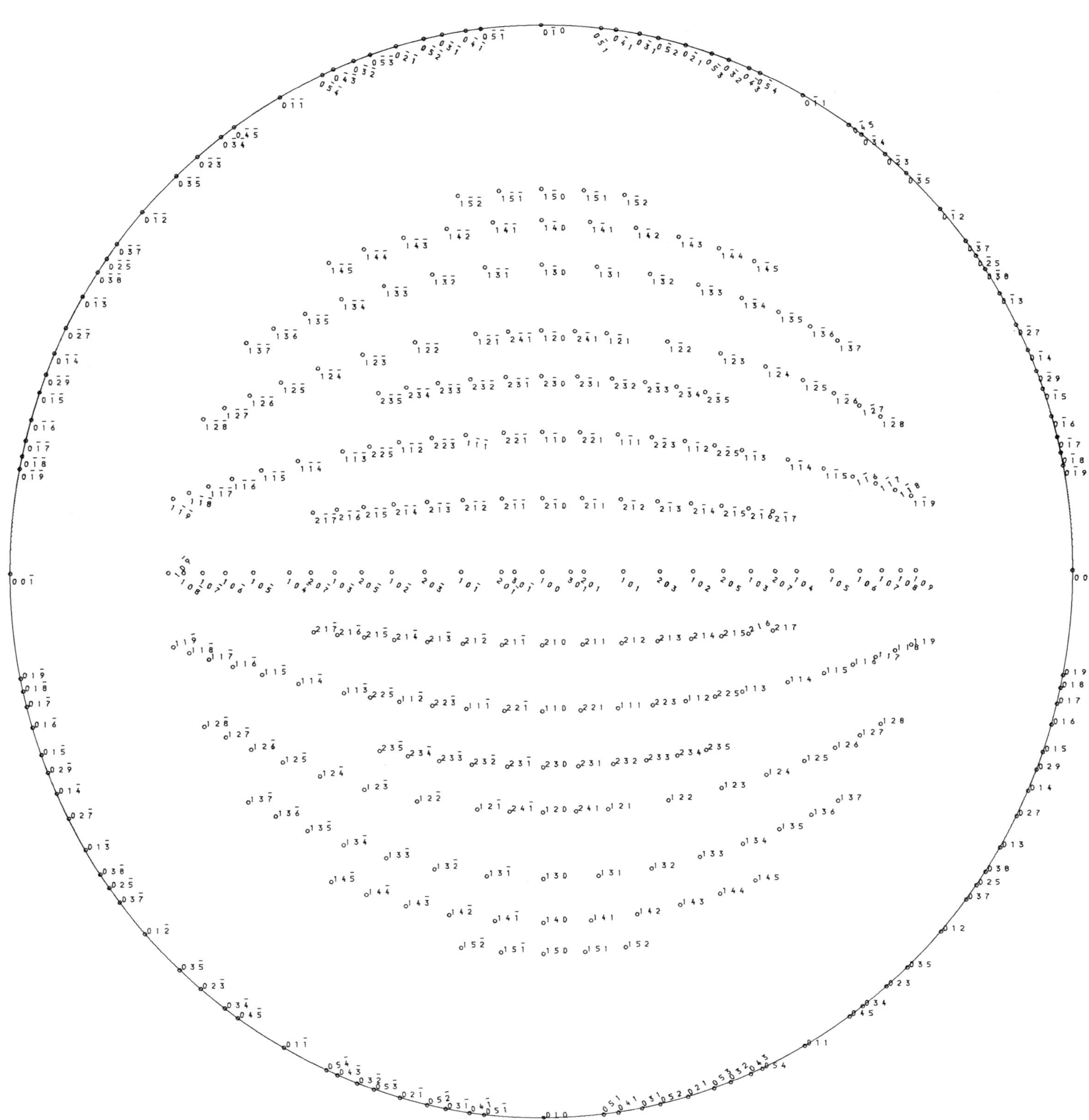

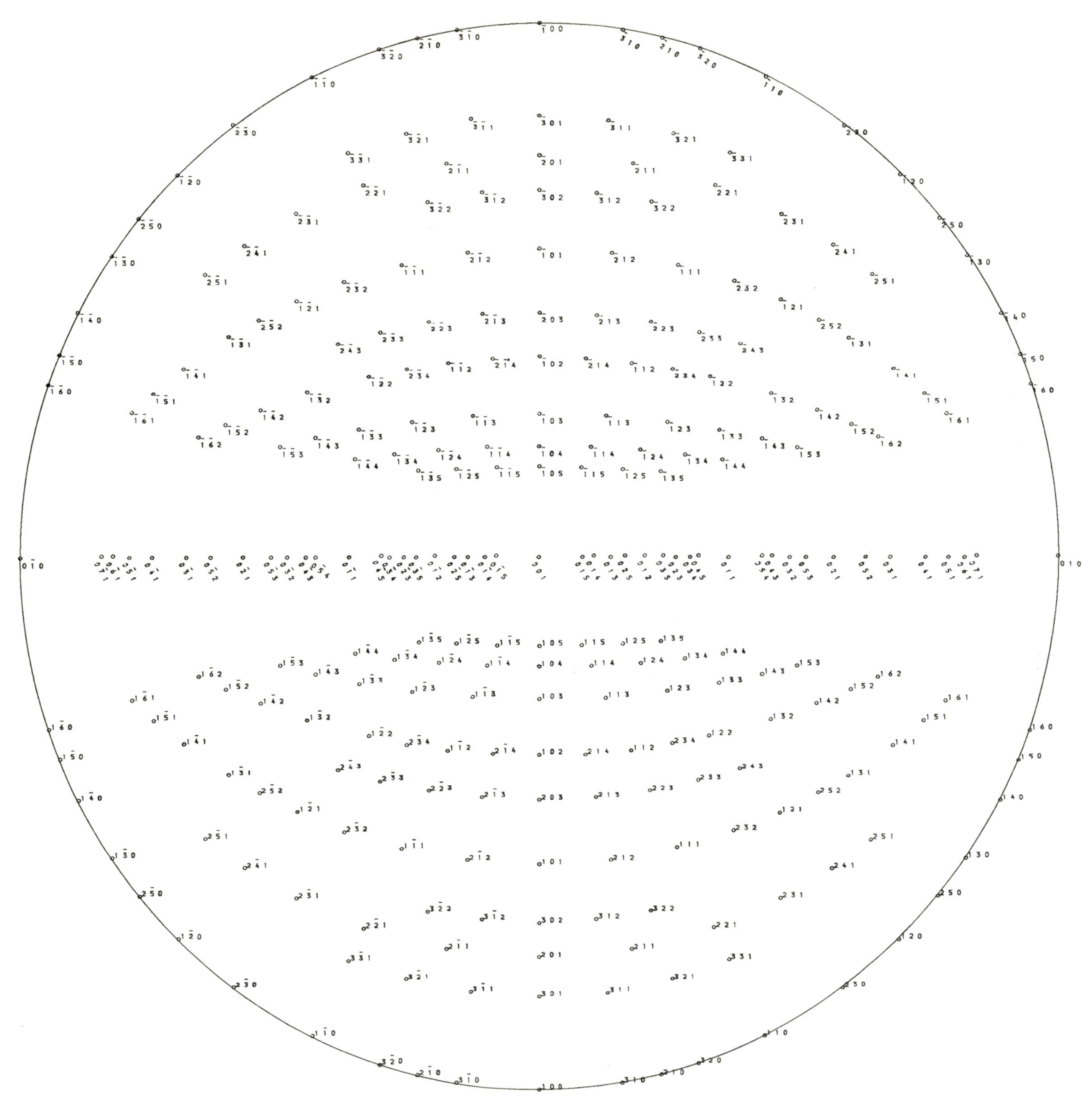

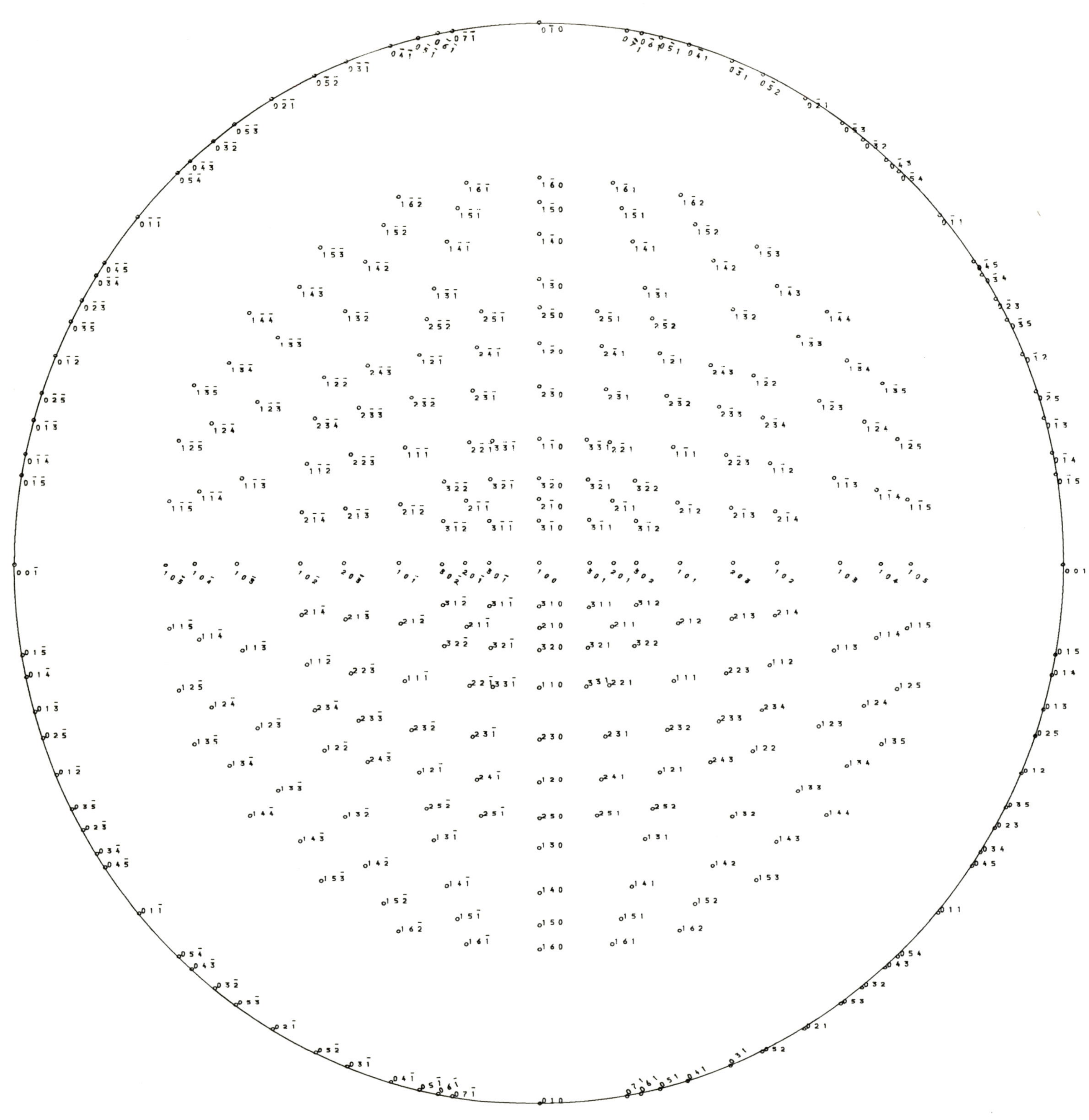

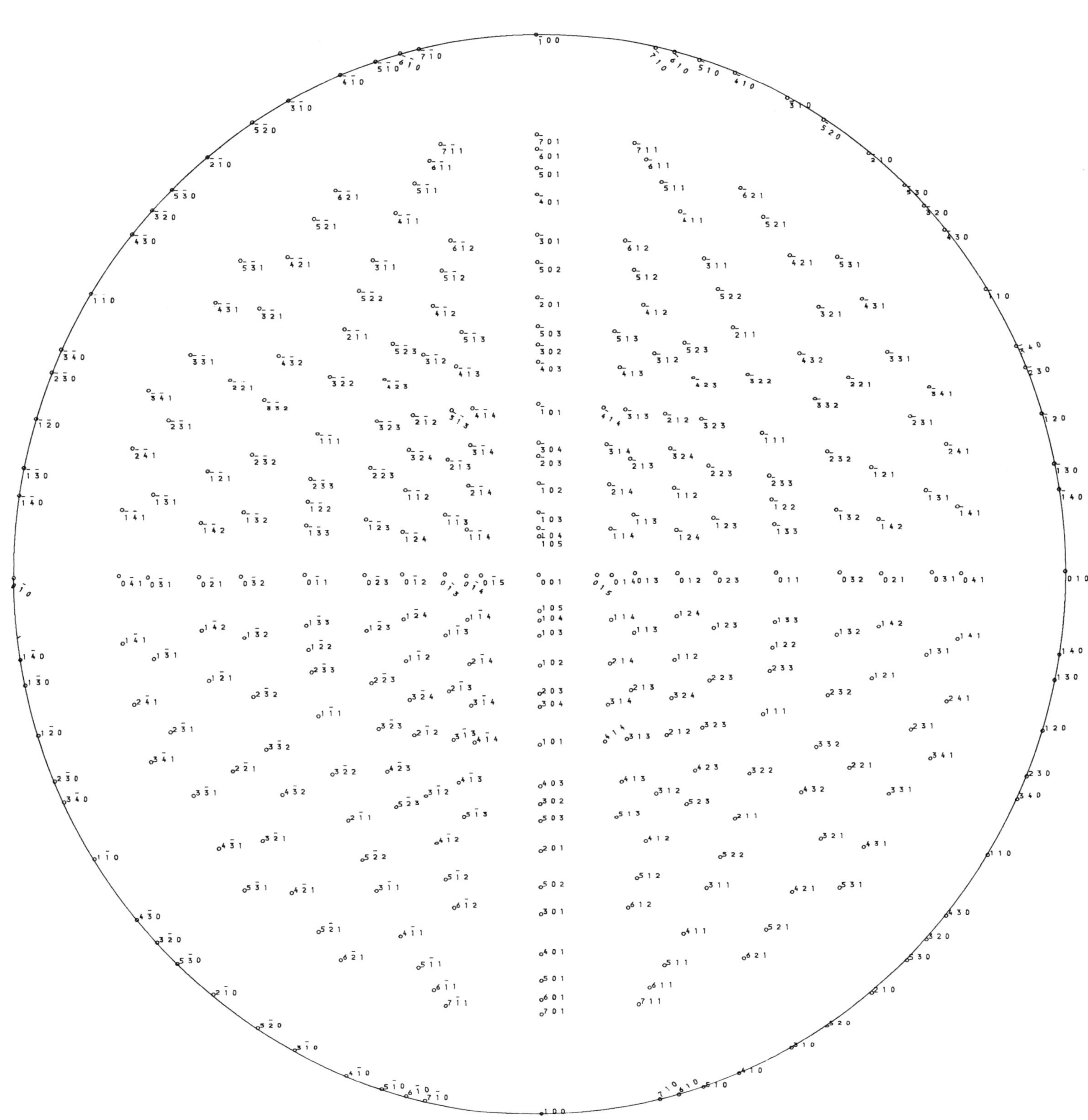

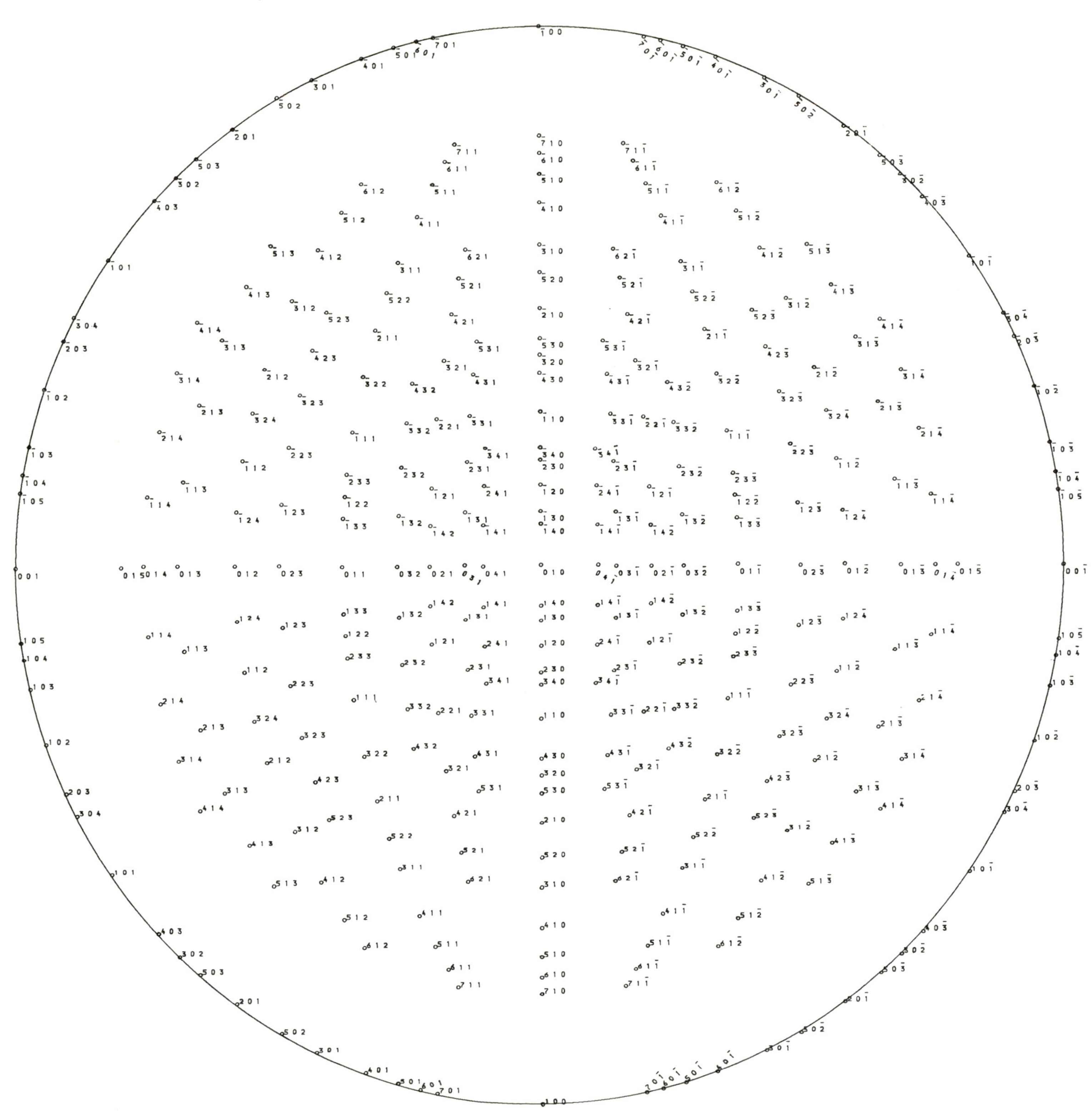

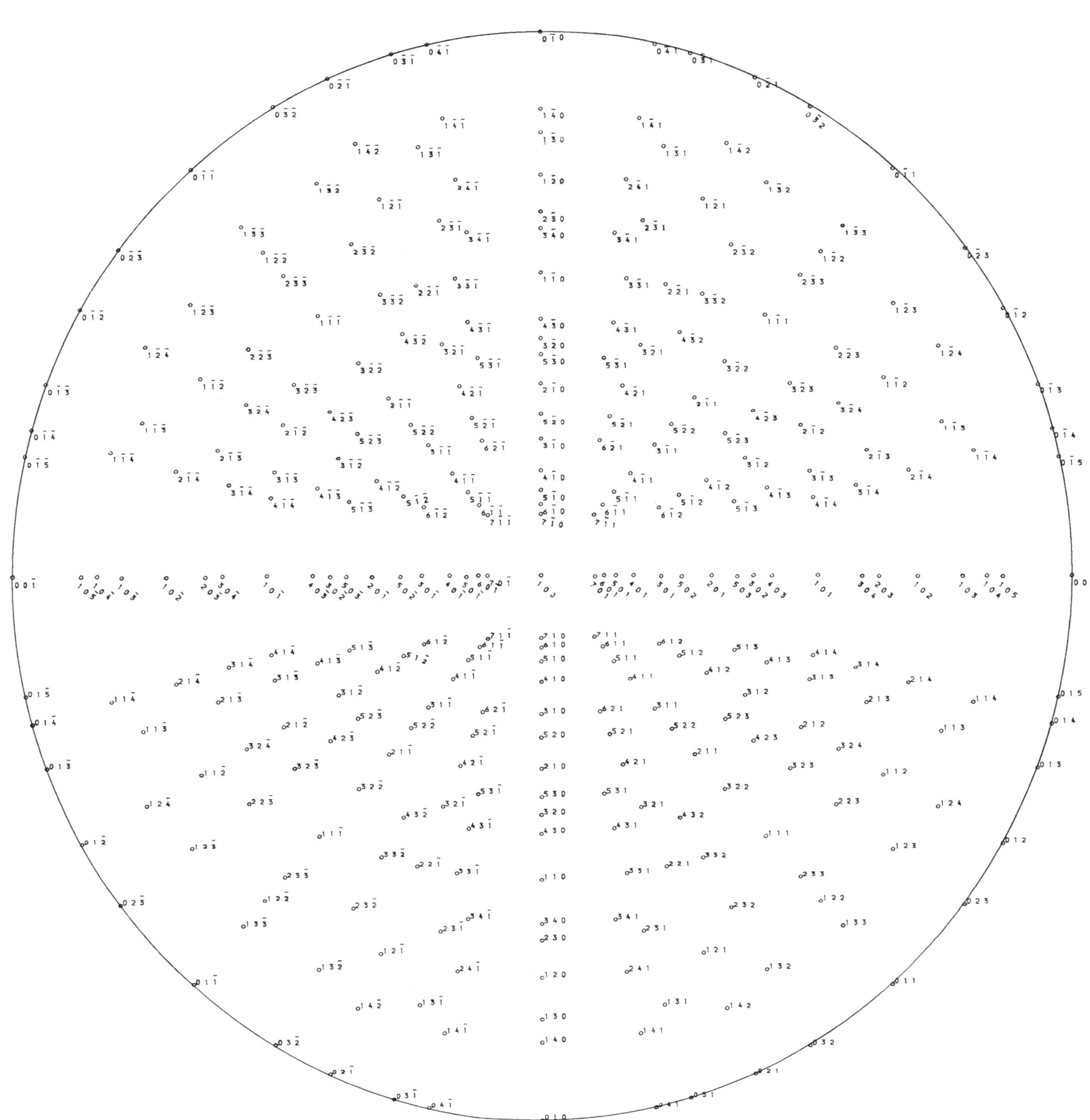

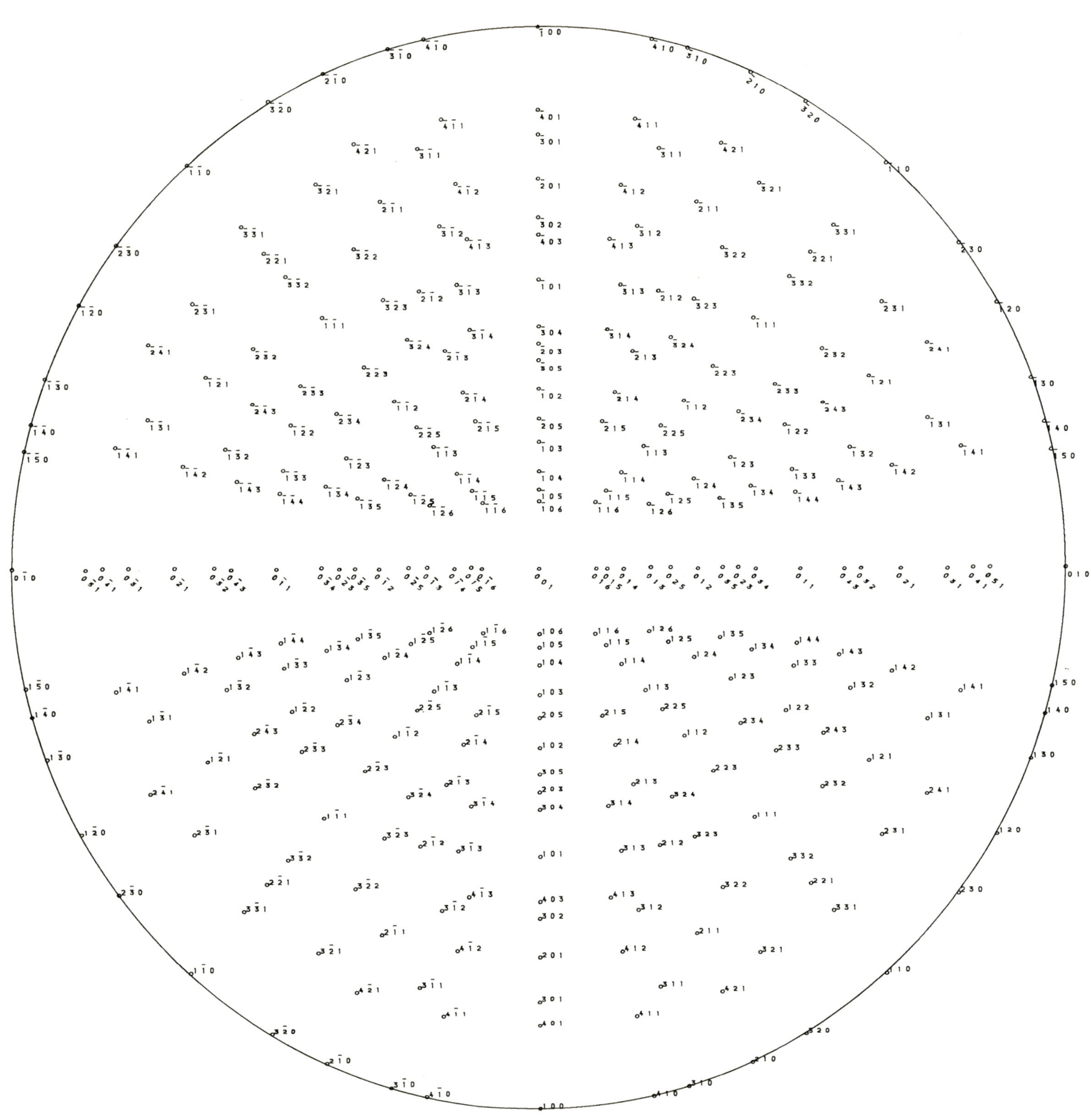

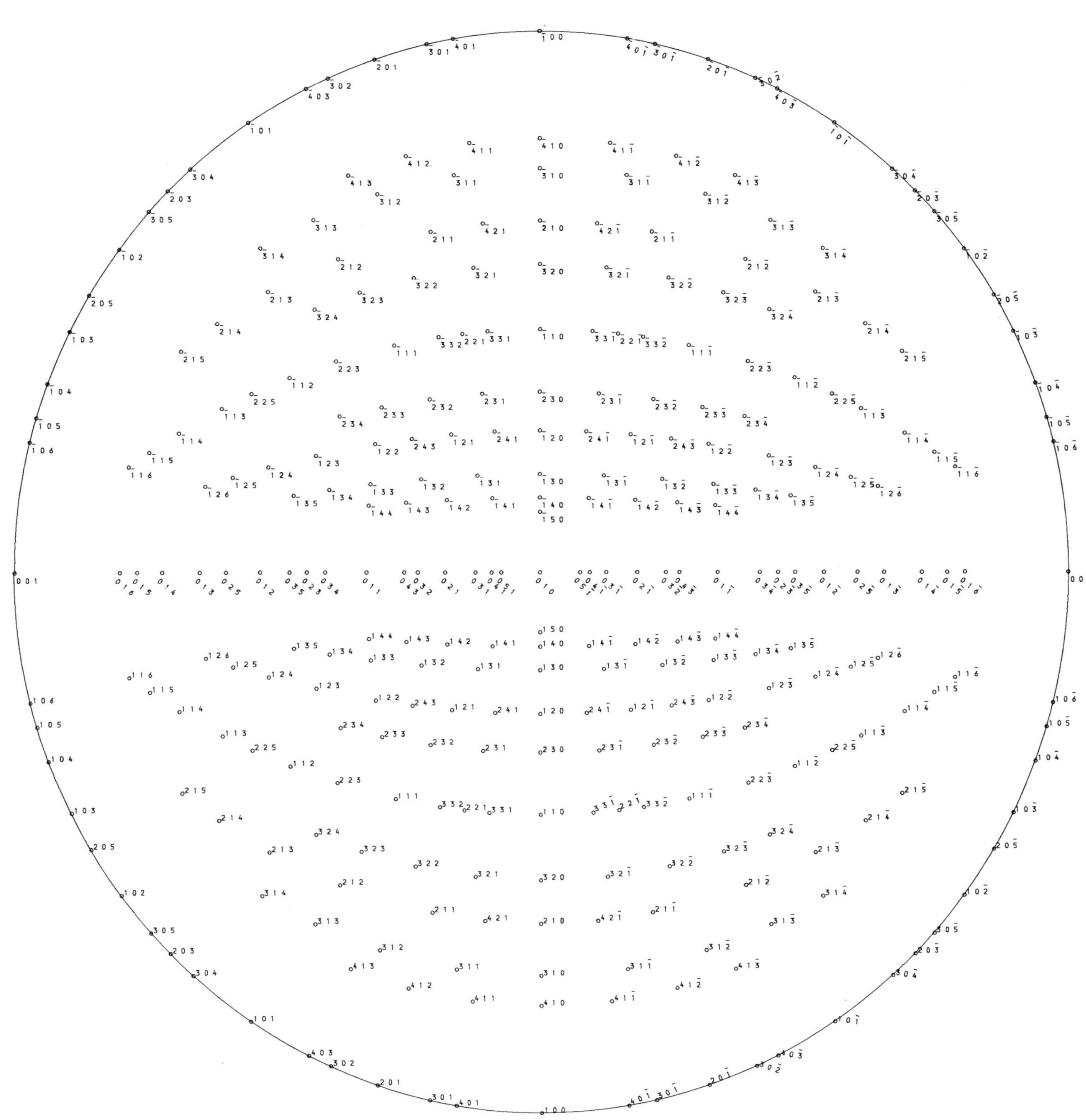

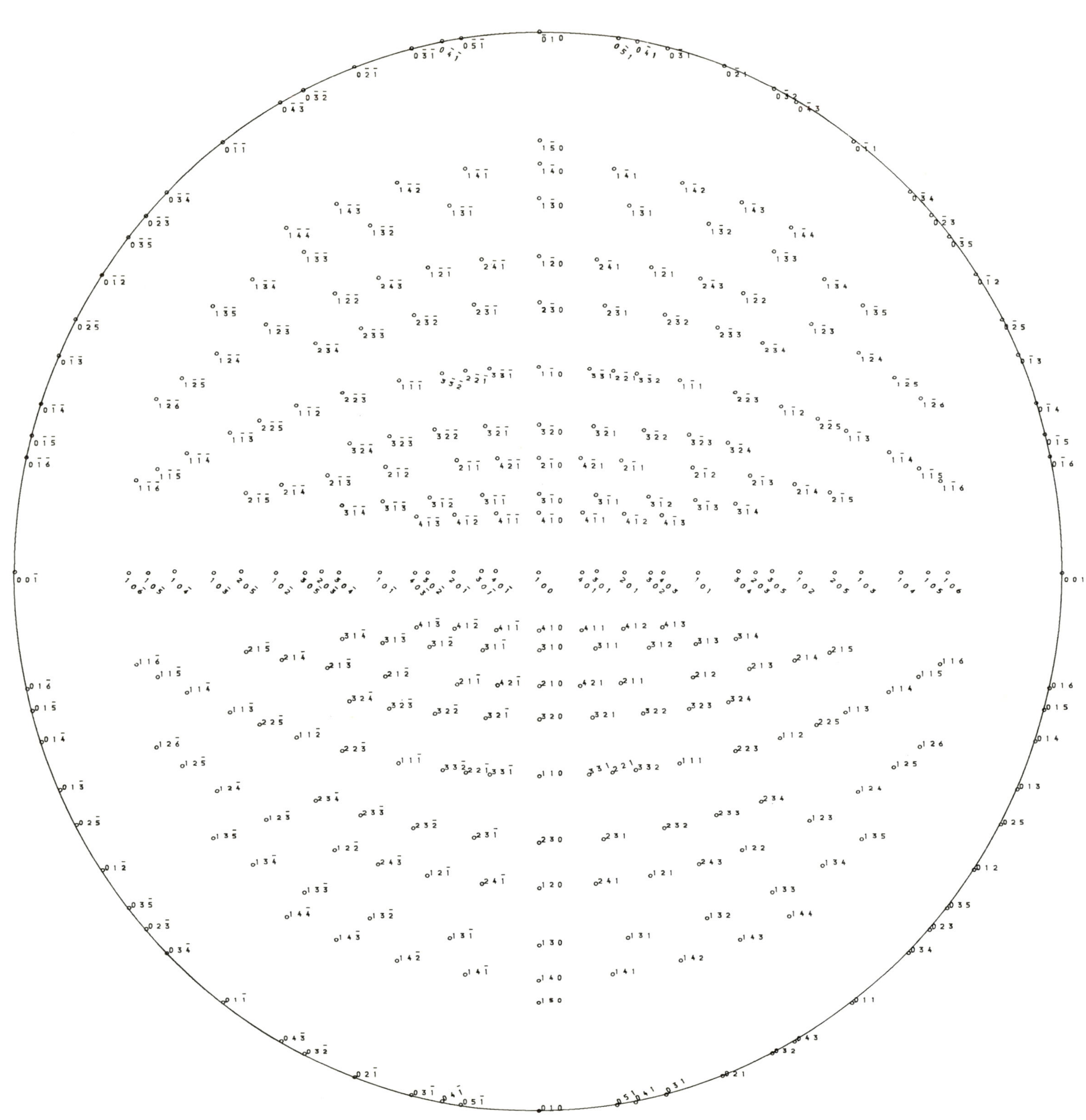

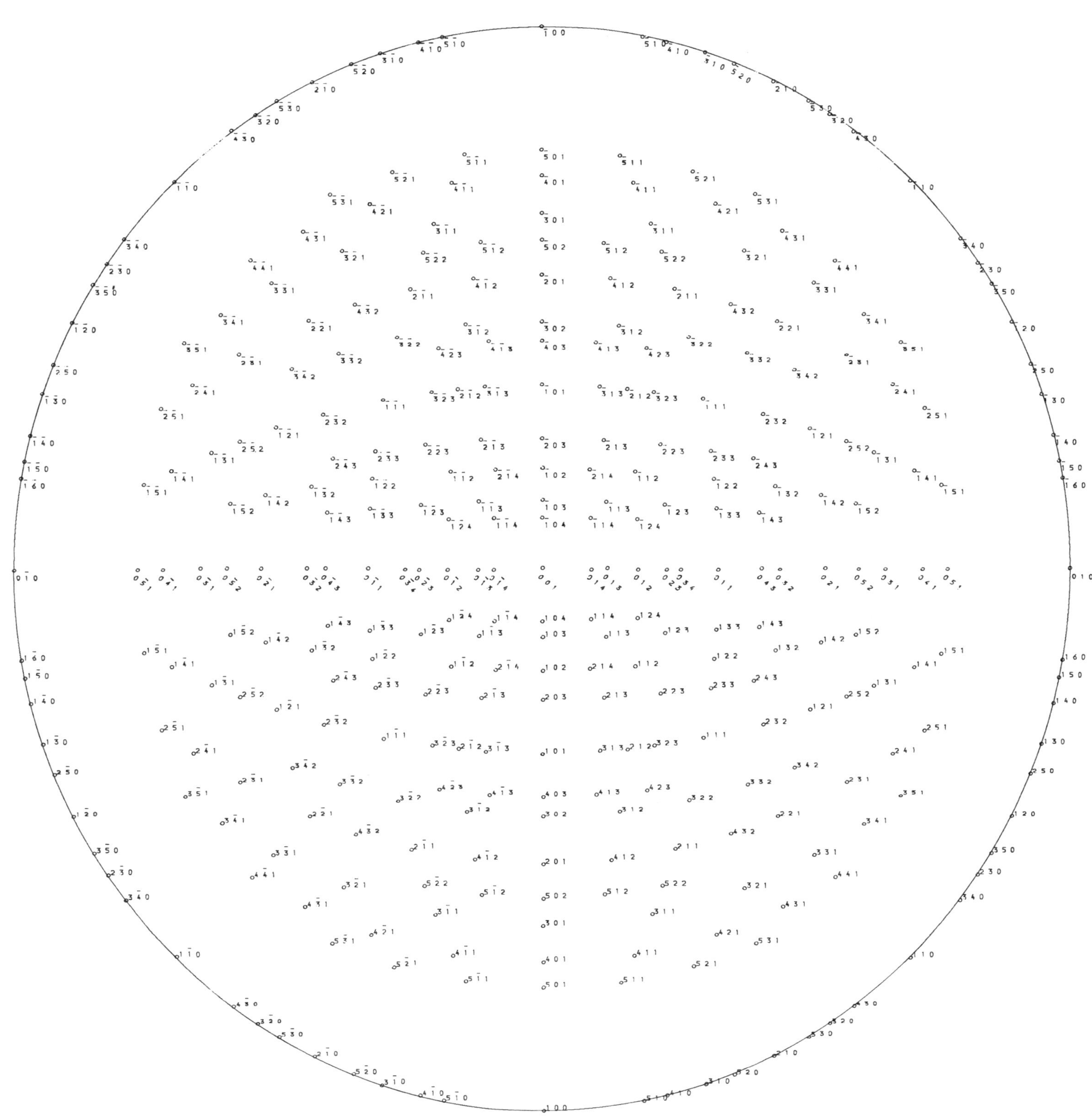

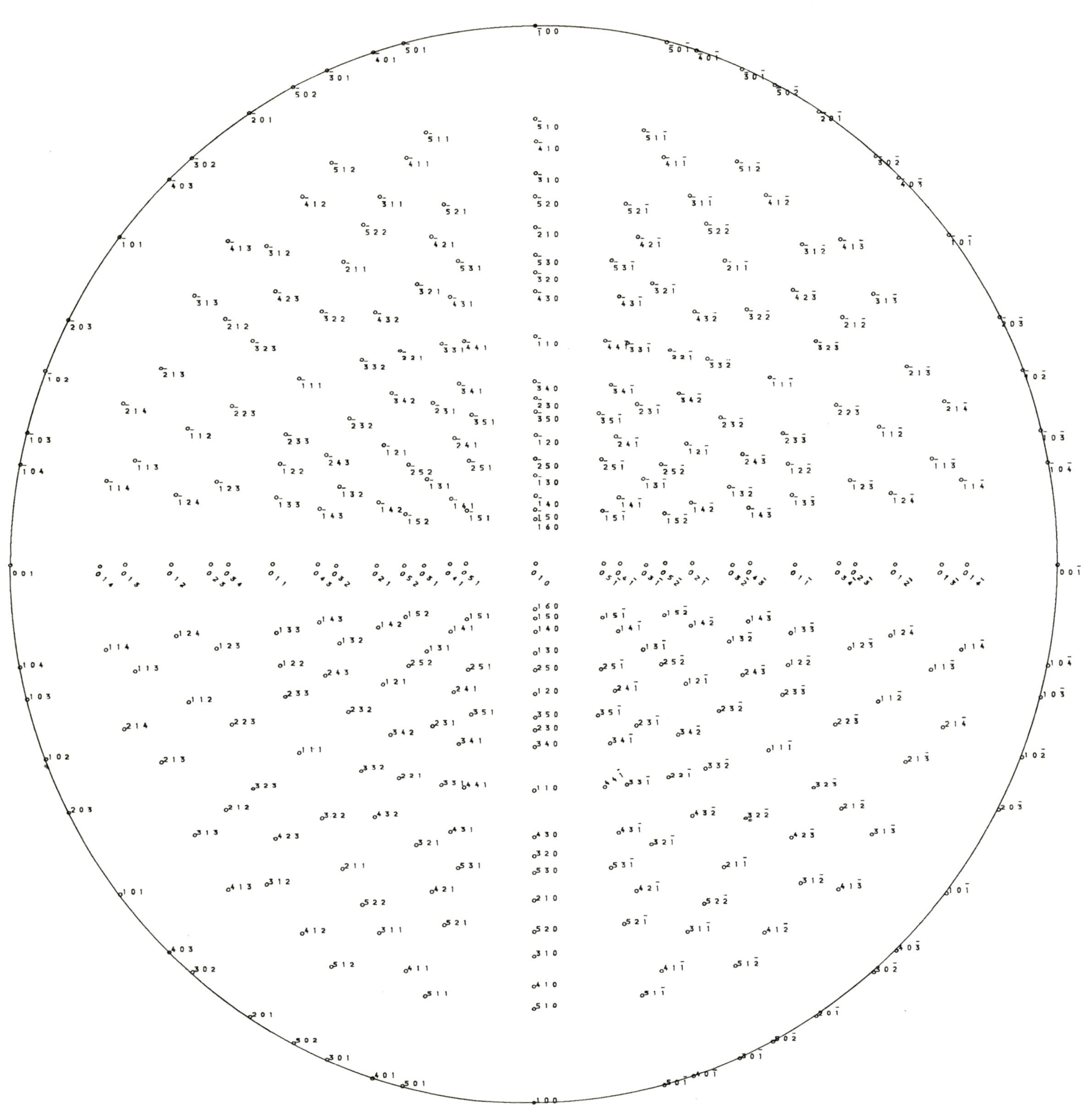

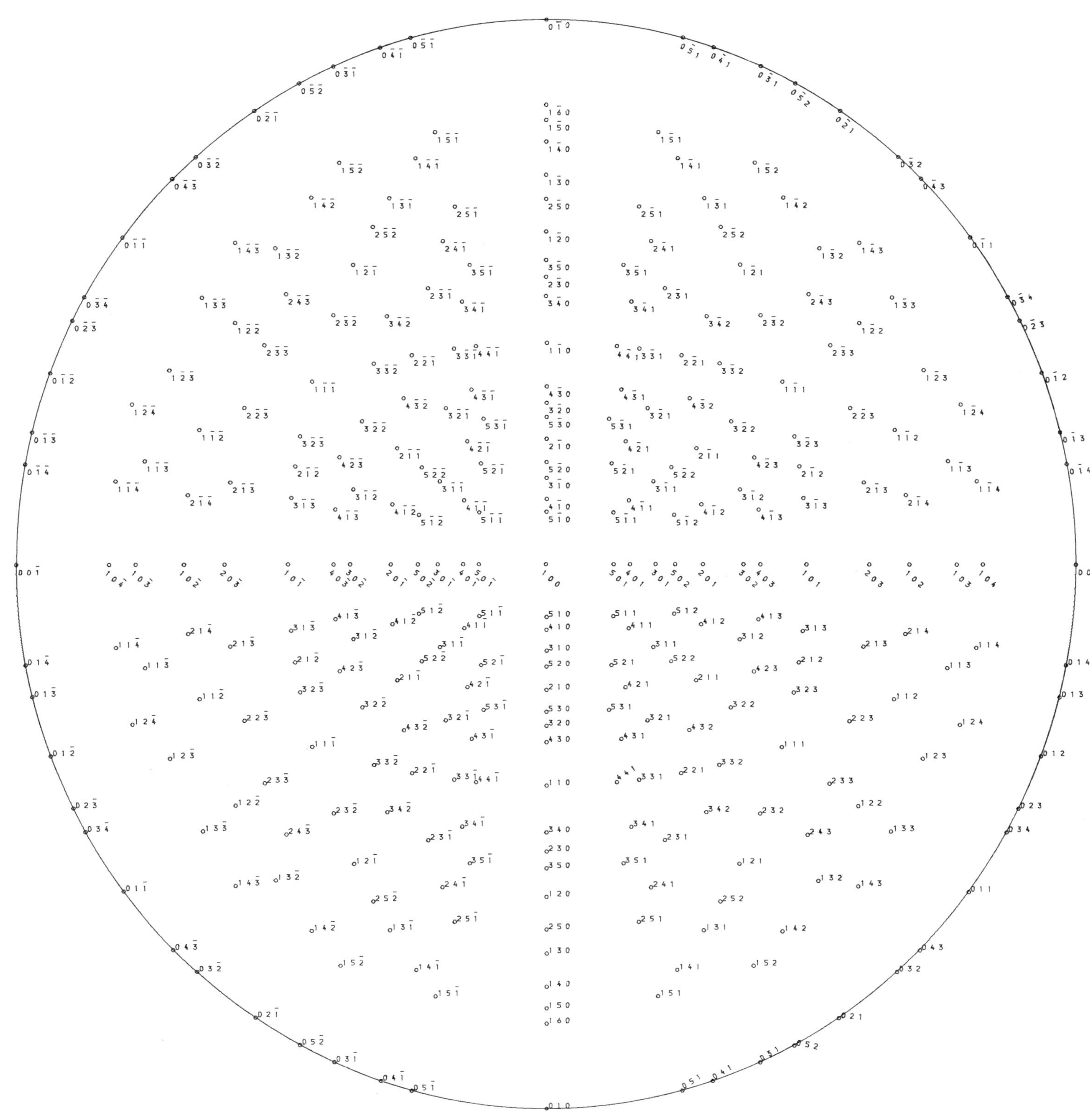

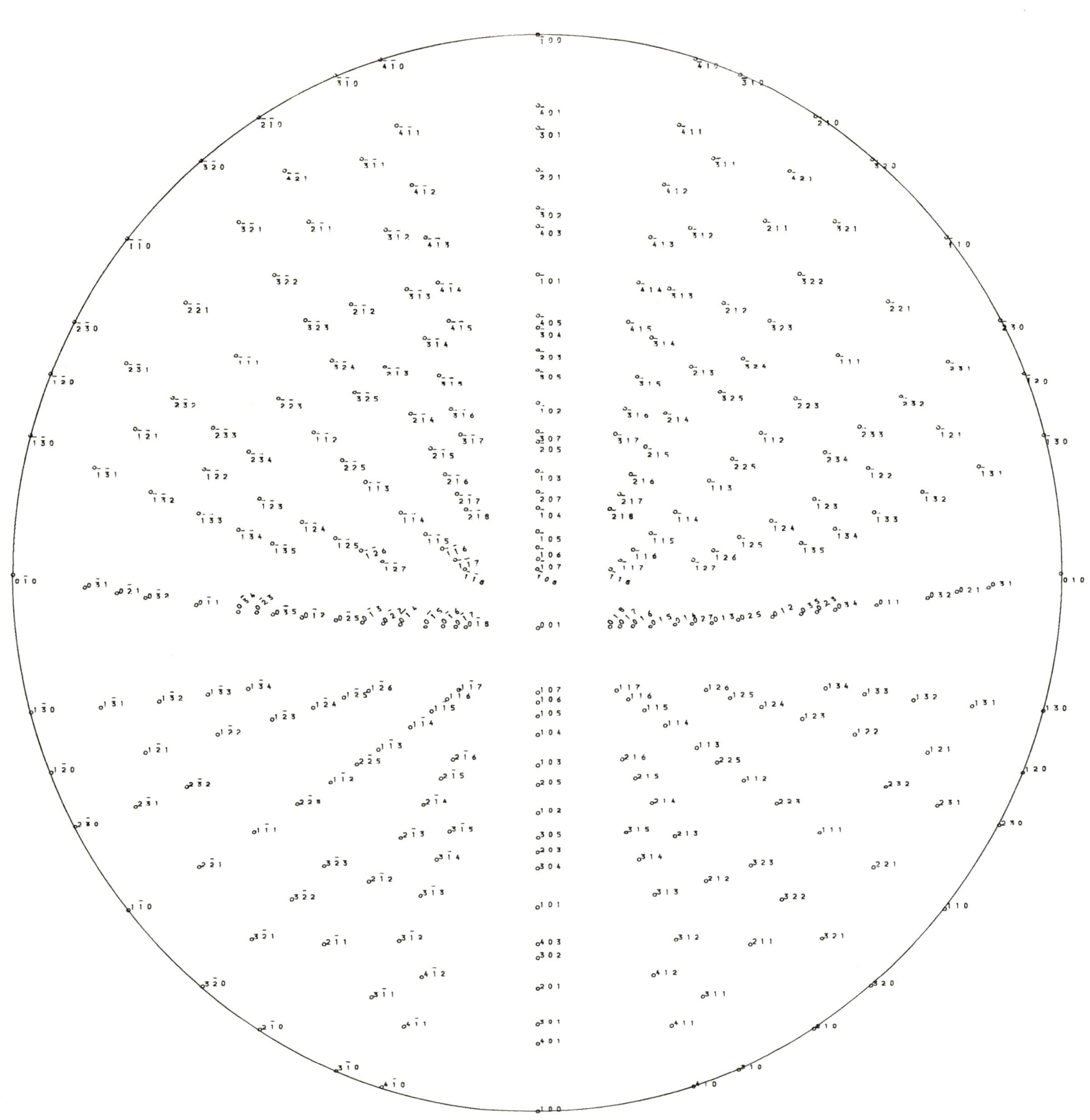

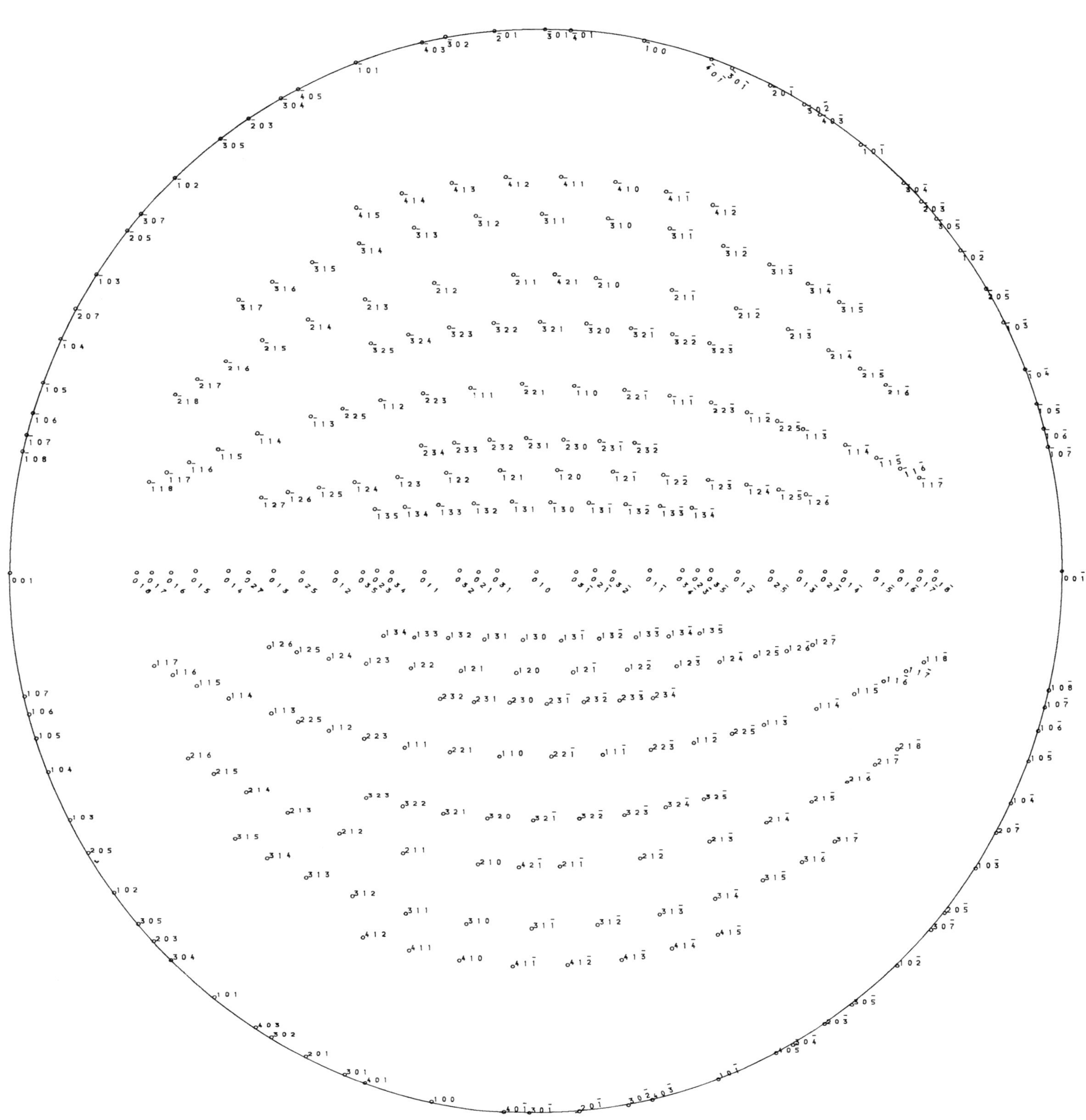

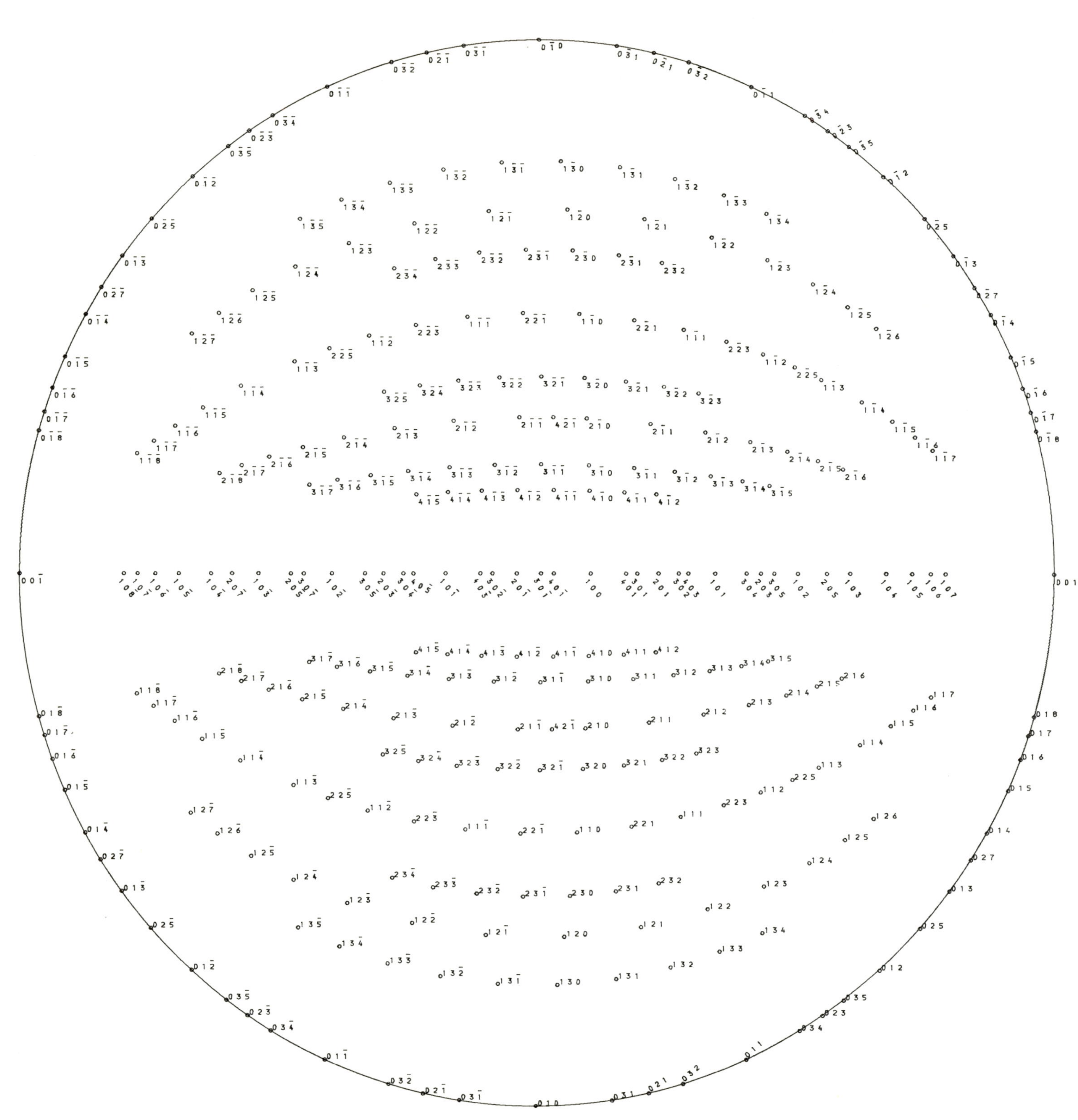

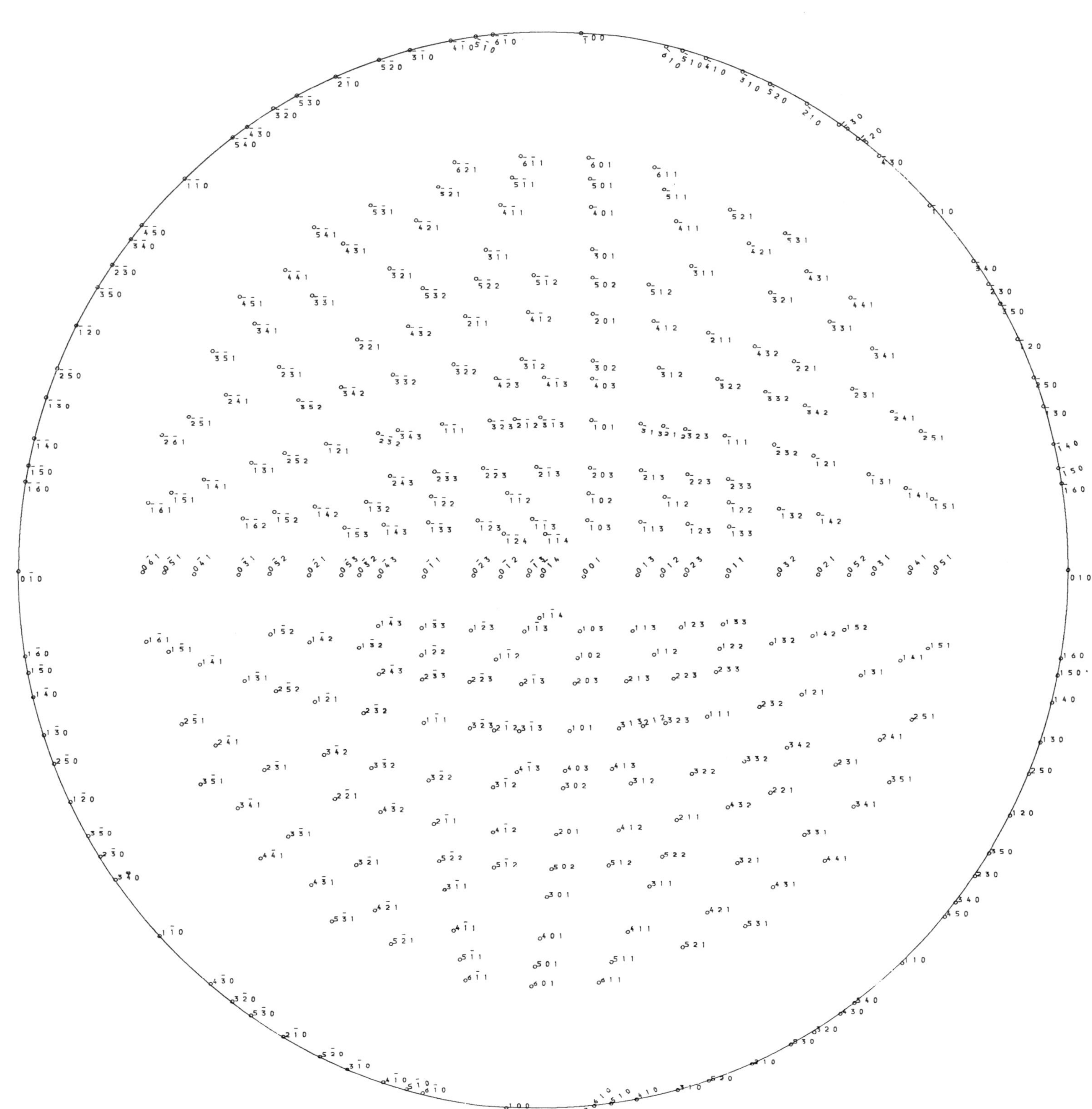

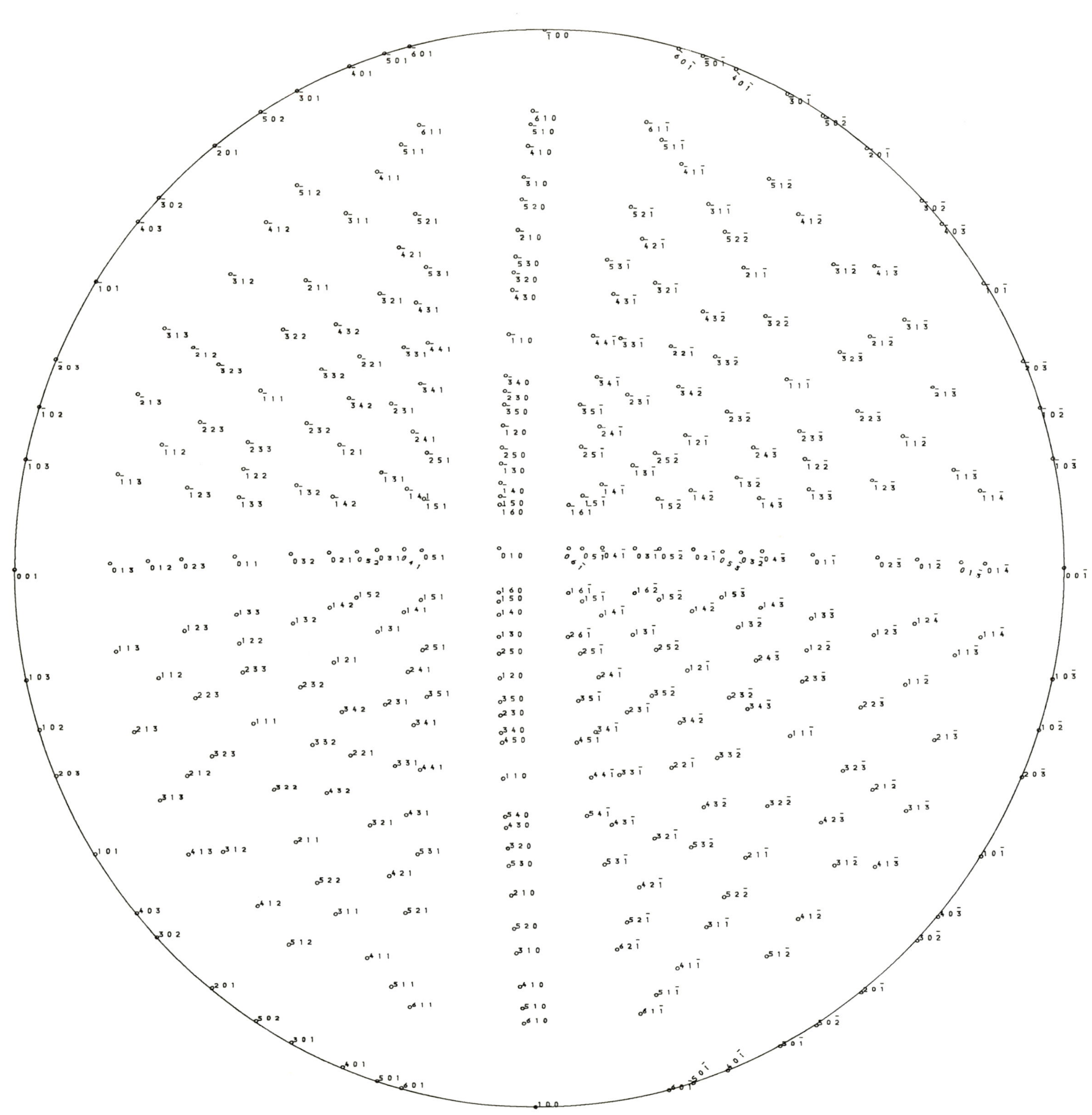

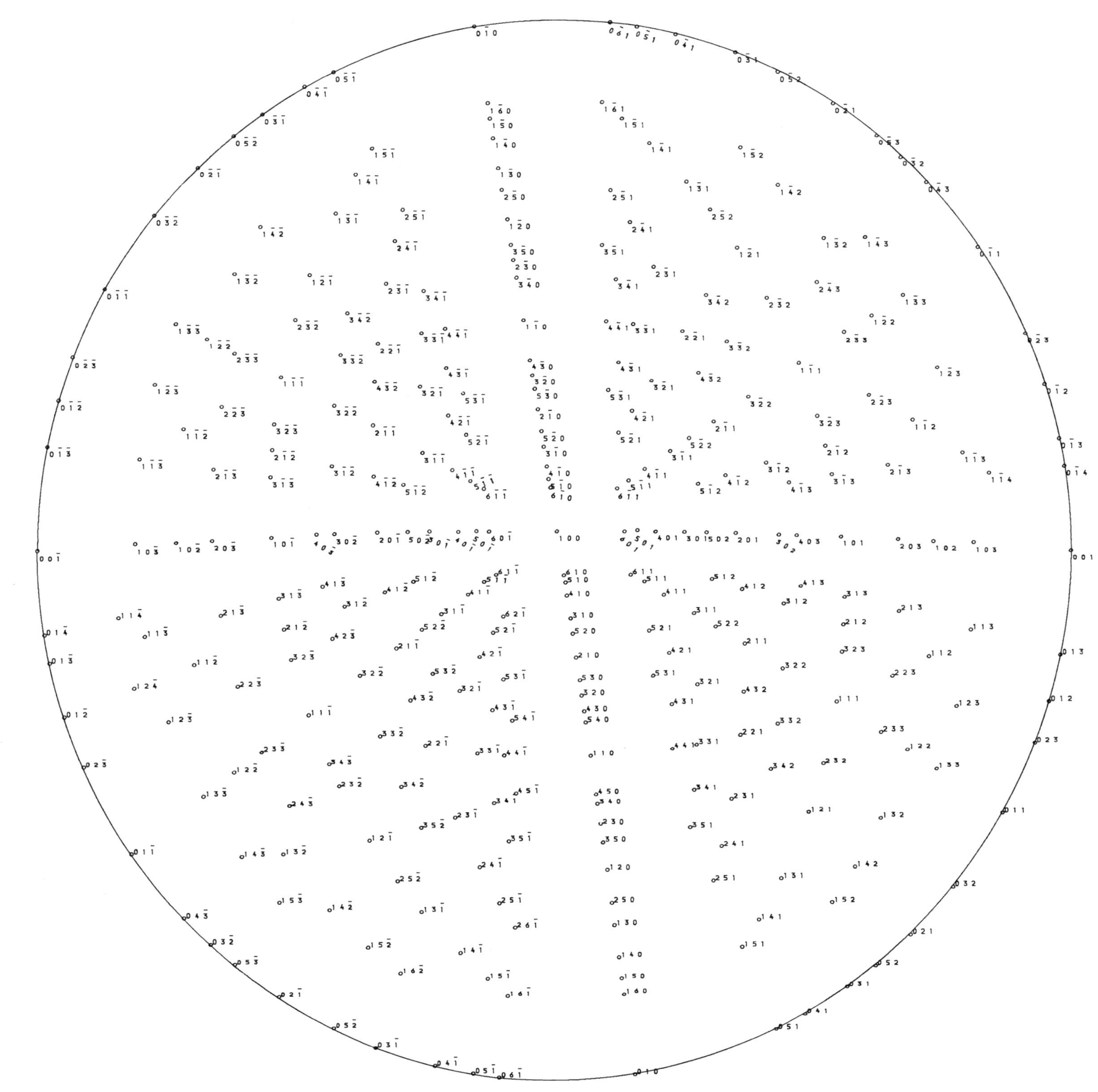